Carl Ritter, William L. Gage

The comparative geography of Palestine and the Sinaitic Peninsula

Carl Ritter, William L. Gage

The comparative geography of Palestine and the Sinaitic Peninsula

ISBN/EAN: 9783742874832

Manufactured in Europe, USA, Canada, Australia, Japa

Cover: Foto ©berggeist007 / pixelio.de

Manufactured and distributed by brebook publishing software (www.brebook.com)

Carl Ritter, William L. Gage

The comparative geography of Palestine and the Sinaitic Peninsula

THE COMPARATIVE GEOGRAPHY

OF

PALESTINE AND THE SINAITIC PENINSULA.

THE COMPARATIVE GEOGRAPHY

OF

PALESTINE

AND THE

SINAITIC PENINSULA.

BY CARL RITTER.

Translated and Adapted to the Use of Biblical Students

BY

WILLIAM L. GAGE.

NEW YORK:
D. APPLETON & COMPANY,
90, 92 & 94 GRAND STREET.
1870.

Entered according to Act of Congress in the year 1866, by

D. APPLETON & CO.,

in the Clerk's Office of the District Court of the District of New York.

CONTENTS OF VOL. IV.

GEOGRAPHY OF PALESTINE.

CHAPTER IV.

	PAGE
JERUSALEM, Ἰερουσαλήμ, Ἱεροσόλυμα, HIEROSOLYMA, THE CITY OF DAVID, THE CITY OF JEHOVAH, THE HOLY CITY, EL-KODS OF THE MOHAMMEDANS, ...	1

DISCUSSION 1. The Central Situation of the City in relation to the World—the Authorities, both ancient and modern, which relate to its Topography, . 1

" 2. The Situation of Jerusalem, and its Division into Hills and Valleys, ... 18

" 3. The Circuit of the present Walls, and the Localities just outside of them, ... 29

" 4. The Interior of the City of Jerusalem: its present Physical Character, and Division into Streets: el-Wadi, or the Street of Mills; the Tyropœon; the Situation of the Baths and Wells on the West Side of the Haram. The Doubtful Situation of Akra; the Antonia; the Serai; the Temple Enclosure on Mount Moriah; and the Mosque of Omar—Kubbet es Sukhrah, . 98

		PAGE
DISCUSSION 5.	The Christian Quarter of Jerusalem, with Golgotha and the Church of the Holy Sepulchre, .	122
,, 6.	The Water Reservoirs and Burial-places in and around Jerusalem, ...	142
,, 7.	The Climate and the Soil, the Plants and the Animals, of Jerusalem, Judæa, and Palestine, .	181
,, 8.	The Inhabitants of Jerusalem—its Population—the Mohammedans—the Oriental and Occidental Christians, and their Subordinate Sects—the Jews, ...	189

NORTHERN JUDÆA.

CHAPTER I.

THE REGION IMMEDIATELY ADJOINING JERUSALEM, . . 213

DISCUSSION 1. Bethany and Abu Dis, on the East of the City—the Wilderness of John the Baptist—Ain Karim, the Convent of St John, and Deir el Masallabeh on the West, ... 213

,, 2. Places directly North of Jerusalem, . . 216

CHAPTER II.

THE MOUNTAIN ROADS, WITH THEIR PASSES WESTWARD TO THE COAST OF THE MEDITERRANEAN: THE PLAIN OF SHARON, AND THE TOWNS OF RAMLEH, LYDDA, AND KEFER SABA (ANTIPATRIS), ... 232

DISCUSSION 1. The Southern Route, by way of Kulonieh, Kuryet el-Enab, and Wadi Aly, . . . 233

,, 2. The Northern Route from Lydda—the Great Caravan Road by way of the Pass of Beth-horon

	PAGE
and el-Jib (Gibeon)—the Branch Road by way of Wady Suleiman, ...	240

DISCURSION 3. The North-West Route from Jerusalem over the Mountains of Ephraim to Kefr Saba, the ancient Antipatris—from Bireh and Jifna to Tibneh, on Wadi Belat—the Burial-place of Joshua—Past Mejdel Yaba, Ras el Ain, to Kefr Saba, ... 243

CHAPTER III.

THE COAST PLAIN FROM THE PLAIN OF PHILISTIA TO THE CARMEL RIDGE, ... 253

Sephela and Sarom, with their Cities and Main Highways—Joppa or Jaffa—Ramleh and the Plain of Sharon—the Eastern or Mountain Road as far as the Plain of Esdraelon—the Western or Coast Road to Cæsarea, and the Promontory of Carmel.

DISCURSION 1. Joppa and Ramleh, ... 253

„ 2. The Plain of Sharon, and the Roads which traverse it: the great Damascus Highway over Mount Carmel to the Plain of Esdraelon, . 265

„ 3. The Eastern or Mountain Road through the Plain of Sharon: the great Caravan Road from Lydda over the Carmel Range to the Plain of Esdraelon, 268

„ 4. The Western or Coast Route through the Plain of Sharon: Kaisariyeh, the ancient Cæsarea Palestinæ—Cæsarea Maritima, originally Stratonis Turris, subsequently Cæsarea Stratonis, . 269

„ 5. Coast Route from Cæsarea to Carmel, by way of Dandora (Tantura, Dor, Dora), and Athlit (Castellum Peregrinorum), . . . 277

SAMARIA, THE CENTRAL PART OF PALESTINE.

CHAPTER IV.

	PAGE
DISCURSION 1. The Nablus Road from Beitin (Bethel) by way of Jefna (Gophna), Sinjil, Seilun (Shiloh), through the Plain of Mukhna to Nablus (Neapolis, Shechem), ...	293
,, 2. The City of Nabulus or Nablus, the ancient Neapolis, the Roman Flavia Neapolis—Shechem at the time of Jacob—Mabortha, the Pass—Gerizim and Ebal, the Mountains of Blessing and of Cursing—the Cuthites, or Samaritans—the Well of Jacob and the Grave of Joseph, . .	302
,, 3. The Road from Nabulus to Sebaste, the ancient Shomron of the Hebrews, the Samaria of the Greeks, the Sebaste (Augusta) of the Romans, and the Usbuste of the Local Population: the Antiquities of the Place, . . .	320
,, 4. Route from Sebaste to the Southern Entrance into the Plain of Esdraelon at Jenin, Ta'anuk, Megiddo, and the Northern Border of Samaria, . .	327

GALILEE, THE MOST NORTHERN DISTRICT OF PALESTINE.

CHAPTER V.

INTRODUCTION, ... 332

Galilee, the Land of the Heathen in the Canaanite Epoch—the Extent of the Territories of Zebulon, Issachar, Asher, and Naphtali at the time of Joshua—the Later Province and Toparchy of Galilee—Upper and Lower Galilee at the period of Josephus, ... 332

CONTENTS.

vii

PAGE

DISCURSION 1. The Southern Portion of Galilee—the Great Plain Esdraelon, or the Jezreel of the Jews—the Brook or River Kishon, ... 343

,, 2. The Mountain Range and Promontory of Carmel, 352

,, 3. The Bay of Acre and the Ports of Haifa (Hepha), and Ako (Akko, St Jean d'Acre), or Ptolemais, 359

,, 4. Nazareth and its Neighbourhood, . . 368

,, 5. The Interior of Galilee—the Upper and the Lower Provinces, the Highlands and the Lowlands, . 378

INDEX OF SUBJECTS, ... 385
INDEX OF TEXTS, ... 401

GEOGRAPHY OF PALESTINE.

CHAPTER IV.

JERUSALEM, 'Ιερουσαλήμ, Ιεροσόλυμα, HIEROSOLYMA, THE CITY OF DAVID, THE CITY OF JEHOVAH, THE HOLY CITY, EL-KODS OF THE MOHAMMEDANS.

DISCURSION I.

THE CENTRAL SITUATION OF THE CITY IN RELATION TO THE WORLD—THE AUTHORITIES, BOTH ANCIENT AND MODERN, WHICH RELATE TO ITS TOPOGRAPHY.

JERUSALEM, built in the heart of Judæa, away from all the great lines of communication which cross the East; separated and protected from the powers lying eastward of it by the Dead Sea, from those at the north and west by almost inaccessible footpaths, and by the Mediterranean Sea, and from those at the south by the broad wastes of desert which stretch away to Egypt; situated on a rocky foundation, destitute of a rich flora, almost without fields, without a river, almost void of springs and any productive soil—has nevertheless gained a place among the great cities of the globe, which, among those of Europe, can only be compared to Rome and Constantinople. And in many respects it has reached a place even higher than they, and affected the world more powerfully; for Jerusalem, as the city of David, as *par eminence* the temple city, has not only had the advantages of age, riches, splendour, trade, luxury, art, and conquest, like the great European capitals; but in the diffusion of its central idea of the unity of a God who must be worshipped in spirit and in truth, in opposition to the idolatries of the rest of the world, it has before, and more eminently since the birth of Christ, carried to the ends of the earth a light which, though having to contend with doubt, falsehood, and superstition, has been enabled to cleave through them all, to dissipate the clouds, and to warm and quicken the mind of man through the whole western world: to affect already that of the East, and to afford assurances that it will do so more and more. What has already been said regarding the central position of the land of Palestine, might be repeated were it necessary, and emphasized still more strongly, regarding Jerusalem, the great world-city, whose geographical, topographical, and historical character we are about to study. Whether the attractive power of Jerusalem,—that force which compelled Assyrians, Babylonians, Egyptians, and all other nations to come up to it (Zech. xiv. 16-18, and Acts ii. 5),—should be ascribed, as Michaelis supposes,[1] to moral grounds, or whether it was its topographical situation, it is alike worthy of study, in all lights and in all its features.

Half a century ago, it would have been impossible to have studied the topography of Jerusalem at all in detail, for no observers had then given it their careful attention; and traditions, legends, and hypotheses had run wild over the whole of the city. During late years, however, much has been done by tracing the ancient outlines; by studying the various styles of architecture; by examining the historical memorials, and discriminating between what is authentic and what is the mere result of middle-age legends, and the fables invented for the ears of pilgrims; and by carefully collating those passages of the Bible and Josephus which relate to Jerusalem,—the main authorities, it may be remarked. To these sources may be added the classics, the fathers of the church, the historians, Mohammedan authors, and the literature created by eastern travel. Doubtless, the ascertaining of many facts, and the placing of them beyond doubt, occasion much difficulty; for the old source of perplexity, the want of

[1] Gesenius, *Comment. zu Isaias* (chap. vii. 1), Pt. i. p. 266.

materials, is compensated by the great number of discoveries, and the divergencies of opinion about them. So great have been the changes by transformation, and by building new structures on the ruins of the old for more than three thousand years, that now it becomes very difficult to come to certain results. The city has outlived a full dozen destroyings and rebuildings; and it is literally true that city lies upon city, ruins above ruins, and such a labyrinth created, that it will be the labour of years to clear up all difficulties and arrive at the simple truth. The masses of rubbish are generally thirty or forty feet in depth before we reach the rock, or the primitive soil where building commenced; and the order of the Turkish Government, forbidding all excavation, and even all entrance into or approach to those localities which are regarded as especially holy, places the greatest difficulties in the way of investigation, even on the part of those who have spent months and years in Jerusalem, to say nothing of those who merely make hasty visits thither. And the places which are open to Christians are so overloaded with traditions and legends coined by the Armenians, Latins, and Greeks, before and after the period of the Crusades, that it is often extremely difficult to arrive at the primitive name where an object is called by several designations, to know whether the Arabic word is formed from a Hebrew or a Syrian root, and whether a later term has or has not driven out an older one, or whether it has merely changed its form. In many of the most central parts of the city, where the Mohammedans mainly live, the streets and squares have remained nameless up to the most recent times.

The difficulties with which we have to contend begin with the very oldest names of the place,—Salem, Jerusalem, Hierosolyma, Ælia Capitolina, el-Kods or Cadytis,—and are not less numerous or less formidable when we come to the designations of the special parts, such as Zion, Moriah, Akra, Bezetha, and their various subdivisions. It is necessary, therefore, at the outset, to state what are the literary authorities with which we have to deal in order to promote an increase of our knowledge.

All the old accounts of the geographical character of the city subsequent to the times of David and Solomon, 1000 B.C., are instructive and valuable in their slightest details; but up to the second conquest of the city and its first destruction, effected at the hands of Nebuchadnezzar the king of Babylon, B.C. 588, we have not information sufficient to allow us to gain any clear conception of the appearance of the place. Up to that date the historical books of the Kings, and those of the prophets, particularly Isaiah and Jeremiah (the Lamentations of the latter more especially), are the only authorities which give us any idea of the nature of the primitive Jerusalem. The return of the people from their seventy years' captivity in Babylonia, and the permission granted by king Cyrus to build the temple again (536 B.C.), although the actual building took place after Cyrus' death and under Darius Hystaspes (515 B.C.), bring us to a new epoch in the geography of Jerusalem,—an epoch characterized, as we learn from the books of Ezra and Nehemiah, by the rebuilding of the walls of the city, the temple, and the royal palace, on the ruins of those which had been destroyed. The effort was made to follow in the re-erection the old wall lines as closely as possible, and to put the new edifices on the site of the old; and we are unable, therefore, to follow the account given by Nehemiah, from the want of more positive knowledge regarding the old city, whose ruined battlements were replaced by the new ones which he describes.

The accounts given by Nehemiah and Ezra would indeed be unintelligible, were it not for the instructive though brief sketch which Josephus gives us of the topography of Jerusalem, and the very full narrative which he has transmitted to us of the siege and capture of the city by Titus in A.D. 70; yet his statements are mainly general in their character, and not given with a view to acquaint his readers with geographical details, but with the incidents which transpired under his eye. On this account, many places which he mentions are alluded to in such a casual and transitory way, and with so little reference to an implied ignorance of the topography on the part of his readers, that we are almost entirely unable to make use of many of his references to names and places. Add to this difficulty still another, namely, that the process of rebuilding had gone on to such an extent under the Syrians, the Maccabees, and the Herods, as to obliterate or change materially the form of many of the old names. And even with this difficulty modified as it has been, especially by the labours of Robinson, yet another appears, viz. that Josephus, writing as he did in Rome, and after the lapse of considerable time, and obliged as he was to depend mainly upon his memory,[1] was led into so many inaccuracies and so many exaggerations, as to render it in many places quite impossible for us to make use of his statements respecting distance.

The historical books of the New Testament, particularly the Acts of the Apostles, throw but little light upon the localities mentioned in the Bible; the profane writers, such as Tacitus, draw almost all their topographical knowledge from Josephus. The accounts prepared still later—such as those relating to the building of Ælia Capitolina by the Emperor Hadrian, A.D. 126; the erecting of church edifices at Jerusalem by the Empress Helena and the Emperor Justinian; those dating from the time of the Crusades, and the establishment of the Christian kings there; the descriptions of the city as it was under its Mohammedan, Arab, and Turkish conquerors—are still more meagre and fragmentary than those which date from even earlier times, and in them error is piled upon error, through a want of knowledge, or the power to sift out truth, or from a naturally superstitious nature, or from a proneness to form hypotheses, or from native superficiality. It is no wonder that it has been so difficult to follow the vein of truth under the heaps of rubbish which have accumulated, and after the most painstaking labour, to attain at last to results which rest upon a secure foundation, and which can be appropriated without hesitation in the interests of science.

It will be necessary to give here a brief summary of the most recent works on the topography of Jerusalem which we shall have occasion to use in connection with those already cited.

[1] Robinson, *Bib. Research.* i. 279, 280; comp. Krafft, pp. 2, 24, 52.

Original Authorities relating to the Topography of Jerusalem.

I. THE SOURCES PRECEDING THE PUBLICATION OF ROBINSON AND ELI SMITH'S WORK IN 1838.

In respect to the earlier period, almost the only complete summary of them is that contained in a lecture delivered at Berlin by my honoured friend, Dr E. G. Schultz,[1] Prussian Consul at Jerusalem. Little more can be done than to rehearse what he has given there, adding to it what he has communicated to me in manuscript at various times. His little work on this subject contains, of course, much that had been stated before; but it also contains no inconsiderable amount of matter never before collected or published. It was hoped that he would be able to gather up the results of his studies, and present them to the world in a more complete form than he has yet been able to do; but a long succession of sickness has hitherto entirely prevented him from accomplishing this task.

1. The list of authorities begins with the Old Testament, to which we must add Josephus' twenty books relating to Jewish antiquities, and the collections of translations and paraphrases found in the old polyglotts. These, as well as the old commentaries, are to be found indicated in the various introductions to the Old Testament.[2]

2. The books of the New Testament contain little which throws light upon the geography of the Holy Land; and that little is made the more unavailable, from the fact that the spiritual import of those writings is such as to subordinate all local peculiarities, and keep them in the background.

3. Josephus' *History of the Jewish War* is the richest treasure of those facts which we consult the pages of the New Testament in vain to find: it brings us down a little on this side of the destruction of the city by Titus.

4. The later Jewish literature relating to Jerusalem has been wrought out from the Talmudists in the most thorough

---

[1] Schultz, *Jerusalem*, a lecture, with map by Kiepert.
[2] De Wette, *Lehrb. d. hebr. judisch. Archäologie*, pp. 5–79.

and available manner by Lightfoot in *Opp. om.* ed. Roterodam. 1686, fol. T. ii., in *Centuria Chorographica Talmudica*, ad S. Matthæum, c. xx.–xli. fol. 185–202.

5. From the Christian epoch, lasting till the capture of Jerusalem by the Arabs, Eusebius and Jerome have left us an inexhaustible treasury of facts relating not alone to the general, but to the special, geography of the Holy Land. The works of some Byzantine ecclesiastical historians may be classed in this group.

6. A new cycle of Christian as well as of Mohammedan writers begins with the time of the Crusades: an elaborate recapitulation of them may be found in Wilken's classic history of those wars.

7. The very valuable *Codice Diplomatico del Sacro Militare Ordine San Giovanni Gerosolimitano, oggi di Malta*, ed. Sebastiano Pauli, della Congregazione della Madre di Dio, Lucca 1773, fol. 2. Tom., contains very rich original documents relating to the topography of Jerusalem as well as of all Palestine.

Schultz, in page 47 of his published work, refers also to some important documents relating to the German Order in Palestine, now to be found in the Königsberg archives. The *Cartulario del Santo Sepulcro*, to be seen in C. Beugnot, *Assises de Jerusalem*, Paris 1842, 2 vols., contains a hitherto unknown description of the city, written in the thirteenth century. Schultz makes an extract in his appendices, pp. 107–120. Eugène de Rosière, *Cartulaire de l'Eglise du Saint Sepulchre de Jerusalem, publié d'après les Manuscrits du Vatican*, Paris 1849, contains in its text a hundred and eighty-five documents relating to the Holy Sepulchre. The *Liber Albus* and *Liber Pactorum*, a collection of Venetian state papers, contain much relating to Jerusalem and Palestine; and the archives of Vienna, which were consulted slightly by Wilken in the preparation of his *History of the Crusades*, have been largely drawn upon by Schultz in collecting his materials.

8. The Arabian geographers Isthakri, Edrisi, Abulfeda, and Ebn Batuta, made accessible to us by Müller, Mordtmann, Jaubert, Mack Guckin de Slame, Reinaud, Lee, and

others; the histories of Sultan Saladin by Bahaeddin, and those of the Egyptian sultans written by Macrizi, and edited by Quatremère; Michaud's *History of the Crusades*, with Reinaud's valuable additions, throw valuable additional light on the topography of Jerusalem. To these authorities may be added the Cadi Mejr ed Din, whose valuable Arabian history, written in 1495, has been drawn upon by von Hammer in the *Mines d'Orient*, T. ii. p. 83, v. p. 161.[1]

9. The comprehensive literature of travel, which does not pass over Jerusalem, is well summed up in Robinson's well-known summary, with which my own, contained elsewhere, may be compared by the reader. It is unnecessary to add, that the admirable volumes of Robinson, as well as other works of recognised value, such as those of Reland, Crome, von Raumer, and Winer, and which have often been quoted in my descriptions of the country taken as a whole, do not overlook the topography of the capital; and although the progress of discovery has been very rapid of late in Jerusalem, we are not to overlook the men who have contributed to this result, and to whom we owe so much. To accept, with a mere courteous expression of thanks, works which may be now in a measure superseded by more recent ones, and to forget our indebtedness to the pioneers in discovery, indicates great vanity, and a lack of true nobleness of mind in those who come after; who, though their results may be more recent, yet could never have gained them without the aid afforded by their predecessors, and who perhaps, after all, have not displayed such talents for investigation, or been rewarded by such happy and unlooked-for discoveries, as some pioneers in the same path have done.

II. THE LATEST AUTHORITIES RELATING TO THE TOPOGRAPHY OF JERUSALEM SINCE ROBINSON AND SMITH.

Among the learned and acute authors who have written recently upon Jerusalem, there are notable instances of great modesty, and also of an envious spirit of detraction from the

[1] The translation, together with another Arabic MS., may be compared in Williams' *Holy City*, vol. i. app. iii. pp. 143-164.

merits of earlier writers. An example of the latter may be found in the voluminous work of G. Williams, formerly chaplain to the English Bishop of Jerusalem, and a man whose position and calling ought to have been a sufficient guarantee that his book would not be characterized by the faults which mar it. The most offensive passages are those in which he attacks the American Robinson; an assault which the latter took very quietly, and without the loss of temper. In the second edition of his work the English chaplain softened his tone, and tried to hush up the affair, letting it go no further; but he broke out in equally unjust invective against a younger explorer, and one who had had no connection whatever with him — Dr Krafft, whom he accused of plagiarism. It was entirely an unjust charge, for the German gave Williams full credit for all that he had really discovered; and showed his own acuteness, thoroughness, and a scholarship even superior, it may be, to Williams, while correcting the Englishman's mistakes. I am rejoiced to say that my own friends, Robinson, Krafft, and Schultz, have always manifested a different spirit; and even Williams must consider the latter a model in the discharge of literary justice. And although I may have occasion, in the course of the following pages, to differ widely from all these gentlemen, and follow my own convictions, yet I cannot forbear alluding to the spirit in which they have laboured, which colours all their controversial writings.

Dr Robinson's classic work[1] has ushered in a new epoch not only for the study of the whole land of Palestine, but for that of the geography and history of Jerusalem; and even now the most salutary influences have begun to flow from the investigations which he made. He has infused fresh youth into a theme of the weightiest import not only for all ecclesiastical life, but interesting in its relation to all the sciences; he has given a new impetus to the study of the Holy Land,—an impetus which has been felt alike on both

[1] Robinson, *Bib. Research.* i. pp. 229-250, embracing the topography; 250-364, the history; 364–418, the other features; 418–431, plans, etc. See also *Later Bib. Research.*

sides of the Atlantic, and has called many strong and bold spirits into the same field. Almost the only scholarly work which preceded Robinson's was that of Dr E. D. Clarke,[1] 1813, in which he took Pococke[2] as a basis, and gave, in addition, the results of his own carefully conducted investigations. Naturally, however, he fell into numerous errors; and it was only in 1838 that Dr Robinson, leaving far behind him the often instinctive and sometimes spirited efforts made by tourists and *dilettanti*, gave his own mind to the task of solving the mysteries which had hitherto perplexed travellers, clearing away the rubbish which had been accumulating for thousands of years, and setting the topography of Jerusalem in a clear light. The magnitude of this undertaking may be best seen after noticing how little so sagacious and faithful a traveller as Niebuhr[3] was able to accomplish when there in 1766. Although he was fully impressed with the importance of studying Jerusalem with great care; although he recognised that it is a city just as interesting to the Christian as to the Jew, and to both the most important of all the cities of the world; yet he did not venture—such was the tyranny of the monks and Turks at that time—to go near enough to the immense stones on the southern slope of the external temple area to gain a full view of them.[4]

Even quite recently, when the eminent geometrician Dr Westphal,[5] and also Dr Parthey (1823), following the efforts of Sieber[6] (1818), tried to improve the existing maps of Jerusalem, by taking measurements of the walls, in order to compare their results with those of Josephus, they were threatened with stones and bullets from the roofs of adjoining houses, and they were compelled to relinquish their under-

[1] E. D. Clarke, *Travels in various Countries of Europe, Asia, and Africa*, vol. iv. pp. 288-394.
[2] Pococke, *Travels*, Ger. ed. Pt. ii. pp. 12-42.
[3] Niebuhr, *Reisen durch Syria und Palästina*, pp. 45-65, with Plates iv. and v.
[4] *Ibid.* p. 141.
[5] Author of the excellent *Agri Romani Tabula cum veterum viarum designatione accuratissima*, Romæ, L. H. Westphal.
[6] F. W. Sieber, *Karte von Jerusalem*, etc.

taking, not without gaining enough new material, however, to warrant them in publishing a map of the city.[1] It was only when the iron sceptre of Ibrahim Pasha was extended over Syria, followed by his subjection of Turks and Moslems, and his generous treatment of Jews and Christians, that the fanatic spirit of the people so far yielded as to allow civilisation to make progress in Palestine, to open Hebron and Jerusalem to strangers as never before, and to allow that thorough course of investigation which has been undertaken and carried on with such success by von Schubert, Robinson, Russegger, Wilson, Gadow, Krafft, Tischendorf, Tobler, and others, during their long sojourn in the Holy Land. Nor have these been all. The American missionaries have done much, the consuls resident in Jerusalem, the members of the newly founded evangelical bishopric there, and many others who have enjoyed the good fortune of being able to prosecute their inquiries for any considerable length of time.

Plans and Maps of the City.

The immediate advantage received from this improvement in the facilities for surveying Jerusalem, was the revision of the map of the city[2] effected by the architect Catherwood, who, in company with his friends Bonomi and Arundale, enjoyed particularly favourable opportunities for studying the city from the roof of the house belonging to the governor of Jerusalem. Here they had a fine panoramic view of the city, and were able not only to take very perfect observations, but they were permitted also to enter the temple area, to visit all the buildings, to use a *camera lucida* in copying their architectural features, and even to examine the foundations on which they stand.[3] A short time before their visit, the curiosity which led some European travellers to enter the sacred enclosure of the mosque had put them in peril of

---

[1] See *Jerusalem und seine nächsten Umgebungen*, von Westphal, vol. i. pp. 385–390.
[2] *Plan of Jerusalem*, by J. Catherwood.
[3] Bartlett's *Walks about the City and Environs of Jerusalem*, pp. 148–168.

death at the hands of the enraged Mohammedans; and yet Catherwood was allowed to spend six weeks of undisturbed quiet in studying the temple on every side, and in making his measurements.

The progress which was the result of these labours, was seen not merely in the map of the mountain on which the temple stood, and its immediate neighbourhood, and in that of the interior of the city, but in many other details. Nevertheless, in all that pertains to the northern and western wall of the city, it falls far behind Westphal's map in exactness, and exhibits the inequalities of surface and the environs of the city in a very imperfect manner: it was, however, an improvement upon the work of Sieber.[1] Yet, as data were then wanting, the measurements of Robinson and Smith not sufficiently well covering the ground to allow the construction of a more perfect map, it remained the basis of those which followed, and indeed of Kiepert's, which accompanies the works of Robinson and Schultz,[2] although with many corrections. Regarding these, Kiepert has spoken fully in the memoir from his own hand, accompanying the maps. The corrections consisted mainly in the far clearer designation of the inequalities of surface in the city, the insertion of several details in the immediate neighbourhood, and of a very valuable profile view of Jerusalem, from W.S.W. to E.N.E., from the upper valley of Hinnom to the highest part of the Mount of Olives.[3] The first of Kiepert's plans was characterized by a rejection of many of the old and wholly useless and unfounded legendary names, but the second contained those of many antiquarian monuments of interest and value. Dr Krafft's map,[4] which appeared in the following year, incorporated all of these; but

[1] *Plan von Jerusalem, entworfen nach Sieber und Catherwood, etc.*, von Kiepert.

[2] *Plan von Jerusalem, nach den Untersuchungen von Dr E. G. Schultz*, geg. von Kiepert.

[3] H. Kiepert, *Memoir zu den Carten, welche Robinson's Palästina begleiten.*

[4] *Plan von Jerusalem nach den Untersuchungen von Krafft, mit Benutzung der Pläne von Robinson und Schultz.*

it contained, besides, the results of his own carefully conducted researches, and was accompanied by a memoir, which made it perfectly intelligible. This was followed by Dr T. Tobler's Sketch of Jerusalem,[1] whose title indicates the special place which the map was intended to fill, namely, to serve as a practical guide to the interior of the city, and to give the correct names of streets, most of which had been spelt in very different ways, and had made the topography of the city almost unintelligible.

The small scale on which most of these maps were executed, the difficulties which lay in the way of access to many parts of the city, and the amount of rubbish and of later buildings which have been allowed to collect, leave many interesting historical places unverified. These difficulties have been increased by the originally hilly character of the city, which naturally led to the erection of the oldest buildings in the valley, from which they gradually receded as the capital expanded. These facts made it necessary to proceed with the greatest care in the task of taking measurements and of making drawings. The first one who entered upon this work with critical minuteness was J. H. Gadow, who spent nine months in Jerusalem in 1847 and 1848, engaged in this undertaking. The result was his map, prepared with the most conscientious care, and containing exact measurements of all the principal objects—the walls of the city, the Haram, and the most important buildings. This map was published in a reduced form by Gadow's friend and companion in travel, Dr Wolff,[2] and the original manuscript was given to the German Oriental Society. We are therefore able to use[3] the whole of the results which he gained. The map of Gadow deviates in some unimportant respects from the earlier ones, especially in the portrayal of the northern and western parts of the walls; but it is completely justified in this by

[1] Dr T. Tobler, *Grundriss von Jerusalem*.
[2] Dr Ph. Wolff, *Reise in das Gelobte Land*.
[3] H. Gadow, *Mittheilungen über die gegenwärtigen Terrainverhaltnissen in und um Jerusalem*, in Z. d. Deutsch. Morgen. Ges. vol. ii. pp. 35—45, 384.

the very accurate map published by the English Government, and prepared by the engineers, Aldrich and Symonds.[1] This work, which is the result of a careful mathematical survey, must be the basis of all future topographical delineations of the city. It is on a scale of four hundred English feet to a mile. It is to be regretted that the masterly model of Jerusalem, executed in relief, in a faithful and elegant manner, by Edwin Smith of Sheffield,[2] and published in 1846, was not based upon the survey of Aldrich and Symonds; yet it should be remarked that the outlines of the city, as Mr Blackburn, the inventor, has portrayed them, are in much closer agreement with Gadow's map than with Catherwood's, and are singularly coincident with those laid down in the royal survey.

Although very great progress has been made in locating the objects of historical interest in Jerusalem, yet it will not seem strange that our course has been so slow in a city which has suffered seventeen conquests, many of which have been accompanied by the destruction of buildings, and the subsequent re-erection of others in their place. These destroyings and rebuildings have been going on for almost 3000 years, and have been effected by the most different people. The complaint which Richardson makes is therefore not without foundation, that it is a Tantalus-like task to try to discover the site of buildings whose names have come down to us, or to find the scene of noted deeds. And Scholtz justly remarks, that in the mass of ruins which have accumulated in the old Jewish capital, it is impossible to identify those which date from different epochs, and to discriminate between them.

The extraordinary difficulties which beset the study of the topography of a great metropolis like Jerusalem, which, unlike Athens and Rome, has not a single authentic monument of its remoter periods to exhibit, must not be forgotten; and keen as have been the intellects, and wide the learning, and strenuous the efforts, of those who have sought to clear

[1] *Plan of the Town and Environs of Jerusalem*, copied from the original drawing of the Survey made in March 1841 by Lieuts. Aldrich and Symonds, in Williams' *Holy City*, i. pp. 9–124.

[2] *Model of Jerusalem*, by Edwin Smith: scale, nine inches to a mile.

up the perplexities which lie in the way, uniform results have by no means been attained; and in many cases I have been unable to form any judgment, so evenly balanced are the arguments offered on both sides of a disputed question, and have therefore been compelled to postpone a decision till the future shall bring more light.

In addition to the records of pilgrimage and travel, among which may be mentioned those of von Schubert and Russegger, which contain measurements of the height above the sea of many objects in Jerusalem, and the classic work of Robinson already referred to, there are the following later and valuable works, all of which I shall have occasion to use in the course of the following pages:—

1. J. Wilson: The Lands of the Bible, Edin. 1847. Vol. i., Jerusalem and its Environs, pp. 406–504; vol. ii. pp. 269-284.

2. W. H. Bartlett: Walks about the City and Environs of Jerusalem; London, 2d ed. pub. in 1850; with very instructive sketches.

3. By the same: A Comparative View of the Situation and Extent of Ancient and Modern Jerusalem; Lond.; four plates, containing very extended panoramic views of the city.

4. David Roberts: La Terre Sainte, Vues et Monuments avec une Descr. histor. Bruxelles 1845. Fol., with twenty-eight plates representing buildings in Jerusalem and its environs.

5. The Christian in Palestine, by Henry Stebbing, D.D., F.R.S.; the drawings recently taken by W. H. Bartlett on the spot, pp. 128-164. Vol. iv., with twenty plates of Jerusalem and its neighbourhood.

6. George Williams: The Holy City; Historical, Topographical, and Antiquarian Notices of Jerusalem. Second ed. London 1849. Vol. ii. contains an Architectural History of the Church of the Holy Sepulchre, by Rev. Robert Willis, M.A., F.R.S., Jacksonian Professor in the University of Cambridge. With an Appendix containing an historical and descriptive memoir illustrative of the Ordnance Survey.

7. Views in the Holy Land, with historical descriptions, by Croly and Brockedon. Four vols., fol. London 1847.

8. Dr Ernst Gustav Schultz, Prussian Consul: Jerusalem, a Lecture read before the Berlin Geographical Society. Berlin 1845. With a map, drawn by H. Kiepert.

9. E. Robinson: Later Researches regarding the Topography of Jerusalem. A supplement to the author's larger work on Palestine. The opposition which Robinson's views encountered[1] from the writers of the two books last quoted called forth from his pen a series of articles, first published in the *Bibliotheca Sacra,* and afterwards collected into the volume whose title is cited above. In the preparation of this work he was led to reconsider his own ground very carefully, and to weigh the value of the opinions of Eli Smith, Wolcott, and others. Tuch, certainly an excellent judge, praised very highly the breadth of his knowledge, his thoroughness, and his exactness, and not less the quiet dignity with which he maintains his position, the last of which is in striking contrast to the tone of his opponent Williams. I agree with this judgment entirely, and must set the impartiality and the love of truth which Robinson displays over against the unworthy attacks of the Englishman, even though I cannot always coincide with my noble and learned friend in all his conclusions. Robinson was unable to take advantage of the work of Krafft, which was issued subsequently to his own. The latter author opened a new path alike divergent from that of Williams and that of Robinson. His work is characterized by great candour, learning, rare acuteness, the closest observation, and practical acquaintance with the whole field. His work I may cite as,

10. The Topography of Jerusalem, by W. Krafft, Bonn 1846, with a map. It may be consulted in connection with the work of his comrade in travel,

11. F. A. Strauss: Sinai und Golgotha. Travels in the East. Third ed. Berlin 1850, pp. 201–342.

12. Dr Ph. Wolff: Travels in the Holy Land. Stuttgard 1849. With a map of Jerusalem (reduced from that of Gadow, his companion). It contains within a very compact compass many a valuable passage and happy sugges-

---

[1] Dr Tuch, in *Z. d. deutsch. Morgenl. Ges.* i. pp. 355, 356.

tion,[1] particularly in relation to the condition of Jerusalem within our own time.

13. Learned and condensed articles relating to the topography of the city, and to subjects which have recently been brought under discussion, may be found in Winer's *Bibl. Realwörterbuch*, under the heads Jerusalem, Antonia, Zion, Golgotha, Temple, etc.

Among the various works which relate to special parts of Jerusalem, such as the Sepulchre, the Temple, the Tombs, etc., I may specify the following:—

1. C. A. Credner: Nicephori Chronographia brevis. Dissert. Gissæ, 1838. 4to. Reges Tribuum Israel: Hebræorum Pontifices summi; Patriarchæ Hierosolymitani; Episcopi Romani, etc.

2. W. R. Wilde: Narrative of a Voyage to Teneriffe, Palestine, etc. Dublin 1840. Vol. ii. pp. 216-400: particularly the passages relating to the city gates, and the lines of the walls at the time of their rebuilding by Nehemiah.

3. James Ferguson, F.R.A.S.: An Essay on the Ancient Topography of Jerusalem, with restored plans of the temple, etc., and plans, sections, and details of the church built by Constantine the Great over the Holy Sepulchre, now known as the Mosque of Omar. London 1847. With many sketches.

4. Dr Joann. Martin. Augustinus Scholtz: Commentatio de Golgothæ et Sanctissimi D. N. Jesu Christi Sepulcri situ. Bonnæ 1825.

5. By the same: Commentatio de Hierosolymæ singularumque illius partium situ et ambitu. Bonnæ 1835.

6. Dr Justus Olshausen: On the Topography of Ancient Jerusalem. Kiel 1833.

7. Otto Thenius: Jerusalem before the Captivity, and its Temple. Liepsig 1849.

8. Alb. Schaffter, V.D.M.: The True Site of the Holy Sepulchre; a Historico-archæological Inquiry. Bern 1849.

9. George Finlay, K.R.G.: On the Site of the Holy Sepulchre; with a Plan of Jerusalem. London 1847.

[1] *Zeitsch. d. deutsch. Morgenl. Gesell.* iv. p. 277.

10. T. Tobler: Scattered Papers in the Ausland, 1847, 1848.

11. K. von Raumer's Contributions to Biblical Geography. 1843. Jerusalem. Pp. 51-63.

12. I ought not to omit in this list the admirable copperplate engraving, from the hand of Halbreiter, eight feet in length, and giving the city and all the neighbouring places within a radius of forty miles. Munich 1850. This map contains eighty different places distinctly entered, and found by means of a key.[1]

13. And, finally, the whole city and the surrounding country are clearly portrayed upon the Relievo Map of Palestine or the Holy Land, illustrating the sacred Scriptures and the researches of modern travellers. Constructed from recent authorities and MS. documents in the Office of the Board of Ordnance. Embossed. London.

DISCURSION II.

THE SITUATION OF JERUSALEM, AND ITS DIVISION INTO HILLS AND VALLEYS.

1. *The Site of the City.*

The present city of Jerusalem lies[2] upon a broad and high plateau, with gently arching elevations, and connected only on the north side with the wide ridge which runs north and south through Palestine, and which forms the watershed of the country. Between the plateau on which the city stands and the great tract at the north there runs no natural ravine worthy of mention: the only barrier is found in a few slight depressions, which serve as the channel of some unimportant brooks, only filled during the rainy season. On all the other sides the hills rise somewhat higher than the plateau on which it stands. In Ps. cxxv. 2, the writer says of the protection which God gives: "As the mountains are round about Jerusalem, so the Lord is round about His people

---

[1] V. Schubert, *Anz.* in *Allgem. Zeitung*, Nos. 213, 91.
[2] Robinson, *Bib. Research.* ii. pp. 258-260.

from henceforth even for ever." The whole of the table-land from which these eminences project is rocky, and rifted by deep gullies or gorges, through which flow no constant streams, but only the intermittent torrents which carry off the waters which fall so abundantly during the heavy rains. The two great depressions which begin on the north-west side of the city, the valley of the Kedron or of Jehoshaphat on the east, and that of Gihon on the west, although displaying at times the most formidable streams, which foam and eddy and roar, yet are not unfrequently entirely dry, excepting in the places where the rain-water is gathered in receptacles prepared for the purpose. Both of those two gorges, which in their course almost skirt the city, unite at a distance of between two and three miles from the place where they appear: this junction is effected on the south-eastern side of Jerusalem. At the same place, a third but less important ravine sets in from the north—the Tyropœon of the ancients. Coming from the heart of the city, it enters the basin formed by the confluence of the other two, near to the pool of Siloah and the King's Gardens. From that point the united gorge pursues its course eastwardly under the name of the Kedron valley, passing the Convent of St Saba, and after traversing a distance measured by a walk of three or four hours, it enters the basin of the Dead Sea. These deep valleys or ravines, in their passage by the city, completely separate it from the adjacent district on three sides—the east, south, and west. They differ from most other gorges of the same character in this, that from a shallow and unimportant beginning they gradually deepen, till in their later course they have the same bold and precipitous sides, which not uncommonly characterize mountain streams at their source.

The city of Jerusalem, surrounded by such steep precipices on the east, west, and south sides, unquestionably owes its security to them; for they effectually shielded it from attack in those three quarters, leaving it exposed only on the north side, where there was an unbroken connection between the rock on which the place was built, and the great ridge or plateau on the north. This exposed quarter was, however, always guarded by the best fortifications which the people could build. In the middle part of the great wall which cuts Jerusalem off from the country lying on the north, there stands the mighty Damascus gate, whose name shows which way it looks, and which alone was the assailable point when the city was surprised and attacked.

The great rock which extends southward, and which is converted by the gorges which encompass it into a kind of peninsula about two thousand feet in height, must have always been, from the very earliest times, the place which was selected as the site of Jerusalem. This tongue of land suggested itself to the Jebusites as a strong situation, and was long occupied by them. David at last grasped the coveted possession, and on it he built the "city of David." And subsequently the influence of this strong position is unmistakeable; for though the city was repeatedly taken and destroyed, yet new walls on the northern side were constantly put in the place of those which had been swept away. How much more transitory would have been the fate of Jerusalem, if it had been built by the sea, or upon a wide fertile plain, like that of Esdraelon! The Psalmist, filled with a consciousness of the wisdom of God in the arrangement of all human affairs, recognises the deep design displayed in this, when he says (cxxii. 3), "Jerusalem is builded as a city that is compact together;" and also (lxxviii. 68, 69), "but chose the tribe of Judah, the Mount Zion which He loved. And He built His sanctuary like high palaces, like the earth which He hath established for ever."

To this elevated plateau, 2300 or 2400 feet above the level of the Mediterranean, and about a thousand more above the surface of the Dead Sea, the city was always confined: it could not extend itself down into the ravines, it could in no way reach beyond them: the only opportunity which it had of enlarging itself was towards the north. To determine just what the limits of those enlargements were, is the most difficult task which we encounter in studying the ancient topography of Jerusalem; for the walls which were built did not depend at all upon any natural and permanent physical conditions of the place, but solely upon the will of man. The changes effected during the lapse of centuries caused old architectural structures to disappear, and give place to later and newer ones, built often upon the ruins of the former. Meanwhile the outlines of the city remained unchanged age after age: the lapse of even a thousand years had no influence upon it. And although Jerusalem was overlooked on three sides by the high hills which were separated from it by the valley of Jehoshaphat and that of Gihon, yet the lack of fire-arms in former days made it inexpugnable on all sides but one. It was assailable by catapults and the stone-hurling machines only on the north; and, consequently, all the assaults which the city endured, from those of Sennacherib and Nebuchadnezzar down to those of the Turks, were directed against the northern wall. The principal point of observation was from the Mount of Olives, and our next step will be to take a survey of the city from that interesting spot.

2. *The View of Jerusalem and its Environs from the Mount of Olives.*

East of the valley of Jehoshaphat rises the Mount of Olives, its western base formed by a steep, high precipice, but its top becoming far less distinctly marked, sloping away in gentle terraces, forming three rounded and by no means sharply defined peaks, the central one of which is the highest. This one rises about 175 feet higher than the highest part of Jerusalem—the old fortress on Mount Zion—the most southerly projection of the rock on which the city stands. The height of the Mount of Olives is given as 2509 feet by von Wildenbruch, 2551 feet by von Schubert, 2249 feet by Symonds. Von Schubert makes it to stand 3860 Paris feet above the Dead Sea; Symonds, 3479 Paris feet. The middle of the three peaks of the Mount of Olives, the highest, is the one on which the Chapel of the Ascension stands; and it is from that point that Halbreiter's panoramic view was taken. The most northerly peak is called in the legends Viri Galilæi. The whole mountain is usually known

by the Arabic word Jebel et Tur, though Edrisi calls it
Jebel Zeitun. Rising as it does 416 feet above the valley
of the Kedron,[1] or, according to some measurements, as much
as 600 feet,[2] it has the appearance of a hill of much greater
pretensions than it really is.

A tolerably large number of olive trees still adorn the
mountain, and give it its name; and at its western base is
the cluster of very ancient ones, known throughout the world
as those supposed by some travellers to indicate the garden
of Gethsemane. A great portion of the mountain consists of
tilled land; even the highest peak of all admits of being
ploughed, and is covered with a growth of barley; yet a
considerable share of the whole surface of the mountain is
rocky, and incapable of cultivation. The place where the
Chapel of the Ascension stands is held by many to be the
place where the Saviour left the earth, and His footsteps are
still shown to the credulous; but the tradition is in direct contradiction to the statement made in the Gospel, " And He led
them out as far as to Bethany; and He lifted up His hands,
and blessed them. And it came to pass, while He blessed
them, He was parted from them," etc. Bethany lies as far
from the summit of the Mount of Olives as the latter is
from Jerusalem. The chapel on the summit[3] is an unattractive building in the interior, built by the Armenians, in the
place of a former one ascribed to the Empress Helena. The
place is a sacred one not only in the eyes of Christian pilgrims, but also of Mohammedans, although they hold that
Jesus ascended directly from the cross to heaven. It may
be said, that false as is the tradition which connected the top
of the Mount of Olives with the ascension of the Saviour,
it is of a piece with the countless fables which have been
woven around Jerusalem. The names which have been
given by the monks to designate places which they make the
scene of biblical events, may not all be so readily shown to be
destitute of a real basis as that given to the church on the

[1] Von Schubert, *Reise in das Morgenland*, ii. p. 521, Note.
[2] J. Wilson, *Lands of the Bible*, i. pp. 416, 482.
[3] Robinson, *Bib. Research*. i. p. 274, and Note, p. 504.

Mount of Olives, but in such a work as this they must be passed over: they contribute nothing to geography, and must be left to those who are content to read the records of those pilgrims who delight to receive such traditions.

The chapel known as that of the Ascension has, then, only this interest and value to us, that from its roof we have the most commanding view of the city, and of a circle whose radius is from thirty to forty miles in length. The prospect which is afforded from that point, Bartlett considers[1] one of the finest in the world, since the whole city can be seen as in a half bird's eye perspective, while yet it is close at hand, and can be inspected in its smallest details. The hills and valley all around, form, says von Schubert, a fine frame to the picture of the sacred city. Eastward, the eye runs from point to point, till at last it rests upon the basin in which lies the Dead Sea; then receding farther, it fixes upon the mountains of Moab, with Pisgah and Nebo prominent among them, although Robinson does not speak of the chain as exhibiting any marked peaks, but as being a long, unbroken, monotonous ridge. The depressed valley of the Jordan can be traced with the eye from its mouth northward, as far as to the site of Jericho and Mount Quarantania; after reaching those points, it is hid by the intervening mountains. Its course is at once detected between Jericho and the sea, by the line of green which runs along its margin, in the strongest contrast with the barren, desolate shores of the Dead Sea, over which hangs a dense hazy cloud of a peculiar hue, suggesting to Wilson the idea of a caldron of molten lead. Robinson saw the Dead Sea so distinctly, that the glance of the sun on the water was distinctly discernible, and so near apparently, that it seemed to be not more than three or four hours distant. The eye can also follow the whole course of the Kedron valley, and the view is limited at the south by the Engeddi hills and the Frank Mountain.

Directly over the Mount of Olives run three footpaths from the city to the village of Bethany. The most northerly

[1] Bartlett, *Walks*, pp. 101-105. Comp. his *Views of the City*, Plate 10; von Schubert, *Reise*, Pt. ii. pp. 520-522.

one of all passes over the highest peak, and is the one most generally taken: the most southern passes over the depression which lies on the south side of the peak. The mountain then changes its name, and rises again, beyond that depression, bearing the title Mount of Offence. It was the Mons Scandali and Mons Offensionis of the early pilgrims, and draws its name from the idolatrous service which Solomon is said to have celebrated upon it in honour of Chemosh, the Moabite god, at the dictation of some of his foreign wives (1 Kings xi. 7). But Robinson was unable to discover that, before the year 1283, any such opinion was current. Notwithstanding the probability that this was the place, and the expression in the books of Kings, that Solomon paid this worship on one of the mountains *before* Jerusalem, indicating one lying towards the east, Brocardus is the first writer who notices the fact, and it seems to have come into general recognition during the Crusades. The Mount of Offence closes the view on the south. The view of Jerusalem from the summit, Bartlett has taken for the frontispiece of his finely executed work: he gives the city, not as it now is, but as it was before the Destruction, while guarded with its three walls.[1]

The footpath to Bethany, now el-Azariyeh of the Arabs, *i.e.* the village of Lazarus, leads along the eastern slope of the Mount of Olives, through a pleasant tract, here and there displaying open corn-fields and fruit-trees, and passes a cluster of whitewashed houses which lie scattered among dark olive trees, which mark the hither bounds of the solitary Jericho desert. On the right stand the fragments of a tree dating from the middle ages; behind this, on a naked hill, some walls, which, when seen near at hand, prove to be the desolate village of modern Bethany. At the entrance to it, the traveller's attention is called to what is said to be the grave of Lazarus.

The most northern peak of the Mount of Olives, Viri Galilæi, is only a quarter of an hour's walk distant from the

[1] Bartlett, *Walks*, etc., pp. 29, 30. See also Bartlett and Bourne, *Comparative View*, Tab. iv.

Mohammedan wely. Upon the summit Robinson measured a base line for the purpose of accurately surveying the city. One result of these measurements was, that the Church of the Holy Sepulchre was found to be just an English mile distant from that place. On the farther side of that northern peak the Mount of Olives bends westward, completely surrounding the upper part of the Kedron valley, which in that portion is filled with gardens and corn-fields.[1] The blending of the mountain with the great rolling plain north of the city here begins, and, without any great natural break, the sacred eminence of which we have spoken passes into the great field of so many battles. Towards this north side, this rolling plain, the Damascus gate, the chief entrance to the city, looks. From it diverge three roads, all of which run more or less northward, one leading north-westward to Beit Hanina, which lies west of the watershed,[2] and sends its waters to the Mediterranean. Two hours' distance from it rises the peak known as the ancient Mizpeh, but now called the Neby Samwil, and supposed to have been the home of the prophet Samuel. It rises above the uniform mountains of Ephraim as a prominent landmark. The second road runs north-eastwardly along the extreme northern base of the Mount of Olives to Anata, Jeba, and Mishmas (Anathoth, Gebah, and Michmash), following the rocky pass between the sharp cliffs, where Jonathan the son of Saul made his valiant opposition to the Philistines, and where the line ran between the tribes of Ephraim and Benjamin[3] (1 Sam. xiv. 4). The third road runs from the Damascus gate due northward past Tuleil el Fulil (Gibeah, the home of Saul) to el-Bireh (Biroth, the city of Benjamin) and Jisna, and is the main road to Nablus and Damascus. From it the mountains of Samaria can be distinctly descried in the distance. Going away from the gate, it is necessary to cross the slight depression formed by the beginning of the Kedron valley, or that on the other side,[4] which develops later into the Valley of Jehoshaphat, and

[1] Gadow, *Mittheil. i.a.l.* iii. p. 38.
[2] Wilson, *Lands of the Bible*, ii. p. 37; Bartlett, *Walks*, p. 104.
[3] Robinson, *Bib. Research.* i. p. 259.   [4] *Ibid.* p. 275.

then to ascend the first rising ground swelling away northward, from the highest point of which the traveller takes his last view of the famous city which he has just left; and arriving at the village of Schafat, only fifty minutes' distance north of the gate, Jerusalem passes entirely out of sight. The rising ground, which in it most elevated portion is passed in twenty-five minutes after leaving the city, is, according to Robinson, unquestionably the Scopus of Josephus, where Cestus coming from Gabaon, the present el-Gib, and afterwards Titus coming from Gophnah, the present Jisna, pitched their camps. From this point Titus had his first view of the city and its splendid temple. On that same place the ravaging hordes of the Assyrians and Chaldæans made their appearance; after them, of the crusaders and the Moslems; and there lay the scene of many a battle and bloody slaughter.

North of Scopus, and a half-hour's distance behind el-Bireh already named, the eminence in the rear of which is Beitin forms the extreme northern barrier in the line of horizon running from the prominent Neby Samwil in the west, and the yet more distant mountains of Samaria in the east. This Beitin indicates the location of the ancient Bethel, *i.e.* house of God, which was as early as the times of Abraham and Jacob a sacred place (Gen. xii. 8, xxviii. 11-19), which was in the time of Joshua a royal Canaanite city (Josh. xii. 16), and afterwards the place where the ark of the covenant was kept (Judg. xx. 26); and a border city between Benjamin and Ephraim. After Symonds' measurements[1] were completed, Beitin (Bethel) was found to be 1767 Paris feet (1883 Eng. feet), and Neby Samwil 2484 Paris feet (2648 Eng. feet), above the sea: the latter was ascertained to be nearly nine hundred feet the higher of the two, which indeed seemed highly probable. The same measurements determined Neby Samwil, the highest point around Jerusalem, the ancient watch-station Mizpeh, to be two hundred and seventy-five feet higher than the Mount of

[1] Rough sketch of a portion of the triangulation of the Southern Dis. of Syria, examined by F. L. Symonds, Lond. 1849.

Olives (2249 Paris feet, 2397 Eng.), several feet less than von Wildenbruch and von Schubert had made it out to be.

The low situation of Beitin, with its ruins, is concealed from view by the low range of hills lying before it, which forms by no means so distinct a landmark when seen from the Mount of Olives as does the neighbouring Neby Samwil.

The view westward from the Mount of Olives takes in the whole length and breadth of the city of Jerusalem; directly beyond it, in the direction of the Mediterranean Sea, the horizon is bounded by a low flat ridge, which passes by an imperceptible transition into the high plain of Rephaim. I do not find any distinct mention of its name, unless it takes the same with that of the valley which begins in it, and deepens in its southward course, close by the city—the valley of Gihon. This low ridge and the plain of Rephaim form a part of the great watershed line, extending from the Wadi Hanina southward to Bethlehem and Hebron, though the slope of the plain is so far inclined westward that its waters run towards the Mediterranean.[1] As the Mons Scandali, *i.e.* the Mount of Offence, closes the view from the Mount of Olives towards the south-east, the Gihon ridge closes it on the west. On the s.s.w. the lofty plain of Rephaim[2] hems in the prospect; the old name having come down little modified by time, for this was the ancient valley of the Rephaim or giants. The plain was broad enough to afford a fine camping-ground for the Philistines when they came out to attack Jerusalem, and they were repeatedly driven from it by David (2 Sam. v. 18–25). The high plain of Rephaim is a celebrated locality, therefore, and extends far westward into that deep and narrow Wadi Werd, which, when joined with the Beit Hanina, extends as far as to Wadi Surar and Nahr Rubin, and offered to the Philistine armies the most direct and available approach to Jerusalem. On an eminence in the more southern part of this depression can be seen from the Mount of Olives the Greek Convent of Elias (Deir Mar Elyas), and in the rear the hill on which stands Bethlehem,

---
[1] H. Gadow, *Mittheil. i.a.l.* iii. p. 37.
[2] Robinson, *Bib. Research.* i. pp. 219, 220, 276.

while the mountains of Hebron are seen in the distance, forming the horizon. Northward the plain of Rephaim is seen, over which runs the aqueduct of Solomon and the road from Bethlehem, which to-day seems to be scarcely less rich than it is painted by Isaiah (xvii. 5), covered as it is with fields of wheat.[1] Here the ruins of former houses are seen, some of them extending very close to the walls of Jerusalem; between them, however, comes the Valley of Hinnom, whose southern border is formed by an unimportant rocky ridge—the northern hem, so to speak, of the plain of Rephaim. This slight elevation, which is directly opposite to Mount Zion, and a little east of the usual road to Bethlehem, bears the name of the Mountain of Evil Counsel. Its base rises steeply from the Valley of Hinnom for twenty or thirty feet, and then displays shelving, rocky sides, in which are several excavations which have served as graves. Higher still it is less steep, and at the top becomes quite flat, and merges itself gradually at the south-east in the plains of Rephaim. South of the Mount of Evil Counsel there begins a small wadi, running eastward parallel with the Valley of Hinnom, but only half so deep, and enters the lower end of the Valley of Jehoshaphat. The eastern face of the mount, where it touches the Kedron valley, is equally high, but not quite so bold as that which is opposite to Mount Zion. The shattered houses which are scattered over this eminence seem to betoken the existence of a former village there: among them seems to be the ruins of a Mohammedan wely, as well as of an ancient Christian church and convent. The village was standing two hundred years ago, and was mentioned by Cotovicus and Doubdan. The monks have asserted since the fifteenth century that this mountain was the one where the house of Caiaphas stood, in which the Jewish priests and scribes took counsel with Judas Iscariot regarding the betrayal of the Saviour (Matt. xxvi. 3, 4; John xi. 47); and from this legend springs the name now generally applied to the eminence.

[1] Gesenius, *Commentar. zu Isaias*, i. p. 559.

DISCURSION III.

THE CIRCUIT OF THE PRESENT WALLS, AND THE LOCALITIES JUST OUTSIDE OF THEM.

We now turn back to the broad and high plateau which runs down between the Kedron and Gihon valleys in the form of a bold tongue of land, and which is in great part covered with the edifices which constitute the city of Jerusalem. We will follow the line of walls which marks the outer side of this tongue of land, beginning with its north-eastern portion, where one may look down into the upper portion of the Kedron valley. From that point the temple mountain (Moriah) reaches as far as to where it falls away with terraced sides to the Valley of Hinnom. Parallel with the deepening Kedron or Jehoshaphat vale the walls run southward, forming at once the eastern barrier of the city and that of the temple area, as appears decisively[1] in Neh. iii. 30–32. Always following the high margin of the rock, it runs southward till it reaches the cross valley of Hinnom, running from east to west, when it too makes a sharp turn and runs westward to the Mount of David, the renowned Zion. Thus the southern face of the great promontory on which the city stands is alike ornamented and defended by this strong defence. At the abrupt steep of Zion the wall turns northward, and follows the line of the Gihon valley as far as to the north-west corner of the city. Thus far we are on sure ground; for the natural character of the place is such, that as it is now it must always have been: the rock has not changed, and there is no opportunity there for hypothesis to gain a foothold. It would have been far different, if, beginning at the north-west corner, we had sought to trace our course across to the north-eastern one. In order to understand the difficulties connected with this subject, however, and to interpret the description of Josephus, and what he has said upon the direction of the walls towards the four cardinal points, it will be necessary to enter upon that careful study of the physical character of

[1] Krafft, *Topographie,* pp. 54, 100, 155.

the city, and the natural inequalities in its surface, in which Gadow[1] has so carefully led the way.

1. *The Eastern Wall of the City and Temple from the north-eastern corner and the Stephen's Gate to the Mosque el Aksa.*

At the north-east corner of the present wall of the city, now shielding mere fields and a hill scantily covered with houses (the Bezetha of Josephus), belonging to the Mohammedan quarter of the city, lie many ruins and architectural fragments hard to identify, but originating, in the opinion of Gadow,[2] at the time of the Crusades. At all events, the place is the same which Josephus states was the location of the great " corner tower," in the neighbourhood of the fuller's field (Isa. vii. 3). Near this place, according to Krafft, must have been the encampment of the Assyrians. A ditch hollowed out from the rock accompanies this northern wall, running eastward; part of it is walled over, but is in a state of decay: it bears the name Birket el Haj,[3] or Pilgrim's Pool. The ditch and the wall seem, from their parallel courses, to have the same antiquity, and run side by side till they reach the valley of Jehoshaphat, where they turn toward the south. The ditch continues its course near the wall for some distance, but at length ceases in the neighbourhood of the Birket Hummam Sitti Marjam, *i.e.* Pool of the Bath of the Virgin Mary. This ditch is to be seen in Robinson's map, but its southern extremity is wrongly termed the Birket el Hejjeh: it should have been given to that portion of it which lies near the northern wall, the name Hejjeh being probably another form for Haj. The pool by the side of the eastern wall, and not far from St Stephen's Gate, is about four times the larger of the two, but it is generally dry. Its name is derived from the fact that its waters were generally

---

[1] Von Raumer, *Pal.* 312, Note 209; Schultz, *Jerusalem,* p. 57; Krafft, *Topographie,* p. 19.

[2] Gadow, *Mitth.* iii. p. 401; comp. Krafft, *Topogr.* pp. 47, 118.

[3] Schultz, *Jerusalem,* p. 37; Krafft, *Topog.* p. 47; Roberts, *The Holy Land,* Book ii.

used to supply a bath within the city bearing the name of the Virgin Mary. Outside of the gate, east of the Kedron, stands an edifice which is called the Church of the Sepulchre of Mary.

The high eastern rim of Mount Moriah, throughout its entire extent from north to south, is reduced to very small proportions by the nearness of the wall to the edge. Room is left merely for footpaths, and every inch of available space is occupied by Mohammedan graves and those of the crusaders. The place being surrounded by the high terrace of the Haram, the ground becomes so holy, that on the day of resurrection and the last judgment, to be held in the valley of Jehoshaphat, special security is obtained [1] by being buried there. Numerous white gravestones of Jews cover the steep slope of the hill as far as to the very bottom; for they too hold that this place is to be the scene of the final judgment, quoting in support of their belief various passages in the third chapter of Joel. In accordance with this view, they interpret the name Jehoshaphat itself, "The Lord judges."

The name Haram, or Haram es Scherif, is applied by the Mohammedans to the place, inaccessible to unbelievers, in whose midst stands the present Mosque of Omar (Kubbet es Sakhah, *i.e.* Dome of the Rock), which occupies the site of the former temple of Solomon [2] on Mount Moriah. Although in its special form—an oblong quadrilateral—the place has suffered many minor changes, yet the description which Josephus gave of the situation of the various parts of the temple can now be understood and verified. The Gate of St Stephen, near which a legend dating from the fourteenth century asserts that the martyrdom was accomplished, has thus derived its modern name.[3] The account

---

[1] Von Schubert, *Reise*, ii. pp. 524, 525; Bartlett, *Walks*, p. 17; Strauss, *Sinai und Golgotha*, p. 269.

[2] Robinson, *Bib. Research*. i. p. 281; Schultz, *Jerusalem*, p. 32; Catherwood, in Bartlett, *Walks*, pp. 143, 148–168; Krafft, *Topog*. pp. 69, 100, etc.

[3] Robinson, *Bib. Research*. ii. p. 262; Schultz, *Jerusalem*, p. 70; Krafft, *Topog*. p. 149.

in Acts vii. 58 is, that "they cast him out of the city, and stoned him." From the time of Arculfus to that of De Suchem, the scene was supposed to be on the north side of the city, and in front of the Damascus gate. From the position of St Stephen's Gate on the north-east side of the Haram, and on the way leading by the Church of Mary's Sepulchre to the Mount of Olives, it is sometimes called by Christians the Bab Sitti Marjam. The fact that, on the outside, four lions are cut in the stone above the roadway, makes it evident that the gate is not of Mohammedan construction, and it probably dates back to the time of the Crusades. At the north-eastern corner of the Haram there are still to be seen colossal remains of a tower, apparently of very ancient origin, and connected with the Birket Israin by a little gate. Krafft holds this to be the "sheep gate" mentioned in Neh. iii. 1, with which he begins his description of the re-erection of the walls on the west side. The name was derived in Nehemiah's time undoubtedly from the sheep-market, into which the creatures were driven from the east, as is still the custom among the Beduins. Robinson looked for the gate thus designated by Nehemiah farther southward.[1] The gate known by the name of St Stephen is sometimes called Bab es Subab,[2] *i.e.* the Gate of the Tribes, and is the only one open towards the east: it is therefore always taken by the pilgrims when on their way to Jericho. The Golden Gate (Porta Aurea of the Crusaders), called by the Arabs in former times Bab er Rachmeh, the Gate of Grace, and now Bab el Daheriyyeh, Gate of Eternity, is no longer opened; but since the time of Omar it has been walled up, in order to prevent access on the eastern side. As it has a depth of seventy feet, its interior[3] was changed into a little mosque. If opened, it would lead to the interior of the temple area, on which account it was walled up by the Saracens, there being a Mohammedan tradition that a new king shall one day pass

---

[1] Robinson, *Bib. Research.* i. p. 277.
[2] Krafft, *Topographie,* p. 48.
[3] Bartlett, *Walks*, p. 17; Wolff, *Reise*, p. 48; Williams, *Holy City*, ii. pp. 313, 355, 358, Note 3; Roberts, *The Holy Land*, Book iii.

through that gate to take command of all the earth. This unquestionably is based upon the Christian idea of the coming of Christ's kingdom, and is the reason probably why sentinels are always posted on the inner side of this gate. At the time of the Crusades it was only opened once a year—on Palm Sunday—to celebrate the entrance of Messiah into Jerusalem, which it was thought had been through this gate. (Matt. xxi. 8; John xii. 13).

Externally the gate displays a double arch of Roman architecture, recognised as such by Pococke, Robinson,[1] and others, and ascribed by them to the Emperor Hadrian. They suppose that it was built at the same time that he erected on the site of Jerusalem the Ælia Hadriana, and replaced the ancient Jewish temple by one dedicated to Jupiter. Jerome states that in his time, A.D. 400, he saw a statue of the god near the equestrian figure of Hadrian (probably the two statues of which the *Itinerar. Hierosol. ad ann.* 333 says: "sunt ibi et statuæ duæ Adriani."[2]). They stood on the same side of the area, where, a little farther to the south, Hadrian's palace was erected, which Jerome also asserts that he had seen. The conjecture of Pococke was confirmed by Krafft's discovery[3] of a Latin inscription with Hadrian's name over the southern wall. The finished style of the carved work of the east gate, as well as that found in a now inaccessible southern gate, near the temple terrace and below the Mosque el Aksa, seems to have the same antiquity, and to date from the same emperor's era.

The inscription, whose first letter is imperfect, has now been made out in full by Schultz, and runs, as we learn from Tuch's *Oriental Journal*, TITO. AEL. HADRIANO. Although, as Tuch[4] remarks, it does not refer to Hadrian the founder of the Ælia Capitolina, but to his successor, Tit. Ael. Hadr. Antoninus Pius, in whose honour it was set up by the ruling governor, yet it remains a lasting monument of

[1] Robinson, *Bib. Research.* i. pp. 296, 297.
[2] *Itin. Anton. Aug.* ed. Parthey et Pinder, p. 279.
[3] Krafft, *Topog.* pp. 40, 73, etc.
[4] *Zeitsch. der Deutsch. Morgenl. Ges.* Pt. iv. pp. 253, 395.

the fact, that at the restoration of the city on the east and south sides as well as on the north, the effort was made to restore the wall of Agrippa, which had been destroyed by Titus, unchanged in its general direction. This Roman restoration has continued, according to Robinson, Schultz, and Krafft, unchanged in its main character throughout the middle ages up to the present day; and even the southern portion of the city, which was shut out by Hadrian's wall, remains shut out by the south wall until now.

Catherwood,[1] who was permitted to make the most minute investigations regarding the exterior of the Golden Gate, as well as of its interior, confirms the Roman character of its external double arch with its capitals, but leaves it undetermined whether the interior wall, eleven feet in thickness, though a walk of columns passes up to the formerly accessible temple terrace, is a remnant of the ancient Jewish temple and city wall, or the work of Hadrian's time. The situation of this gate now walled up may have lain tolerably near the middle of the chief (east) entrance to Solomon's temple, as it does at present in relation to the mosque. Yet, as Robinson remarked,[2] it lay a little northward of the eastern entrance to the temple; and Gadow's very accurate observations place the inner side of this now half-destroyed gateway rather on the north side of the little plateau surrounding the Mosque of Omar,—a position which is confirmed by Symonds' map of the city. It is possible that that situation may exactly indicate the site of the ancient east gate, which was provided with a vestibule, and regarded with reverence as the chief portal. The watchman of this gate, mentioned by Nehemiah (iii. 29) by name, held, according to 2 Chron. xxxi. 14, a more honourable position than the other Levites who tended the gates. The most sacred gifts were entrusted to him; and, indeed, the ancient temple gate on the east side was covered with gold, from which circumstance it is probable that

[1] Bartlett, *Walks*, pp. 158-160, Note. Comp. Roberts, *The Holy Land*, Book iii.; Ferguson, *Essay*, p. 94.
[2] Robinson, *Bib. Research.* i. p. 284; Gadow, in *Zeitsch. i.a.L* iii. p. 45.

CIRCUIT OF THE PRESENT WALLS.

the name Porta Aurea, used by the crusaders, takes its origin.[1]

The nearer this portion of the eastern wall of the city comes to the south-eastern corner of the Haram or temple area, the nearer does it approach the steep declivity of the Valley of Jehoshaphat. It is here that the oldest and most colossal stones are found, at the extreme south-eastern corner: the distance of the wall from the verge is scarcely ten steps. It is here that Gadow has, in the course of his very accurate investigations, traced the remains of an arch which he compares with that discovered by Robinson, and entered by him upon his map under the title of the Ancient Bridge. According to his view, the latter belonged to that row of mighty arches[2] which once spanned the Tyropœon, the hollow between Mount Moriah and Mount Zion. Bartlett has devoted two views to this bridge, constructing it theoretically in its ancient shape and proportions. The remains of the arch on the east side of the Haram is, according to Gadow's measurement, less colossal than that on the west side, yet still of no inconsiderable size.[3] The two upright stones are about eleven feet, the transverse ones about sixteen feet in length. The great depth of the Kedron valley—at least a hundred and fifty feet—makes it entirely impossible that there should ever have been a bridge spanning the valley;[4] and it is probable that this arch formed the support for a flight of steps leading to the fountains and gardens below.

At the same corner Gadow discovered the remains of an ancient cistern: the traces of the cement cling to the oldest stones of the wall, and may be seen nine feet above the present pile of rubbish. The discovery was first made in consequence of excavation. It, as well as the fragment of the arch above, entirely escaped the eye of the English surveyors. This south-east corner of the temple area, which forms at the

---

[1] Krafft, *Topographie*, p. 155.
[2] See copy of this in Bartlett, *Walks*, p. 135; also frontispiece in Bartlett and Bourne, *Ancient and Modern Jerusalem*.
[3] Robinson, i. pp. 219, 286-289.
[4] Robinson, *Bib. Research*. i. pp. 232, 271.

same time the south-east corner of the city wall, belongs to the most remarkable and the most ancient remains of Jerusalem, even if I do not go further, and say, of all existing architectural monuments.

This wall is made up in its lowest layers of very large hewn stones, which, according to Robinson,[1] were placed there as far back as in the time of Solomon. The upper portion of the wall is indeed of modern construction; while the colossal blocks—which, however, are not uniformly found in the lowest layer—seem in many cases to lie below the rubbish, but in this south-east corner lie particularly exposed, and are almost universally supposed to form a part of the primitive foundation. The appearance is as if in every part of the wall an original structure lies beneath, and that in later times new walls have been superimposed upon the ruins of the more ancient ones. The line of separation between the old and the new is therefore always discernible, but not always regular. The ancient blocks are found in some cases much higher up than in others; here and there the rents are filled in with rude masonry; sometimes the whole wall is modern, and the base of it hidden by the rubbish. Robinson found on the east as well as upon the south side of this wall, squares of from seventeen to nineteen feet in length, three and four feet in width, and one even seven and a half feet high. Near the St Stephen's Gate he found one stone twenty-four feet long, six feet broad, and three feet high. Nor is it alone the colossal character of single stones in this wall which excites the spectator's astonishment: the extent of the structure is not less remarkable, extending as it does along the temple area a distance of nearly fourteen hundred German feet, which, when we add the parts which have been filled in during modern times, coincides with the measurement of this side of the temple wall given by Josephus.

The south wall of the Haram runs westward from the high south-east corner, and in its middle part is somewhat concealed from view by the adjacent Mosque el Aksa, which

[1] Robinson, *Bib. Research.* i. pp. 232, 237, 238.

runs out towards the south. Its whole lineal extent can be traced, however, to the south-west corner, although this lies within the city. From this point, the wall which forms the western boundary of the temple area takes its regular course northward; but its west side is not to be plainly traced, since it is concealed for the most of the way by the buildings which are erected against it.

The south wall of the Haram, Gadow found on measurement to be eight hundred and sixty feet in length. It extends across the entire breadth of the temple area, but it does not seem to correspond so closely with the proportions assigned by Josephus[1] as does the length of the eastern wall. Yet it must be said that here, as in all other cases, it is very difficult to ascertain the exact distance implied in ancient measurements. Gadow remarks that, despite this want of concord in the two accounts, the theory is untenable that a part of the wall is of modern construction, and has been joined to the old portion; for there is a uniformity in the style throughout the entire structure which disproves such a theory. The stones are all of them three feet long and three feet high. On the west side, a few feet from the s.w. corner, there are to be found the remains of the very oldest portion of all; colossal stones[2] in close connection with the fragment of an arch already alluded to, which Robinson discovered, and which he supposed to be a part of a bridge once spanning the Tyropœon and leading to the Xystus. The lower stone in the south-west corner of the wall has, according to Gadow's measurements, the great length of twenty-nine and a half feet. In a direct line with this is the Jews' Wailing Place, whose colossal walls have been made familiar to readers by manifold descriptions and drawings. In consequence of their peculiar construction, with bevelled edges and polished surface, they resemble in their appearance the colossal blocks incorporated in the east wall of the Haram; but they are entirely unlike any Roman or Saracenic remains, and probably date back as far as to the age of Solomon. I have already referred to the occurrence

[1] Robinson, *Bib. Research.* i. pp. 291, 292.
[2] Bartlett, *Walks*, p. 140, Tab. xix.; comp. Krafft, *Topographie*, 113.

of similar architectural forms in the walls of Hebron which surround the grave of Abraham. These at Jerusalem are the stones which were set there in the reign of Solomon or his successor, and are referred to by Josephus as " fixed for all coming time." On this west side, near the so-called southwest corner, a piece seemed to Robinson,[1] at his first visit, to have been detached from its place, and to be threatening to fall. On repeated visits, he discovered that, although the stones projected in such a way as to indicate that the wall was rent and might shortly fall, yet that it was built so, and that the blocks were in their natural and primitive position. The outer surface, he discovered, formed a regular curve, and the arrangement of the stones was such as to form the commencement of an arch. This, he conjectured, formed a part of a bridge which once spanned the Tyropœon, and led to Mount Zion. This monument was a proof, in the opinion of the discoverer, of the antiquity of the whole wall from whose side it springs. He also supposed that it was the γέφυρα of Josephus which led from the court of the temple to the Ξυστός, *i.e.* the open terrace before the Asmonæan Palace, and which connected Moriah and Zion. It served as an avenue of escape for the remnant of the Jewish defenders of the city after being driven from the temple; and after they had passed over it to Mount Zion, they destroyed the bridge, cutting off the approach of the Romans, and enabling them for a short time to prolong their defence. It is indeed singular that the existence of such a structure should have been so speedily forgotten; but that is no more so, than that it should have been so easily destroyed by the retreating Jews.

The approximate length of this ancient Gephyra, according to Robinson's measurement of the breadth of the Tyropœon, was three hundred and fifty feet. The traces of the arch, according to the same authority, are plainly discernible, and extend fifty-one feet along the wall. The stones may be seen in three layers, occupying their original position. Each one is above five feet four inches thick. Some of them are very

[1] Robinson, *Bib. Research.* i. pp. 237, 286. [See also *Later Researches.*]

long, one of them being over twenty feet, and another over twenty-four. Robinson's very careful measurements are fully confirmed by his successors; yet Wolff,[1] in his travels, adds a remark which, if it should be confirmed, would give an entirely different character and import to this arch. It is evident that it would have required many similar ones to have sustained a bridge across the Tyropœon, but Robinson searched in vain on the slope of Mount Zion opposite for any trace of a similar abutment; yet this part may have been covered with the accumulations of centuries. There has been no lack of objections to Robinson's hypothesis, founded mainly upon the various views regarding the course of the Tyropœon, and the arrangement of the interior of the city. Wilson[2] propounds the conjecture that this fragment of an arch has some connection with the subterranean fragments of the Mosque el Aksa.

Williams, Schultz, and Krafft supposed the location of the ancient Xystus to have been farther north, and regard the Gephyra of Josephus not as a bridge, but an earth wall like that which now passes the house of the cadi,[3] whose position corresponds to the ancient location of the terrace connected with the Xystus, and which to-day serves as a means of communication between Mount Zion and the Haram terrace. It forms the present street of David, and is in part built upon rubbish collected in the Tyropœon. The colossal arch lies altogether too far down the valley, says Krafft, to have formed a part of a connection between the two opposite mountain slopes. The steep declivity on the eastern side of Zion presents one face at least thirty feet in height; and the end of the bridge must have rested on a site even higher than that, in order to give free access to the hill. Krafft supposes it much more probable that this arch served as the foundation for a flight of steps leading down to the bottom of the declivity, and that there was a similar flight on the opposite side. The accounts of Jeremiah and Josephus hint not obscurely at the existence

[1] Wolff, *Reise in das Gelobte Land*, p. 67.
[2] Wilson, *Lands of the Bible*, i. p. 468.
[3] Schultz, *Jerusalem*, p. 28; Krafft, *Topogr.* pp. 15, 60-62, 94, etc.

of several stone stairways, and they are found even now in various places on the west side of the Haram. Their existence appears to afford a key to the character of the arch discovered by Gadow, outside of the eastern wall, and already referred to. Tobler made an interesting discovery, in 1846, of some subterranean vaults near the Mckhemeh, or Court of Justice, which he supposed to have some connection with the arch discovered by Robinson. Should that discovery be fully confirmed, it will do much to settle this disputed question. Williams agrees with Wilson in considering the arch the work of Saracens, and as dating back to a period subsequent to the age of Justinian. Still, in spite of all objections which have been made to Robinson's theory, and into which I cannot go, von Raumer, in his latest edition, adheres to the view originally promulgated by the American discoverer. In this Bartlett coincides. The grounds on which von Raumer bases his judgment, regarding this as well as many other disputed points, are fully detailed in the special discussion on Jerusalem, in his classic work,[1] to which I will briefly refer the reader, instead of going over the ground myself. It is evident that there will long be differences of opinion regarding a locality so little known as Jerusalem, just as there have been in regard to a city much nearer us—the great Italian capital. Where the most acute and learned men have failed to come to a unanimity of mind on the spot itself, those of us who must follow the labyrinth by means of only secondary helps may not claim an authoritative power of decision. For my own part, I assume no umpire's place, excepting in matters of a purely geographical character, and do not attempt to pronounce authoritatively upon matters of which only they who have studied the history and archæology of the entire subject can be competent judges. I wish to express my indebtedness to the brief though admirable work of one of the latest travellers in the Holy Land—Philip Wolff. From him I have gathered much of what will follow; and to him also I am largely indebted for the method of treatment. I should not omit to state, that in his impartial account he has in-

[1] Von Raumer, *Pal.* pp. 251–321, 393, etc.

corporated several new and important facts; and, in particular, a valuable description of the temple wall and its two southern corners.[1] Using his work somewhat exhaustively, I shall be compelled to pass more hastily than I could wish over the larger and more comprehensive works of those who have entered deeply into the discussion of the antiquities of Jerusalem.

2. *The Southern Wall of the City from the Mosque el Aksa to the Zion Gate, and the south-west corner of Mount Zion.*

Near the Mosque el Aksa, which breaks the southern wall of the temple area near the middle point, the present city wall leaves that of the temple, and runs for a short distance directly southward, before turning in a direct angle westward, towards the Tyropœon and Mount Zion. This mosque, which is at the most southern part of the temple area, seems to derive its name el-Aksa, *i.e.* the outermost, from the fact that, of the three especially hallowed mosques—namely, those of Mecca, Medina, and Jerusalem—it is the most northerly one, and therefore the one farthest away. It does not lie exactly at the middle of the southern wall of the temple area, but is three hundred and thirty-seven feet west of the south-east corner. East of the city wall at this point, the direct descent to the Kedron valley is a hundred and fifty feet, and the ascent to the highest point within the temple area is a hundred, making an entire altitude of two hundred and fifty feet, and giving that corner of the city the imposing aspect for which it is celebrated. The south wall of the city, measured on the top of Ophel, is sixty feet high; and the corner which it makes with the southern wall of the temple area forms a tolerably square tract, which serves as the garden of the mosque.[2] This place was once evidently covered deeply with rubbish; for the surface of the ground is fifty feet lower inside of the wall than it is on the outside.

In the innermost corner of this angle stands the mosque

[1] Ph. Wolff, *Reise in das Gelobte Land*, pp. 64–68.
[2] Robinson, *Bib. Research.* i. p. 285.

of el-Aksa,[1] already named, which, with its neighbouring buildings, was first portrayed in Symonds' plan, and described by Catherwood, as the interior was then accessible. The main structure seems to owe its existence to Justinian, while much of the adjoining architecture is of Mohammedan origin.[2] It is two hundred and eighty feet in length from north to south; it has a fine dome over the central part of the nave, and has on each side three minor naves. The breadth of the whole is one hundred and eighty feet, and the roof is supported by thirty or forty pillars of various materials, partly of Saracenic, partly of Roman workmanship. On the west side there is a second mosque, two hundred feet in length, called Abu Bekr; and connected with this, still another one running northward, the Mosque of Moghrebin, *i.e.* of the Africans. This is utterly unimportant, and has probably been erected since the end of the fifteenth century. These structures are mainly interesting from the fact that they are supported by massive and extensive subterranean arches, by means of which the surface of the whole enclosure, formerly a shelving one, is levelled up on the south, as we have it to-day. From the north façade of the Mosque el Aksa, built in fine Norman Gothic style, formerly inaccessible to Christians, and only described by Ali Bey from hearsay, a double vault conducts us out of the inner area of the Haram under the mosque, and ends at a double gate in the southern wall of the city, which is adorned with Corinthian columns, but is at present walled up. It is considered to have been the gate to the temple of Jupiter erected by Hadrian, and that the ascent had to be made through this passage. On both sides, east and west, these mighty subterranean structures continue, extending with their countless pillars and arches for hundreds of feet, and perhaps underlying the whole breadth of the southern area; for there are passages leading down in the western mosque, that of the Moghrebin, near the south-

---

[1] Ferguson, *Essay*, Plate ii.: Interior of Mosque el Aksa.
[2] Comp. Catherwood, in Bartlett, *Walks*, pp. 155-163; Ferguson, *Essay*, p. 139; Williams, *Holy City*, ii. pp. 301-313; Ferguson, Tab. iv. and v.

west corner, and also near the opposite or south-east corner. The Turks hold these entrances to have been parts of Solomon's temple. In the middle ages these subterranean vaults were known; and Felix Fabri,[1] who obtained access to them through the shattered wall of the city, thought them to be the stables of Solomon, and says that six hundred horses could have been kept in them; a comparison which may have arisen from the fact that the Knights Templar, who lived on this side of the city, may have appropriated them to this purpose. The whole of these wonderful subterranean passages have been in a measure examined and sketched by Catherwood. After exploring the whole of the area thus filled, he ascertained that the vaults and arches are supported by fifteen rows of square pillars, underlying the whole southern part of the temple enclosure. In many cases, the roots of the cypresses and olive trees growing above have pierced through, and may be seen overhead. The subterranean passages extend three hundred feet towards the west, one to two hundred feet northward over an uneven bottom, and the pillars vary from ten to twenty-five feet in height. Some of these are four and a half feet in diameter, very well hewn, often made of large bevelled stones, and executed in the Roman style. The whole place is only accessible through some holes in the outer wall, and therefore so faintly lighted that the study of its architecture is very incompletely effected; and it will be difficult to pronounce with any degree of certainty upon the age in which these works were made, and in what style, till a more perfect exploration shall be undertaken. At present there is the greatest diversity of opinions regarding the time when they were constructed, and the nationality of the builders. It is known from Josephus that the southern side of the temple area had gates which opened to the interior, and that there stood a princely hall which Herod the Great built close to the south wall, at the same time that he enlarged the temple area.[2] Josephus[3] says of this palace of Herod, that it was the "most remark-

---

[1] Robinson, *Bib. Research.* i. p. 302.
[2] Wilson, *Lands of the Bible*, i. pp. 468–472.
[3] Krafft, *Topogr.* pp. 62, 73, 74.

able work that ever the sun shone on," magnificent and
luxurious, like all that monarch's buildings. It seems to have
been a triple hall, extending from the Kedron valley to the
Tyropœon, and displayed four rows of pillars, of which those
at the sides were lower than those in the middle, like those
to be seen in the Mosque el Aksa. It may be possible that
Herod made use of older arches and vaults which, as Josephus suggests, were in connection with the temple of Solomon,
and were of use in constructing his splendid work, which
stood so high that it made those giddy who looked down into
the Kedron valley. It may have formed a part of his plan
in erecting that building, to enlarge the temple enclosure on
the south, and level it.

After the destruction effected by Titus, Hadrian erected
his new edifices on the southern side of the temple enclosure;
especially a large gate, probably the same one which Catherwood observed from the inner side of the area, and which was
traced on the outside by Wolcott.

Shortly before the middle of the sixth century, the Emperor
Justinian[1] erected a magnificent church in Jerusalem, in
honour of the Virgin Mary. The description of it given by
Procopius borders on the fabulous. Only at the southern end
of the temple enclosure can its situation be determined; and
after the invasion of the Arabs it was transformed into the
building which afterwards became, with supplementary additions, the Mosque el Aksa.[2] At the period of the Crusades,
this structure, with the numerous ones adjoining it, was
reckoned as the portico of Solomon's temple. It was afterwards called the Palatium, and was the first residence of the
Christian kings of Jerusalem. It afterwards became the
dwelling and guard-house of the Knights Templar; but the
historians and pilgrims of that time have left no clear description of it. The difficulties of investigating and of judging
are here very formidable, and many points can be regarded as
by no means settled.

[1] Robinson, *Bib. Research.* i. pp. 296–300, 384; Wolcott, *Bib. Sacra*,
i. 1843, pp. 17, 18; Rödiger, *Rec. in Allg. Lit. Z.* 1842, No. 110.

[2] Roberts, *The Holy Land*, Book iii.

Among the latter may be reckoned, by way of illustration, a subterranean passage,[1] beginning close to the eastern declivity of Ophel, and terminating very near Mary's Fountain, on the western side of the city. It is built of great hewn stones, and seems to have been an ancient sewer, coming from the city in the direction of the temple enclosure. It has been brought to light in recent times by being used by the fellahs and Beduins of the neighbourhood as a means of access to the city. According to Schultz, it runs as far as the western wall of the Haram; but precisely where it terminates is yet unknown. Gadow has located its mouth in his manuscript map of the city. Tobler, who has located it a little farther south-west of Mary's Fountain, is probably the first who has been bold enough to explore this passage. He passed six hundred and twenty-two feet up its length, as far as the neighbourhood of the pillars which run around the temple enclosure, but the masses of rubbish prevented his further progress. Another passage begins a hundred paces (according to Robinson three hundred) south of the Dung Gate, and was traced by Tobler for a considerable distance far within the wall, and near the west end of the temple bridge, in the quarter el-Mugharibeh. This subterranean canal was also used in 1834 by the fellahin to enter the city secretly for purposes of robbery.

These and similar hidden passages leading out of the ancient city are unquestionably the ones which Titus sought to wall up at the time of the great siege. We learn from Dio Cassius, that before he attempted to cut off those channels, the Jews were in the habit of making frequent sallies, particularly in the neighbourhood of the fountain of Siloah. When the Romans took the city, they found many bodies lying in those passages, the remains of those who had probably been smothered there. It was from such a subterranean asylum that Simon,[2] with several dependants and stone hewers whom he wished to use as helpers in his plan, made

---

[1] Schultz, *Jerusalem*, p. 41; Tobler, in *Ausland*, Jan. 22, 1848, No. 19, p. 74; Robinson, *Bib. Research*. i. p. 265.

[2] Krafft, *Topographie*, pp. 83, 84.

his exit into the temple enclosure. He had remained concealed till provisions had entirely failed, and then, rendered desperate by extremity, he wrapped himself in a purple mantle, attempting to terrify the guard, and so escape. He was taken, however, and carried to Rome to grace the conqueror's triumph.

The present southern wall of the city, which on the side of the el-Aksa mosque is wholly modern, does not allow what is beyond to be seen at all, viewed from the outside, so entirely are the traces of the southern gate of the temple enclosure [1] concealed. Yet Gadow claims to have traced, on both sides of a building touching the el-Aksa mosque, the remains of an ancient arch, observed by no previous traveller. It appeared to be composed of ancient fragments, since the portion of the arch inside of the wall displayed finer workmanship and a different style of ornament than that outside of the wall. If these fragments were the decorations of an ancient gate, it must have led directly into those lower rooms of el-Aksa which have been considered by some as its mere foundations. A window ten feet in height allowed Gadow to discover, still beyond, a cruciform Basilica, with columns and low vaults, which he supposed must have been the unaltered lower storey of the church of Justinian. Before the erection of the external city wall—that is, before 1536—there must have been free access afforded by the open arches and the confused masses of fragments; and that this was the case is evident from the language of pilgrims, Felix Fabri among the rest. The high point of land extending southward [2] beyond the wall, and considered now to have been the ancient Ophel, falls away in gentle terraces, the upper ones of which embrace the entire breadth. This projection, whose southern extremity has been so much modified by art, is by no means so bold an object as its neighbour Mount Zion; and yet it rises to a considerable height above the lowest part of the Kedron valley, and gives one a conception of the depth of that valley. The eastern declivity of Ophel is steep and

---

[1] Robinson, *Bib. Research.* i. p. 262; Gadow, MS. plan.
[2] Gadow, in *Zeitsch. i.a.l.* iii. p. 40.

hard to climb,[1] if one tries to ascend it from Mary's Fountain lying at its base. The bottom of the Tyropœon, on the west side of Ophel, lies higher than those of the Kedron and Hinnom valleys in their lower course. It falls away in a gradual succession of terraces, planted with olive trees, like Mount Moriah itself; and at its point of junction with the valleys last named, it does not glide into them by an insensible transition, but displays a terrace of considerable elevation even there. The eastern wall of the Tyropœon is not so steep as that on the west leading up to Mount Zion. Gadow, after the most careful search, was unable to detect any traces of ancient walls crossing these terraces, and leading from Ophel[2] to Zion. The Ophel ridge, it will be seen by the reader, parts the Kedron and Tyropœon valleys. Its highest portion might be covered with houses, and cultivated; its steeper slopes remain mostly naked, or sustain scattered groups of olive trees, between which ascend steep pathways. At the south-east corner of the city wall, Ophel was a hundred feet lower, according to Robinson's measurement, than the upper wall of the temple enclosure; while the end of this rocky tongue over the Pool of Siloah is about forty or fifty feet above the water in it. The extent of Ophel from east to west—that is from one edge to the other—is about three hundred feet. There is not a single building upon the whole of this tongue of land, although, as we learn from Josephus, its entire extent as far as Siloah was formerly reckoned as belonging to the city. When the new wall was built by Jeremiah, it extended from Zion across the Tyropœon as far as to the Pool of Siloah, which furnished water to the King's Garden; from that point it turned northward, taking an ascending course as far as to the wall Ophel, whence it ran to the east gate (Neh. iii. 15-28). In ver. 26, we are told that "the Nethinims dwelt in Ophel, unto the place over against the water gate toward the east, and the tower that lieth out." From this passage, and from the circumstance that Ophel is not reckoned by Josephus among the hills of

[1] Robinson, *Bib. Research.* i. pp. 232, 311 et sq.
[2] *Ibid.* i. pp. 231, 261.

Jerusalem, and in almost all the eight places where the name occurs in his history it refers to buildings, and not to an eminence of land, it has been thought that the name[1] has been improperly applied to the tract of rising ground, but that at the outset at least it referred to edifices of some kind. According to Krafft's observation,[2] Ophel was a walled fortress, begun by Jotham the son of Uzziah (2 Chron. xxvii. 3; 2 Kings xv. 35); and it was referred to by Isaiah (xxxii. 14): "Because the palaces shall be forsaken; the multitude of the city shall be left; the *forts* and towers shall be for dens for ever, a joy of wild asses, a pasture of flocks."

This Ophel or Ophla, if we speak of it as an edifice built on the tongue of land sometimes known by the same name, served as a place of refuge, and, together with the archives, the citadel, and the town-house, was fired at the time of the taking of the lower part of the city before the hostile ranks of the Romans succeeded in forcing their way to Zion and the upper city.

Following these data, and the tortuous course of the southern wall indicated by Josephus, Krafft differs from the most of his predecessors, and agrees with Robinson, that the present southern wall followed an entirely different course from the ancient one, striking the temple enclosure not at el-Aksa, but at its south-east corner, and turning at a sharp angle and running due north, forming the eastern wall. Manasseh completed the building of the Ophel works in which the Nethinim, *i.e.* "outcasts," lived. This part of the city was made over to the "hewers of wood and drawers of water," possibly to the descendants of the Gibeonites, certainly to all vassals, and those who were taken captive by David and other kings. The eastern and southern slope of Ophel became a place of the greatest importance in connection with the history and topography of Jerusalem, since close by it are the two most profuse supplies of water,—the wells of Mary and of Siloah.

It will be impossible to enter into a special description of

[1] Schultz, *Jerusalem*, p. 59; Williams, *Holy City*, ii. p. 365, Note 7.
[2] Krafft, *Topographie*, pp. 23, 118, 154.

the Tyropœon, the so-called valley of the cheese-makers, in the course of this examination of the walls which pass around the city. To speak of the original character of this place, and the changes which have come over it in the course of the repeated construction and destruction which it has experienced, lies outside of my present purpose. According to Gadow's expression, the southern portion of Jerusalem, thrusting itself out in a southerly direction, may be compared to a peninsula, partially cleft, and lying between two deep valleys. The definiteness which the mouth of the Tyropœon displays at the place of its junction with the valley of Hinnom is confirmed by all observers; but regarding the commencement of this valley, whether it is in the west or in the north, there is a great and still unsettled difference of views, of which I cannot speak till we come to consider the northern side of the city. It is in the Tyropœon that the present wall, composed as it is of stones of all forms,[1] sizes, and ages, displays the many changes which have been witnessed during the long past. It is surmounted by a single palm, which stands upon a high terrace. Westward, towards the eastern slope of Mount Zion, is the gate known as Bab el Mugharibeh, *i.e.* the African[2] Gate. By the crusaders, Franks, and monks, it is known as the Dung Gate. This term corresponds by no means with the gate of the same name in the wall erected by Nehemiah: that lay outside the present wall, somewhere along the south foot of Mount Zion, and below the garden of the Armenian Convent (Neh. iii. 14).[3] In the monkish legends it is confounded with a position farther north. Both probably received their name from the piles of ordure which lay outside. The gate in the Tyropœon is at present closed, in consequence of the disturbances in modern times. The explorer can, however, climb up by means of steps to the battlements of the wall, and pass over this gate to the quarter of Mount Zion.

[1] Bartlett, *Walks*, etc., p. 17.
[2] Robinson, *Bib. Research.* i. pp. 238, 262.
[3] *Ibid.* i. p. 319; Schultz, *Jerusalem*, p. 58; Krafft, *Topog.* p. 151; v. Schubert, *Reise*, iii. p. 544; Gadow, p. 44.

The portion of the city before the Bab el Mugharibeh is at the present time covered with heaps of rubbish thirty or forty feet high, overgrown with a prickly and unapproachable hedge of cactus, which in the very lowest places is higher than the neighbouring garden of el-Aksa. From this cause, —as well as from the refusal of permission to wander freely over this tract at the south-west corner of the Haram, and from the existence there of the miserable Turkish quarter, stretching southward into the hollow between Zion and Moriah, and clustered in narrow streets around the Mosque el Mugharibeh, from which the vicinity and the gate take their name, Hareth el Mugharibeh or African Quarter, and Bab el Mugharibeh or African Gate,—this part of the city is examined with much more difficulty than other parts. All the houses built in that quarter stand on ground which belongs to the Mohammedan schools of the adjoining Haram, and which is leased to the occupants; a circumstance which accounts for the want of any regular streets and edifices of value. This abandoned quarter, in consequence of its nearness to the Haram, has been called the quarter of the black watchmen of the Haram, and carried Gadow's thoughts back to the Nethinim, the servants of the ancient temple, who had their dwellings in this neighbourhood in the time of Solomon and of Nehemiah. Through narrow and crooked streets of the African quarter, the wanderer finds his way eastward, until he comes out upon the place already referred to, where are still seen the titanic stones which date back to the time of Solomon.[1] The spot is now known as el-Ebra, or the Jews' Wailing Place, and has been repeatedly represented in pictures. It lies a hundred paces north of the remains of the colossal arch discovered by Robinson. The Jews visit it every Friday to offer up their prayers. They bend themselves to the earth, and lament the fate of the nation, in the very scene where that nation's existence passed away, and where their ancestors poured out their blood. It is well known that, at the time of their insurrection under

---

[1] Krafft, *Topographie*, p. 113; Bartlett, *Walks*, p. 140, Tab. xix.; Robinson, *Bib. Research.* i. pp. 237, 238.

Hadrian, they were driven from the city; under Constantine they were permitted to approach near enough to view the spot where their sacred temple once stood; and at last it was permitted to them, in consideration of the payment of money, to come once a year into the city, on the day which commemorated the sacking of the city by Titus.

In the twelfth century, in Benjamin of Tudela's time,[1] the place now frequented was the usual spot where the Jews met to pray. They held that it was the vestibule of the Holy of holies in the ancient temple; and they have continued to keep the place under their control up to the present day, by making continual sacrifices to the Turkish Government. The place is not much exposed to the view of the fanatical Moslems, and in its secluded neighbourhood Bartlett found old greyheaded Jews reading their prayer-books, and the women clad in long veils walking alongside the walls, kissing the stones, looking through the chinks, and repeating their prayers with great devotion; but he perceived no weeping and wailing. On the contrary, the people, though sunk in poverty almost to the extent of needing to beg alms, seemed to pride themselves, as of old, upon their being a chosen race. South of this wailing place there is, according to Gadow, a narrow street or lane leading up a steep ascent to the Haram; between this and the ruin of Robinson's broken arch there is a stairway of eighteen or twenty steps leading to the court of the el-Mugharibeh mosque, which lies within the circuit of the Haram, and on a level with the enclosure. Only from this one place on the west side can it be now seen that Moriah is strictly called a mount. Gadow, on attempting to ascend this staircase, was driven back with vile epithets; but Tobler was able in 1846 to make some valuable observations, which will be spoken of on a subsequent page.

We now leave the closed African's Gate (Bab el Mugharibeh), the Dung Gate of the pilgrims' legends, and leaving the Tyropœon, pass westward. After a short distance our course is checked by the steep ascent leading to Mount Zion.

---

[1] The *Itinerary* of Rabbi Benjamin of Tudela, vol. i. p. 70; Bartlett, *Walks*, p. 139.

Robinson estimated[1] the bold escarpment which faces the south-west corner of the temple enclosure to be from twenty to thirty feet high: he found it in the same condition in which it had been left, it would seem, at Josephus' time; but the adjacent valley was well-nigh filled with rubbish.

Over the high ridge of Zion, nearly 2500 feet in width, and close by the beginning of its southern declivity, the present wall runs in a zig-zag course, making many angles towards the south-west. At length it turns at a sharp corner, and pursues a direct course northward.

This wall, wholly of Saracenic construction, compared by Bartlett with those which surround some of the old cities of England—York, for example—passes from the Tyropœon straight up the steep wall[2] of Mount Zion, whose top, at the time of the taking of the city by Titus, was covered with houses, and formed one quarter of Jerusalem. At the time of Josephus the wall appears not to have crossed the highest part of the mount, but to have passed around its base. The south-east slope of Zion, down which there was, both at the time of Nehemiah (iii. 15) and of Josephus,[3] a flight of steps leading from the "city of David," as well as the south-west slope down which another flight led to the Birket es Sultan of the Arabs, the Lower Pool of Gihon, according to the usual designation,[4] was more gentle and rounding than that which lay west of the Valley of Gihon and south of the Valley of Hinnom. The deeper the later ravine sinks in its eastward course, the more bold and precipitous becomes the southern slope, with its wave-like knolls, between which winds the aqueduct from Bethlehem, sometimes called the Aqueduct of Pilate. The whole exterior part of Zion, where it rises from the valley, bears the impress of being formed by an accumulation of rubbish. At its bare eastern base may be seen an oval cistern, not narrowed at the top as they usually

---

[1] Robinson, *Bib. Research.* i. pp. 231, 264.
[2] Bartlett, *Walks*, etc., pp. 15, 17–22, Tab. i.; also frontispiece to Bartlett and Bourne, *Modern Jerusalem*.
[3] Krafft, *Topographie*, pp. 61, 152.
[4] Gadow, *i.a.l.* iii. p. 40.

CIRCUIT OF THE PRESENT WALLS. 53

are found, but covered with terra cotta: of it only half is preserved. South-west of it there is a spacious room hewn out of the rock: its length is forty feet. On Symonds' plan of the city it is designated as the Cave of St Peter. At the west foot of the mount, and near the south-east corner of the Birket es Sultan, there is the ruin of a structure of doubtful use, partly hewn out of the rock and partly made of masonry: it is connected with some ancient cisterns which are called on Symonds' plan the Bath of Tiberius (Hammam Tabariyeh),[1] but have been designated by Tobler as Bir el Jehudi, and also the Palace of David. A little farther north, and on the west slope of Zion, directly above the Aqueduct of Pilate, which crosses the valley on low arches,[2] there is another fragment of old masonry, forming a narrow wall, which passes round the whole southern extremity of the mountain, and comes to its termination at the city wall, where it crosses the Tyropœon. The whole side which slopes towards the Valley of Hinnom is traversed by several footpaths, which wind up to the Zion Gate, the southern entrance to the city. This slope is dotted with clusters of olive trees, which have secured for themselves a footing at the various landings; and here and there is space enough for a bit of land to be cultivated. Towards the east there are few traces of former habitation; but on the south-west side of the so-called Nebi Daud, as far as to the Hammam Tabariyeh, Schultz[3] found unmistakeable traces of former masonry and connected cisterns, showing beyond a doubt the site of the ancient wall of the city of David. Near the Hammam Tabariyeh must be sought the location of the Dung Gate and the Valley Gate mentioned in Neh. iii. 13, 14. In that neighbourhood Gadow[4] observed two immense piles of rubbish, of great antiquity, which rise like artificial walls, and which seem to have some connection with a former investment of the city. The complete disappearance of the dwellings of man from

---

[1] Schultz, *Jerusalem*, p. 27.
[2] Robinson, *Bib. Research.* i. p. 264.
[3] Schultz, *Jerusalem*, p. 58; Krafft, *Topographie*, p. 151.
[4] Gadow, *i.a.l.* iii. p. 41.

this part of the city, the quarter anciently known as the city of David, is one of the inexplicable phenomena connected with this place of wonders. The top of Mount Zion is a large plateau, excluded from the city by the present wall. In its centre stands the Nebi Daud, or the Grave of David, the possession of which is in the hands of the Mohammedans. A little farther north, near the wall, is the Armenian church, with the house of Caiaphas; and between the two is the place which the monks point out to credulous pilgrims as the spot where the cock crowed, and where Peter wept. The Armenian church stands almost due south-west of the Zion Gate: west, south-west, and south of the church are the graves[1] of Christians, each denomination lying by itself, —the Armenians next to their sanctuary, at the south the Greeks, in the middle the Latins. South of the Nebi Daud is the spot connected with the new bishopric, and consecrated by Bishop Gobat. This is surrounded by a wall. As Strauss says,[2] Mount Zion has become a resting-place for the dead. Footpaths lead obliquely down the mountain-side to the Valley of Hinnom; but there is no road which connects the southern gate of Jerusalem with the adjacent country.[3]

This Nebi Daud, with a mosque and the adjacent Armenian church, occupies one of the most memorable localities in ancient[4] Jewish history; yet it is probable that its present appearance gives us little token of what it was at the time of David, so much has it been changed during the lapse of centuries. At the time of Robinson's[5] visit, Ibrahim Pasha had erected his dwelling there. The mosque, which is the object of Moslem pilgrimages, and the reputed grave of David, also in the possession of the Mohammedans, are not accessible to Christians, and we have but partial descriptions of them. The monks' legends tell us, that over the grave of David is the apartment used for the first celebration of the Lord's

---

[1] Robinson, *Bib. Research.* i. p. 228.
[2] F. A. Strauss, *Sinai und Golgotha*, p. 250.
[3] Robinson, *Bib. Research.* ii. p. 19.
[4] Krafft, *Topographie*, p. 168.
[5] Robinson, *Bib. Research.* i. pp. 241, 243, 262.

Supper; but regarding it we know nothing further. There is a large stone hall, fifty or sixty feet long, and thirty broad, furnished with an altar, where Christians repeat their prayers, and sometimes celebrate mass: near it is a second and larger apartment, in which the Mohammedans go through their devotions. The building was once a Christian church. It is called by Cyril in the fourth century the Church of the Apostles, and at that time was held to be older than the buildings of Constantine. The *Itinerar. Hierosol.*, written in 333, designates it as the house of Caiaphas,[1] and states that at that time the Pillar of Scourging stood at its gate. From this circumstance, Tobler supposes that the oldest Via Dolorosa ran from Zion northward to the Holy Sepulchre, and not from the present Chapel of Scourging, near the seraglio of the governor, westward. In opposition to a former view, which Tobler considers the one held by the crusaders, Krafft[2] asks whether the official residence of the high priest, where the Sanhedrim met, is not rather to be looked for in the north-east of the temple enclosure, near the Roman Palatium of Pilate, which was unquestionably on the highest part of Akra, and coincident with the ancient Antonia, and the present seraglio of the Turkish governor. According to this, the ancient Via Dolorosa would coincide with the one now visited by pilgrims, running from east to west. According to Robinson's researches,[3] we are indebted to Marinus Sanutus for the first full description of the Via Dolorosa, and for the prominence which it has received as the object of pilgrimages.

During the middle ages, the above-mentioned Church of the Apostles was called the Cœnaculum, and the legends attribute to it all kinds of remarkable attributes. During the time of the crusaders, a Franciscan convent was built here. From Maundeville and De Suchem we learn that at their time this church was still in the possession of Latin

---

[1] *Itin. Burdig.* ed. Parthey, p. 279; Tobler, in *Ausland*, 1848, No. 21, pp. 71-82.
[2] Krafft, *Topog.* pp. 63, 165, 166.
[3] Robinson, *Bib. Research.* i. pp. 233, 252.

monks. A hundred years later, H. Tucher of Nuremberg found it transformed into a mosque. For a century a Minorite convent was established in the lower storey, but in 1561 it was dispossessed, and purchased the present Latin Convent of St Salvadore at the north-western corner of the city, in which most pilgrims find reception.

The little Armenian church, which lies a short distance north of Nebi Daud, is said to occupy the site of the house of Caiaphas the high priest: in its court may be seen the burial-places of the Armenian patriarchs of Jerusalem; and many legends are connected with the spot. There seems, however, to be nothing authentic connected with the place that is older than the fourteenth century. On the other hand, the statements which locate the grave of David there are of the greatest antiquity; and there is no reason for doubting that in this very place, on the top of the mountain, and in the centre of the city known by his own name, and within the wall known to have encompassed it, the bones of the great singer king were interred. We have the express statement in 1 Kings ii. 10, "So David slept with his fathers, and was buried in the city of David." In like manner, Solomon,[1] Rehoboam, Abiah, Asa, Jehoshaphat, Ahaziah, Amaziah, Jotham, and Josiah, were all buried in the royal vaults; and up to the time of Josephus the spot was called the graves of David, of the sons of David, of the kings of Judah, or simply of the kings. Every one has his own special vault. Those who died of unclean diseases, like Jehoram, Joash, and Uzziah, were buried, not with their fathers, but in an adjoining field; while the idolatrous Ahaz was laid in the suburbs of the city, away from the bones of his ancestors (2 Chron. xxviii. 27). Of the priest Jehoiada we are told (2 Chron. xxiv. 16), that "they buried him in the city of David among the kings, because he had done good in Israel, both toward God and toward his house." At the first conquest of Jerusalem, effected by Nebuchadnezzar B.C. 588, the kings' graves were not disturbed, since Nehemiah speaks (iii. 15, 16) of the steps leading up from the valley to the city of David

[1] Krafft, *Topographie*, pp. 205-211.

at the time of the rebuilding of the wall. This serves as a landmark to us in locating the wall, and is referred to by Josephus.[1] The reason that the tombs of the kings were spared does not appear, since it would seem natural that the treasures which used to be buried with them would have proved a strong temptation; but it may be that the graves, like those of the Egyptian kings at Thebes, passed unnoticed. They were also unknown to the Syrian and Babylonian conquerors; for, according to Josephus, it was the high priest Hyrcanus, son and successor of Simon Maccabæus, who, in order to effect the raising of the siege, plundered the graves of the kings, and took from that of David alone 3000 talents, or more than half a million of pounds sterling, using the money to bribe the enemy to withdraw. Herod the Great followed his example, but committed his robberies by night. He found no money, however, only articles of royal adornment and jewels; and when he tried to press farther in and reach the graves of David and Solomon, there broke forth a flame which caused the death of two of his followers, and caused him to desist from his sacrilegious attempt. (May not the collected gases have kindled on being touched with fire?) In order to atone for the deed, he erected a costly marble monument close by. Josephus informs us that neither Hyrcanus nor Herod went so far as to the coffins themselves, but that they were kept back by a mechanical contrivance which prevented any one from going beyond a certain point. He gives no further particulars; but it would seem that some such appliances were used as those found at Gadara in the ancient Hauran, where traces of swinging doors are discovered. Remains of the same are also seen in the tomb of Helena, north of Jerusalem.

From the words of Peter concerning the outpouring of the Holy Ghost (Acts ii. 29, "Let me freely speak unto you of the patriarch David, that he is both dead and buried, and his sepulchre is with us unto this day"), it is plain that there was then universal agreement regarding the grave of the shepherd-king. Otto Thenius and Krafft have shown, from

[1] Krafft, *Topographie*, p. 152.

the legends of subsequent centuries, that the account of these graves is covered with obscurity, but that thus much is certain, that the graves of the ancient Jewish monarchs must be sought, if not directly beneath the Cœnaculum, at least in its neighbourhood, and on the summit of Mount Zion.

Passing from the outside of the wall through the gate into the city, at the right lies the Jews' quarter, full of dirt[1] and filth, displaying the greatest poverty and the lowest social condition, full of crooked lanes and piles of rubbish, and displaying a single great building—the synagogue of the Sephardim. The most of the houses in the Jews' quarter stand upon the slope that descends eastward to the Tyropœon. Close to the Zion gate, and at the right hand, are the pitiable hovels of the lepers; on the left hand is the Christian quarter, Hareth el Nussarah, extending to the Armenian Convent and its garden, and embracing the neighbouring small convent, Deir el Zeituneh, belonging to the Jacobite Syrians and Armenian Christians, both being frequently called Monophysites. This spot, lying at the culmination of Mount Zion, affords the best view of the south of Jerusalem. A confused picture it is, as one looks towards the Jews' quarter, and the eye wanders over so many perished walls, so much rubbish and ruin; and the mind is filled with sadness and gloomy thoughts as it traces the history of this once famous city. What a change since the time when Jehovah had His throne there, and inspired the words of Ps. xlviii. 11–15: "Let Mount Zion rejoice, let the daughters of Judah be glad, because of Thy judgments. Walk about Zion, and go round about her: tell the towers thereof. Mark ye well her bulwarks, consider her palaces; that ye may tell it to the generation following. For this God is our God for ever and ever: He will be our guide even unto death." But how soon was this proud Jerusalem, as Micah the Morasthite prophesied (Jer. xxvi. 18), compelled to become heaps of stones, and Zion a ploughed field! (Micah iii. 12.)

The pitiful hovels close by the gate first attract attention. They stand behind hedges of thorn, prickly cactus plants, and

[1] Bartlett, *Walks*, etc., p. 80.

rubbish, and are only inhabited by a class of people who are called lepers. Whether their disease was the leprosy of the Bible, or some other, Robinson could not judge: its symptoms seemed to him to agree with those of elephantiasis. At all events, they are miserable creatures,[1] pitiful outcasts from society, marrying and perpetuating their disease among their children, who grow up to manhood with the appearance of health, when the disease suddenly breaks out, and they seldom outlive their fortieth or fiftieth year. Tobler,[2] who devoted a great deal of attention to these unfortunate creatures, terms their hovels Biut el Masakin, *i.e.* huts of the lepers. He found sixteen of these pitiful structures inhabited by about thirty persons. They were mostly supported by alms and casual gifts, and were under the supervision of a sheikh who had the disease himself. Although outside of these people the complaint is known among the Jews and Mohammedans of the city, yet, according to Tobler, it does not spread; and the hovels at the Zion gate serve not only for the lepers of the city, but those of the neighbourhood of Jerusalem. The Byzantines, as early as the time of the Empress Eudocia, contributed to the sustenance of the poor lepers there: in like manner did the Franks during the Crusades, caring also for the unfortunate creatures similarly afflicted at Damascus and at Jaffa. The Turkish Government is, however, entirely indifferent to their distress, and offers no help to temper their misery, despite the numerous examples which Europeans, both Christians and Jews,[3] have set before them in this thing.

3. *The Western Wall from the south-west corner northward to the Latin Convent St Salvador and Kasr Jalud.*

Only the northern portion of Mount Zion is comprised within the modern wall, and this portion is mainly taken up by the Jewish quarter and the great Armenian Convent at

[1] Robinson, *Bib. Research.* i. p. 243; Bartlett, *Walks*, p. 75.
[2] *Ausland*, 1844, No. 115, pp. 459, 460.
[3] Tobler, *über Aertzte, Apotheken und Krankenhauser in Jerusalem*, Nos. 114, 115.

the south-west corner of the city. Viewed from the lower Valley of Jehoshaphat, this part of the mountain,[1] in consequence of its steep descent to the deep Hinnom valley, seems the highest point of Jerusalem. Robinson estimates the height of the summit of Zion as three hundred feet above the lowest point in the valley just mentioned. But far less does its height appear viewed from the upper part of the Valley of Gihon; and still farther north, Zion is considerably overtopped by the Latin Convent. The northern portion of the mountain is so covered with walls and edifices, that it is difficult to determine its original extent. Robinson defined its northern border as a line running south of the street leading from the Jaffa gate directly eastward, and considers the depression occupied by this street as the primitive location of the ancient Tyropœon, which, in his view, ran around the northern foot of Mount Zion, and then turning in a right angle toward the south, took its course toward the Valley of Hinnom. More recent topographers dispute this point, and consider the street running eastward from the Jaffa gate to be no primitive valley, but hold that the Tyropœon extended due north through the heart of the city as far as the Damascus gate. This difference lies at the foundation of the great discussion which has arisen in the effort to interpret Josephus' description of the localities in this city at the time of its capture by Titus. Robinson adduces as the main support of his theory, the fact that, if one looks southward at any point along the street running eastward from the Jaffa gate, his eye meets a steep though not a high slope, upon whose rim it is possible to overlook the roofs of all the houses that occupy that pitiful depression. Krafft says,[2] in this connection, that the northern edge of Zion runs parallel with the street leading eastward from the Jaffa gate to the Haram; and that it is sensibly elevated above the almost contiguous terrace of the elevation on which the Church of the Holy Sepulchre stands. He says that no valley begins here, and that Robinson only meant that the

---

[1] Robinson, *Bib. Researches*, i. p. 264.
[2] Krafft, *Topographie*, p. 4; Schultz, *Jerusalem*, p. 54.

east side of Zion rises steeply above the depression which
runs through the whole city from the Damascus gate at the
north to below Ain Silwan (Siloah), that is, the Tyropœon,
where it joins the Valley of Ben Hinnom. Robinson himself noticed that the cross valley along the north foot of
Mount Zion, which he held to be the upper Tyropœon, had
for eighteen hundred years been filled with rubbish, yet he
believed[1] he could trace its primitive course. That the mass
of houses and the accumulation of so many ages have made
investigations there extremely difficult, is self-evident; and on
this account the greater credit is to be given to the labours
of a recent observer,[2] the impartial Wolff, whose efforts in
tracing the ancient localities within the city have been
thorough beyond all precedent. A remark made by him
seems to me strictly just, that before we attain to certainty
regarding the topography of ancient Jerusalem, it will be
necessary to make surveys and excavations in far greater
measure than has thus far been done. Von Wildenbruch,[3]
together with Wolcott the missionary and Johns the architect, while preparing for the foundation of the new evangelical
church, have discovered the remains of aqueducts forty feet
below the present surface. These remains bear in their
workmanship the traces of the greatest antiquity. In reaching them, the excavators passed first through ten feet of
earth, then through ten feet of rubbish, then through ten
feet more of earth, and lastly through ten feet more of architectural remains,—circumstances which must make us cautious
against too hastily adopting theories drawn from the present
appearance of the surface.

Waving the discussion of these points for the present, we
turn back to the southern portion of Mount Zion, whose gate
is formed by a stately square tower, although, unlike the
Jaffa gate, no road passes through it to the adjacent country.
The greater portion of the western part of Mount Zion from

[1] Robinson, *Bib. Research.* i. p. 281.
[2] Dr Phil. Wolff, *Reise in das Gelobte Land*, p. 74.
[3] Von Wildenbruch, in *Monatsber. der Berl. Geogr. Ges.* Pt. iv. p. 143.

the gate to the south-west angle of the wall is occupied by the Armenian Convent, whose size is so great that it is able to accommodate from two to three thousand pilgrims at a time. It is the only one in Jerusalem whose appearance can be called stately; it is supplied with a good façade, is well paved around, and partly hid by fine trees: and the well-conditioned monks give proof, says Bartlett,[1] that they live in peace with themselves and the world. Guests of position are entertained in accordance with their rank, but those who are held in less consideration are received into the Latin Convent. A massive gateway leads to the spacious court of the Armenian Convent, which is kept very neatly, and surrounded by picturesque edifices.

The Armenians are the most prominent religious sect in Jerusalem, and through their industry they have acquired considerable wealth. Their convent, which is named after St James, because he is believed to have been beheaded here by Herod, receives liberal donations from believers in foreign lands: their church is profusely decorated, and their gardens, which surround the whole building, are filled with rare trees, and afford a fine view towards the west, south, and east. The spot is one of the most lovely in all Syria. At Easter the richest and most crowded bazaar in the city calls together Armenian pilgrims in crowds from all provinces of the Turkish Empire.

The western wall of the city, with its many projecting bastions, called Abraj Ghuzeh, *i.e.* the Towers of Gaza, runs directly north and south, and lifts its head directly above the adjacent west valley of Gihon, passing the modern Turkish barracks, and ariving at length at the great fortress of the city, el-Kalah, or the Castle of David. This, with its numerous towers, one of them the well-known Hippicus, occupies a conspicuous position. It seems to date from the period[2] of the occupation by the Romans, who would scarcely have overlooked a situation so favourable for the purposes of for-

[1] Bartlett, *Walks about the City*, p. 78; Olin, *Travels*, vol. ii. pp. 304-306.
[2] Wilson, *Lands of the Bible*, i. p. 432.

tification. The preparation had already been made by Herod, and Titus wished to retain the old fort as a trophy of his victory.

On the north side of the citadel, and directly adjacent, the wall is broken by the Jaffa gate, called also by pilgrims the Bethlehem gate, because here the two roads meet, one of which runs southward to Bethlehem, and the other westward to Jaffa. Outside of the gate the roads part, and there is frequently to be seen the meeting of pilgrims, monks, and pedestrians, receiving the greeting of their friends who have come out of the city to welcome them. This Bab el Chalil,[1] as it is called by the Arabs, or Hebron gate, is a massive square structure, to which access is given on the east; but within there is a sharp angle, and the exit is towards the north. According to an Arabic inscription over the entrance, the present walls of the citadel were erected by command of Sultan Suleiman in 1542. They appear to take the place of the walls of the middle ages, which were destroyed several times during the Crusades, but always rebuilt. On this side of the city the re-erection could scarcely be effected at any other place than the one which is occupied, and the old materials have doubtless been used in making the new walls. So secure did the position seem in the earliest times, that the Jebusites, the first possessors of whom we have record, defied David with scorn when he came up against their stronghold: "Thou wilt not come in hither; the blind and lame shall drive thee back"[2] (2 Sam. v. 6-9; 1 Chron. xi. 5-8). Here David subsequently took up his residence, called it after his own name, surrounded it with bastions, and gradually enlarged its size.

The exact position of the fort called Millo[3] is not given in the Old Testament, and it is held by some to have been at the north-west corner of Zion, and by Schultz and others at the north-east angle. The geographical limits of Zion, too,

[1] Robinson, *Bib. Research.* ii. p. 17.
[2] I render the quotation literally from Luther's German translation: the meaning in the English Bible is doubtful, if not even absurd.—ED.
[3] Winer, *Bibl. Realw.* ii. pp. 96, 735.

are not easily defined, since its position is not indicated in the Old Testament, and Josephus never uses the word, but always speaks of the "upper city." The account of the kings' graves in the "city of David" shows that its limits extended southward beyond the present walls. It lay, therefore, at the south and south-west parts of the territory bounded by the great natural ravines; and Moriah had then no connection with it, as we learn from 1 Chron. xxi. 18, and 2 Sam. xxiv. 18-25, in the account of David's purchase of the thrashing-floor of Araunah the Jebusite, for the purpose of erecting there an altar to Jehovah, and subsequently a temple. The name Moriah, too, comes as rarely into use in the Old Testament as that of Zion: in the theocratic language of the prophets and psalmists, it is employed to indicate the whole city of Jerusalem as the hallowed dwelling-place of Jehovah, and very rarely the mountain on which the temple stood. Such passages as that which opens the forty-eighth psalm are rare: "Great is the Lord, and greatly to be praised in the city of our God, in the mountain of His holiness. Beautiful for situation, the joy of the whole earth, is Mount Zion, on the sides of the north, the city of the great King."

With far greater certainty are we able to designate the north-west corner of Zion in the present citadel[1] el-Khalil, the position of which in the period before the destruction by Titus is manifested by remains of very massive proportions. The present citadel is an irregular assemblage of square towers, which are defended on the side towards the city by a wall, on the outside of which there is a deep ditch and an escarpment, which appears to be Roman in its character: it probably dates from the time of Hadrian. Robinson's description is the most detailed, and I shall adopt it as my guide. The massive exterior works comprise so much space, that if they were cleared of their rubbish they would hold thousands of soldiers. At the capture of Jerusalem in 1099, this fortress was the strongest which the Saracens possessed, and it was the last which was surrendered. Wil-

---

[1] Robinson, *Bib. Research.* i. pp. 306-310, 316, 376; Bartlett, *Walks*, etc., p. 85; Roberts, *The Holy Land*, Books ii. and iv.

liam of Tyre speaks of it as the Tower or the Citadel of David, and says that it was made of hewn stone, and of great strength. When, in 1219, the wall around it was destroyed by the Mohammedans, according to Wilken, the fortress was spared. The work indeed was regarded so strong as to be indestructible.

Since the year 1522 this stronghold has been familiarly known by the name of the Castle of the Pisanese,[1] because, as we learn from Adrichomius, during the time that the crusaders held the city, Christians from Pisa built it. In the interior the most conspicuous object is the tower at the north-west, on which the Turkish flag is usually seen waving: its great antiquity secures for it this honour. The upper part is modern, but the lower portion is composed of huge quarried stones with bevelled edges, and unquestionably retaining their original position. This is the Tower of David *par eminence*: it is no other than the Hippicus of Herod,[2] which was spared by Titus, as Josephus asserts, in order to remain as a perpetual testimony of the tremendous difficulties which the Roman conquerors overcame. Josephus, who has given a full description of it, says that at that tower, in the north-western part of the city, each one of the three walls had their commencement. It was erected by Herod, and was named in honour of his friend Hippicus, who had fallen in a battle with the Parthians. The tower was rectangular, every side twenty-five Jewish ells long and thirty high. Above this massive portion was a cistern, twenty ells high, with rooms twenty-five high, together with breastworks two ells, and pinnacles three ells high. The altitude of the whole structure was eighty ells: the stones which composed it were of great size, twenty ells long, ten broad, and five high, and were of marble. Although these details were all given from memory, and may perhaps have been a little coloured, yet they agree in the main with the portion which remains, which is really of massive proportions, and has stood unchanged during all the vicissitudes through which Jerusalem has passed. Both Schultz and von

---

[1] Adrichomius, *Theatr. Terræ Sct.* p. 156.
[2] Bartlett, *Walks*, p. 85, Tab. vi.

Schubert recognise in this Pisan castle the ancient fortress of David, and the Tower of Hippicus built by Herod. Robinson's careful measurement confirmed the same view; for the appearance of the massive bevelled stones at the base, so similar to those found in the walls of the Haram, at the Jews' Wailing Place, at the old broken arch, and also at Hebron, led him to the irresistible conclusion, that all these works date back at least to the time of Herod, and perhaps still earlier. The Tower of Hippicus is quadrangular, yet not perfectly square: its eastern side is fifty-six feet four inches long, its southern seventy feet three inches. The height of the ancient portion is forty feet; and were it not partially buried in rubbish, it would probably reach fifty feet. The stones are not removed from their old places: they remind one at once of the walls referred to above; yet they are smaller in their dimensions, they being nine, ten, and in some instances twelve feet long. Though bevelled at the edges, yet in the middle part the stones are often left rough, which gives the whole structure a much less finished look than it would otherwise present. The present entrance is on the west side, but it is half-way up the side: the old portion has no accessible opening. Josephus alludes to the existence of two other towers built by Herod in the same form, but of still more gigantic dimensions: one of them he named Phasaelus, in honour of a friend, and the other Mariamne, in honour of his mistress. They stood below the Hippicus, close by the first old wall, which ran north of Zion to the temple. The edge of the eminence on which they were placed was thirty Jewish ells above the Tyropœon, which gave the towers a very commanding aspect. The royal castle and palace of Herod were connected with the Hippicus and the other two towers; the whole were very strongly fortified, and fitted up with great splendour. Josephus surpasses himself in his description of the magnificent halls, gardens, and sculptures. Nothing is left of all this but the basis of the Hippicus: what Titus spared was subsequently razed by Hadrian, who wanted to use the stones elsewhere, for the purposes of his Colonia Ælia Capitolina.

CIRCUIT OF THE PRESENT WALLS. 67

Still a fourth tower, opposite the Hippicus and the other towers at the north, and at the north-west corner of the third or outer wall of the city, is described by Josephus as eight-sided, and as being seventy Jewish ells in height. From its summit it is asserted that, after the sun had gone down, Arabia and the Dead Sea could be descried. It must, in order to have afforded such a prospect, have stood upon that prominent elevation of land that lies north-north-west of the present north-west corner of the city. About seven hundred feet from the modern wall, and upon a ridge even higher than the summit of Zion, Robinson believed that he could trace foundations which seemed to him to indicate the former existence[1] there of towers or fortifications stretching northward for a distance of six hundred and fifty feet, and playing a very prominent part in the siege of Jerusalem. Robinson thinks that it could not have been the Psephinos; but, taken in connection with other traces, he suspects that it must have had some connection with it. Schultz[2] was of the same opinion, and believed that he could trace with much confidence the old defences of Agrippa past these apparent remains of the Psephinos, and relics of cisterns, as far as to the grave of Helena. In the abundant relics of former architectural objects, he thinks he can follow the course of the outer wall with far more certainty than from the imperfect descriptions which have come down to us. But in case that distant line of masonry had been the outer wall, the area of the city would have been doubled, for which the number of population at that time would give no warrant; and the defence of so long a line would have required so large a body of men, that it would have been vulnerable at many points, and could have offered but a feeble resistance. According to Krafft's subsequent more thorough investigations, these supposed remains of towers and fortifications prove to be only the lower portions of cisterns which were once there. They have no solid foundations, and are all above ground.[3]

[1] Robinson, *Bib. Research.* i. p. 314.
[2] Schultz, *Jerusalem*, pp. 62, 63.
[3] Krafft, *Topographie*, pp. 37-39.

We must therefore hold to the view, that the western wall of the ancient city, of which we have no full description, did not run so far to the north as the point suspected by Robinson, but terminated at the present north-west corner, in the neighbourhood of the Latin Convent of St Salvador and the so-called citadel of Goliath (Kasr Jalud). Near the latter lies an old wall, now broken, but probably connected anciently with the Kasr, beneath which there is to be seen a broad open arch, the outer west wall of which still shows traces of the great stones with which the mighty structure was once covered.

These gigantic remains[1] at the present north-west corner of the city are the traces of the last great structure in Jerusalem on the north before the destruction effected by Titus, *i.e.* the third wall built by Herod Agrippa, ten or twelve years after the death of Christ, for the protection of the new town. He probably took as the starting-point the princely Psephinos Tower, to which the subsequent wall of Hadrian was probably contiguous.

Josephus speaks of it as the most remarkable structure of the whole third wall. Robinson recognises the antiquity of this historical monument, although he does not use it to confirm his theory,—a fact which does much to recommend the accuracy and trustworthiness of his observations.[2] These works, he remarks, appear to have been built upon the ruins of a yet older wall, possibly that of Hadrian or of Agrippa; for at the south-western corner, near the ground, there are three layers of great bevelled and hewn stones which run diagonally into the mass, in such a way that it may be seen that they were there before the town and the bastion were built: they were probably the remains, he thinks, of the old third wall. These three layers of gigantic bevelled stones are, according to Krafft, the relics of a former external covering, and the unquestionable traces of the octagonal form of the Psephinus, from which, according to Josephus, the wall ran directly eastward. Even the name of this remnant, fifteen

[1] Schultz, *Jerusalem*, p. 95; Krafft, *Topog.* pp. 40-42.
[2] Robinson, *Bib. Research.* i. pp. 314, 318.

or twenty feet high as it is, and composed of small stones, knit very firmly together by the mortar, confirms the suspicion that it was the noble Psephinos, since the word ψήφινος signifies "made of small stones." In the middle ages, during the Crusades, Tancred encamped in its neighbourhood, and hence it was called Tancred's Tower. Brocardus, who visited it in 1283, and who describes its location and surroundings in much the same way as Josephus does, terms it, in consequence of its elevated position, Neblosa.

Krafft seeks to justify his position in opposition to that of Schultz and his predecessors, who advocated the theory of an extension of the wall a long distance north of the city, by showing that all natural protection which might have been afforded by the nature of the land was wanting there; while in Josephus' account of the eastward direction of the wall from the Psephinos tower on, there are localities still discernible in the immediate neighbourhood of the present wall, and of valuable service in identifying the site of the ancient one. Josephus says that the wall lay along the line of graves opposite that of Helena (on the south), and then ran down towards the city, past the royal caverns (Herod's cave, now the grotto of Jeremiah): it then turned towards the Potters' Field, and finally united itself to the ancient wall at the Kedron valley. These are indeed only indications, and may allow us to attain to probability, but they do not ensure certainty. We turn back now to the region outside of the western wall, before the Jaffa gate and the Castle of David, where runs the Valley of Gihon from north to south, with its two pools,[1] commonly known as the upper and the lower, or the larger and the smaller. The valley runs first southward, or rather south-eastward, to the Jaffa gate, and then directly southward, till it reaches the southern extremity of Zion, and turns eastward towards the Valley of Jehoshaphat. Robinson did not fail to remark that it was strictly an upper valley of Hinnom, and perhaps should be called Gihon only in its upper portion; a name which

[1] Robinson, *Bib. Research.* i. p. 328, etc.; Wilson, *Lands of the Bible*, i. 493.

is often[1] met with to indicate a watercourse, and whose original application to this valley is a subject of doubt. The word Gihon[2] (to spring, to jet forth) occurs very early in the Old Testament, and Josephus says that Solomon was crowned at the fountain of Gihon. At the time of king Hezekiah, who constructed water-works for the protection of the city in case of attack by an enemy, we are told (2 Chron. xxxii. 30; 2 Kings xx. 20) that "this same Hezekiah also stopped the upper watercourse of Gihon, and brought it straight down to the west side of the city of David." But it by no means appears from this passage which of the various works around Jerusalem is meant, and whether it was that which is met as one comes out of the city, near the Jaffa gate. Isaiah (xxii. 9–11) and Sirach (xlviii. 17) describe the destruction of the watercourses at the approach of the Assyrians: they speak of an old pool outside of the city, whose waters were conducted within two walls, and collected within a new or lower pool. The historical circumstances connected with these accounts make it highly probable that this Gihon spring, and the old pool, which received its name in contradistinction to the new one within the city, lay on the north side, in the neighbourhood of the Damascus gate, and not on the west, although the later legend has wrongly connected the reservoir in the present Gihon or upper Hinnom valley with the water-works built by Hezekiah, and has given the name upper and lower pool to the reservoir lying, according to Robinson, seven hundred paces west of Kasr Jalud. The Arabs call the upper pool Birket el Mamilla, the lower one outside of the city Birket es Sultan. Another one, below the upper pool, but within the walls and north of the citadel, and at present surrounded by houses, has received from pilgrims the name of Hezekiah's[3] Pool: the Arabs, however, call it Birket el Hammam, or el-Batrak, a corruption of the name given it by the crusaders, who called it Lacus Patriarchæ, because it

---

[1] Winer, *Bibl. Realw.* i. p. 428.
[2] Krafft, *Topographie*, pp. 119–124.
[3] Bartlett, *Walks, etc.*, p. 89, Tab. vii.

furnished the baths of the patriarchs with water. The name given by pilgrims is by no means an ancient one: neither Brocardus in 1283 nor the crusaders employed it. Quaresmius (1616 to 1625) appears to be the first[1] who insisted that the pool called in his time the Piscina Sancti Sepulchri was the one which Isaiah (xxii. 9) speaks of as Hezekiah's pool, the water of which that king brought into the city by an aqueduct running north-westward.

Traces of high antiquity in the upper one of the two pools, *i.e.* the one known by the Arabs as Mamilla, show that it sent a branch southward to the Gihon, as now its surplus waters flow towards the Jaffa gate (the valley gate of Nehemiah), before which was the Dragon[2] Fountain, which Nehemiah rode past while he examined the west side of the city by night, in order to make his plans for the restoration of the walls of the city (Neh. ii. 13, iii. 13).

The name Birket el Mamilla is derived from a church in the neighbourhood, long since destroyed, called after St Mamilla. The legend states that, at the time of the Persian invasion under Chosroes II. in 614, the bodies of twelve thousand slaughtered Christians were preserved in its neighbourhood. At the time of the Crusades there were the graves of Christians there, but later the Moslems laid out one of their burial-grounds there. Robinson found the pool very dry, yet full in the rainy season. Its length from west to south-east he ascertained to be three hundred and sixteen feet, its breadth two hundred feet, its depth about twenty feet. Its walls are composed of small stones overlaid with mortar. Its position indicates the beginning[3] of the Valley of Hinnom, called more fully by Jeremiah (xix. 2, 6) the Valley of the the son of Hinnom, or Ben Hinnom, which the Greeks contracted into Geënna: from this springs the Gehinnom of the moderns, or, as Edrisi and others write it, the Wadi Jehennam. Its cradle is surrounded by the gentle hills of

---

[1] Quaresmius, *Elucidat. T. Sanct.* T. ii. fol. 717.

[2] Krafft, *Topog.* pp. 124, 186; Robinson, *Bib. Research.* i. 319, 326; Williams, *Holy City*, p. 302; Tobler, *Ausland*, 1849, No. 20, p. 78.

[3] Robinson, *Bib. Research.* i. p. 272.

the watershed which must be passed by one going to Jaffa. The rain-water collects in the valley, and runs south over a stony bottom, fifty to a hundred paces wide, and forming a ravine or gorge fifty feet deep, extending to the south-west corner of Mount Zion, where it becomes a true valley of Hinnom: the name Valley of Gihon is nowhere applied to the upper portion in the Old Testament.

Tobler says, that at the Mamilla pool there is at present no spring of fresh water, but that it sends its surplus in the winter through a canal into the city. He asserts that he has often been refreshed by drinking at the little waterfall which is made by the issuing of the water at the south-west corner of the pool. Gadow[1] found the length of the Mamilla basin to be a hundred and twenty-five paces, and its breadth eighty paces. At its west side he found a spacious cavern in the rock, which appeared to him to be an ancient grave. He noticed also that an aqueduct, originally covered, but afterwards left open, led thence to the Jaffa gate, which therefore must lie lower. It passes its south side and then enters the city; but he did not learn whether it ends at the Patriarchs' Bath (Pool of Hezekiah), as is generally supposed.

For centuries the name Hezekiah's Pool has been in use: it lies east of the Jaffa gate, on the west side of the street which runs northward to the Church of the Holy Sepulchre, on which account it used to be called by the monks the Pool of the Holy Sepulchre. Now it is generally known as the Birket el Hammam, from the circumstance that its water is commonly applied to bathing purposes. Sometimes the name is given in a fuller form, Birket Hammam el Batrak, *i.e.* the Bath of the Patriarchs,[2] because its water was used in the patriarchs' ablutions, according to the popular notion. Standing on the house of Mr Whiting, the American missionary, Bartlett was enabled to make a very perfect sketch of this basin, surrounded as it is by other houses, whose flat roofs and little domes he has retained in his picture, adding the scattered fig and palm trees which stand here and there, and on the left the top of the Church of the Holy Sepulchre. At the

[1] Gadow, p. 125.   [2] Krafft, *Topogr.* pp. 124, 125.

right the view extends down the Tyropœon valley. The massive rectangular building before the church belonged to the Knights of St John at the time of the Latin kingdom of Jerusalem. Eastward the eye runs past the Mosque of Omar to the Mount of Olives. According to Gadow, this pool is only full during the rainy season: it is eighty to a hundred paces in length, and fifty or sixty in breadth; up to a height of ten feet from the bottom, its walls are covered with cement in a good state of preservation.

As this pool was between walls, and was moreover between the walls of the upper and lower city described by Josephus, Schultz[1] believed that he had found in it a confirmation of the existence in that neighbourhood of the fortifications erected by Hezekiah, referred to in Isa. xxii. 9–11, in the words where the prophet appeals to the timorous, and those who have forgotten their God, at the time of the Assyrian invasion: "Ye have seen also the breaches of the city of David, that they are many; and ye gathered together the waters of the lower pool: and ye have numbered the houses of Jerusalem, and the houses have ye broken down to fortify the wall. Ye made also a ditch between the two walls for the water of the old pool: but ye have not looked unto the maker thereof, neither had respect unto him that fashioned it long ago."

Unquestionably this pool must have lain between the two walls (the first and the second of Josephus' account, though Robinson locates it within the second), and not within the city, whither it was necessary to conduct the water in order to supply the besieged, and to cut it off from the besiegers. It lay therefore outside, and failed accordingly in fulfilling the purpose of its builders; for the third wall, which now throws it within the city limits, was built long subsequently. Notwithstanding, the statement made in the passage quoted above, 2 Chron. xxxii. 30, "This same Hezekiah stopped the upper watercourse of Gihon, and brought it straight down to the west side of the city of David," seems to correspond closely to the present position of the Mamilla Pool, and that of the

[1] Schultz, *Jerusalem*, p. 83.

patriarchs regarded as the Pool of Hezekiah; on which account Robinson[1] was inclined to hold the latter as unquestionably Hezekiah's Pool, there being in his view no other way of explaining the use of the words "upper and lower pools" than by supposing that these were the ones indicated. Yet, aside from the failure of the object in view when the lower one was built,—namely, the supply of the city in time of siege,—the plain meaning of the words in 2 Kings xx. 20 seems to militate against this hypothesis, where we are told that Hezekiah "made a pool, and a conduit, and brought water into the city." It is not *to* the city, but *into* the city. With this the passage in Ecclus. xlviii. 19 agrees: "He fortified his city, and brought in water into the midst thereof; he digged the hard rock with iron, and made wells for waters." From this it would seem that the expression "the pool between two walls" may be referred to some other locality in the valley between Zion and Moriah, and that although Josephus, in describing the Patriarchs' Pool as lying "between two walls," has called it Amygdalon, or the Almond Pool, and given no indication of any connection of it with the aqueduct built by Hezekiah.

This Birket el Hammam, according to Robinson (whose estimate differs somewhat from that of Gadow), is about two hundred and forty feet long, a hundred and forty-four broad, but of insignificant depth. The bottom is made of rock, and is covered with mortar, and levelled. In the month of May it was found to be only half full, and could scarcely be supplied during the whole of summer from a source like the Mamilla Pool, which is itself so likely to be dry. At the northern end stands the Coptic Convent, at whose erection antique walls with bevelled edges were exhumed, making it probable that the pool once extended northward as far as the city wall. The expression employed in Isa. xxii. 9, regarding the collection of waters in "the lower pool," is no less applicable to the Birket el Hammam than to another and much larger basin below the Jaffa gate, near the turning of the Gihon valley, and its point of immergence in the Hinnom

[1] Robinson, *Bib. Research.* i. pp. 326-329.

valley. The latter pool cannot possibly be conceived as ever within the city. Although it is usually called by pilgrims the Lower Pool of Gihon,[1] from the circumstance that it derives its water from the upper pool, yet it is commonly designated by the Arabs Birket es Sultan. Pilgrims sometimes, however, give it the name of the Beersheba or the Bathsheba Pool, a title applied usually and more strictly to an insignificant little basin or ditch just at the left on entering the Jaffa gate. It has, however, been recently filled up, at the instigation of the French consul.

The large reservoir just mentioned, although from all appearance of great antiquity,[2] and possibly identical in situation with that mentioned in Neh. iii. 16 as opposite to the sepulchre of David, has received the name of the Sultan's Pool, from the circumstance that it was built contemporaneously with the erection of the wall on Mount Zion by Sultan Suleiman, between 1520 and 1526, as is testified by an Arabic inscription. The rocky walls of the Gihon valley form two sides of the pool: some flat stones are set up against these rocky faces, and two walls of hewn stone are laid directly across the valley to serve as the ends. Over the southernmost of these two the road which comes from Bethlehem runs, room being left for a fountain now dry, and the Arabic inscription above mentioned. As this great cistern is filled only a portion of the time with rain-water, it is by no means a superfluous task to supply the city by means of aqueducts, the best known of which is the one previously described, which comes from the pools of Solomon, and is conducted over nine stone arches, resting on the northern wall of this pool of Gihon. An inscription on this bridge of arches indicates that it was built by the Egyptian Sultan Mohammed ibn Kelavun, who reigned between 1294 and 1314. But this was unquestionably only the restoration of a much older structure, traces of which Wilson found on the Mount Zion side, in immense hewn stones, which convinced him that there was at an early period a gigantic watercourse

[1] Robinson, *Bib. Research.* i. pp. 24, 327; Krafft, *Topogr.* p. 185.
[2] Wilson, *Lands of the Bible,* i. p. 495.

there.[1] The city walls are a hundred feet higher than the pool, whose dimensions reach the remarkable figures of 592 feet in length, 245 to 275 in breadth, and 35 to 42 in depth. The aqueduct, which is known in modern times as that of Pontius Pilate, had fallen very much out of repair, but within recent years it has been restored by the Turkish governor, as a very important public work. Its course gives with tolerable accuracy the level of the ravine, from the lower pool on, where it crosses on the arches of masonry from the foot of the Mountain of Evil Counsel to the southern base of Mount Zion and the junction of the Gihon, Hinnom, and Tyropœon valleys. From this point Tobler believes that he has traced this Etham aqueduct, whose further course in the Tyropœon was formerly unknown, within the city wall from the Dung Gate northward to the Suk Bab es Sinesleh, *i.e.* to the residence of the kadi and the Mekhemeh, where it begins to supply the city with water.

We now turn, with Gadow,[2] northward from the Jaffa gate, and follow the Gihon valley, covered as it is with fine groups of olive trees, and during the rainy season with patches of grain and cucumbers. The city wall bears uniformly in a north-westerly direction, and the road runs parallel with it, though at a distance of some forty paces from it. It lies about eight feet below the lowest layer of stones in the wall, and follows the course of a kind of platform between it and the wall. The sides of this platform are in many cases regularly walled up: in others there are openings in it, as if once for the purpose of holding water. At the north-west corner of the city these subordinate walls rise to the height of four or five feet: they lie at a distance of ten or fifteen feet from the city wall, whose irregularities they follow to a certain extent, as well as does the course of the ditch, which here and there has a depth of from eight to fifteen feet.

At the well-defined north-west corner of the city, near the Mohammedan wely, there are to be seen in the wall, at a

[1] Wilson, *Lands of the Bible*, i. p. 494; Gadow, iii. p. 38; Tobler, in *Ausland*, 1848, Nos. 19, 22, p. 73.
[2] Gadow, iii. p. 41.

height of four or five feet from the earth, two immense capitals adorned with foliage, and set into the masonry in a reversed position. A couple of similar ones may be seen in the inner wall of the same wely. Here the ditch ceases. An old wall runs across it, its height the same as the depth of the fosse, and bearing traces of having once served as the basis of an ancient aqueduct. It is at this point, just in the neighbourhood of the Goliath citadel (Kasr Jalud) and the Latin Convent, the principal home of the pilgrims who visit Jerusalem, that the wall suddenly loses its direction north-westward, and turns sharply towards the east, thence to run across to the upper Kedron valley, where we started in our examination.

4. *The Northern Wall from Kasr Jalud and the Latin Convent eastward, past the Damascus Gate, to the north-east corner, near the St Stephen's Gate.*

The northern side of the city offers a surface so little remarkable for objects of special prominence, that just on this very ground must be sought the reason why there must continue to be so much uncertainty regarding ancient and modern Jerusalem, till future measurements shall lead us to more settled data than we at present possess. Even in the ground plan of the city drawn by Catherwood, elsewhere so valuable, and followed by Kiepert, Robinson, Schultz, and others, the deviations from the later results, gained in the official survey of Symonds and Aldrich (the great bowing convex of the northern wall being shown to be a mere gentle curve), are so marked, that it may be doubted whether an equally careful survey of the ground level, and the preparation of vertical sections, would not show as marked changes, or display as signal errors. The greatest difficulties lie in our way on the northern side of the city, and the acumen of observers has been exhausted in the effort to master them. The absence of prominent monuments on the scenes of great battles, the lack of great fortifications and ancient walls of defence, the masses of rubbish which have been heaped into ancient springs and natural watercourses, as well without as within the city, make it exceedingly difficult to follow the descriptions of ancient authors, such as those of the Old Testament and Josephus, clear as they may have been at the time, and intelligible to those for whom they were prepared. Even when there was a great deal of definiteness, they are hard to follow, and are capable of a diversity of interpretations; but in the case of Josephus the difficulty is much increased, from the fact that he wrote from memory, at a place and at a time different from those where and when the occurrences took place which he narrates. We must grant, therefore, that there is a large field for the play of conjecture: its labyrinthine walks we cannot wholly escape, but must confidently leave to a more advanced study of the place itself, based on future measurements and excavations, and to a more thorough examination of ancient documents, the solution of questions which are now unsettled.[1]

From the Mamilla Pool in the Gihon valley, Robinson[2] wandered along the north-western portion of the wall, passing a great terebinth or butm tree, and then crossed a rolling plain composed of hard limestone rock. The landscape was a desolate one to look upon: here and there olive trees and patches of arable land, but no vines and no fig trees: these only flourish where the land is lower, east-north-east of the Damascus gate. He then turned to the left, and crossed to the grotto of Jeremiah, which lies on the south side of a round hill, being entered through a perpendicular wall. It is surrounded by a small unwalled garden. The land gradually rises between the Damascus and Jaffa gates, and the highest part of the whole city is that which is found in the north-west corner around the Kasr Jalud and the Latin Convent. The roof of the latter commands the whole town,[3] even the highest part of Mount Zion, and affords one of the most extensive, and at the same time one of the most charming prospects in Jerusalem. This high position was near being perilous[4] to the

[1] Scholz, *Reise*, p. 166.
[2] Robinson, *Bib. Research.* i. pp. 233, 234, 238.
[3] See Parthey's MS. for a detailed view of these.
[4] Quaresmius, *Elucid. Terræ Sanctæ*, ii. p. 52.

convent, for in 1600 some hostile Turks in the city endeavoured to excite the jealousy of the governor of the city, and induce him to destroy the convent, because it overtopped the adjacent citadel. Between the grotto of Jeremiah and the city there is no valley, though outside of the wall the land rises somewhat higher than even within the walls at the north-west corner. From the Kasr Jalud to the edge[1] of the Valley of Jehoshaphat, Robinson estimated the greatest breadth of the city at 3060 feet, so that the Damascus gate lies not far from the middle. The whole northern wall presents an imposing appearance, with many sharp angles, towers, and battlements. The portion erected by Sultan Suleiman in 1542, and often repaired, is mostly made of large stones laid in mortar; its clefts and holes have become the refuge of countless lizards which may be seen crawling out. Stones with carved edges, dating from the Roman times, are mingled with those of more modern date. There is only one gate open on the north side of the city, the Bab el Amud, *i.e.* the Pillared Gate of the Native Born. From the fact that the road to Damascus passes through it, it is generally called, however, the Damascus gate. The exterior of this stately gate is from thirty to sixty feet in height. Outside there are ditches excavated in the rock; but as they do not run all the way along the wall, they would not be of much service in the way of defence. On the breastworks and behind the ramparts there is a line of wall, to which stairs ascend at various places. North-west of the Damascus gate there are a couple of gates which are walled up.

The remarks which Robinson makes regarding the topography of the northern side of the city are noteworthy. In one passage he says that "the surface of the high projecting mass of land on which Jerusalem stands sinks steeply towards the east, where it is bordered by the edge of the Valley of Jehoshaphat;" and further on he says, that "from the north side of the present[2] Damascus gate there runs a hollow or shallow wadi southward directly through the city, on the west side of which are the ancient hills of Acra and Zion, while

[1] Robinson, *Bib. Research*. i. p. 260.   [2] *Ibid.* i. p. 259.

on the east there are the less elevated Bezetha and Moriah." Between Acra and Zion there is a second depression or shallow wadi, which, however, is very readily traceable, running from the neighbourhood of the Jaffa gate, forming a cross valley whose course is east and west, and forming a junction with the other. The course of the united valleys is then southward, through a deeper valley bed, as far as to the Fountain of Siloam and the Valley of Jehoshaphat: this is the ancient Tyropœon. The verification of these two shallow wadis running in such different directions, the existence of both of which Robinson acknowledges, while supposing that the one running eastward from the Jaffa gate is the ancient upper Tyropœon, is at the basis of the great Jerusalem controversy, which has been so long and so ably disputed. On the one side we have Robinson as the great authority, supporting his views with masterly ability, and restating them even more emphatically in his supplementary pages than in the old edition of the *Biblical Researches*. Opposed to his views we have among those who first appeared, Williams and Schultz, subsequently Krafft and Gadow. The questions involved are of no slight moment; for on the course of the upper portion of the Tyropœon, whether northward towards the Damascus gate or westward towards the Jaffa gate, depends the interpretation of Josephus' account of the ancient divisions of the city, the position of the beleaguering camp, the location of the early Christian edifices, and the history of the great changes which the city has experienced. Regarding the controversy, I can only say here that Robinson has displayed, in a truly worthy spirit and manner, extensive learning and rare acuteness; and even where I cannot accept his conclusions, I am forced to admit that his book must be regarded as a model in the field of polemics. It was not to be expected, of course, that he could answer the arguments which have only subsequently been put forth by Krafft.

The cross valley, which Robinson says it is still perfectly easy to trace, is confidently affirmed by his opponents to have no real existence. At present it is only to be traced in the course of a street that lies low, the depression running from the Damascus gate southward. On the contrary, not only has it a hydrographical character of its own, indicating that it was once a shallow wadi, but it is even still called el-Wadi by the Jerusalemites, and has this to plead in its behalf, that it runs in a parallel course with the larger valleys on the east and west, those of Kedron and Gihon,—an arrangement which breaks the tongue of land extending southward into two minor and distinct projections. But both hypothetical Tyropœa are filled with vast quantities of rubbish, and so covered with crooked streets and houses, that it has always been exceedingly difficult to gain a clear conception of their topographical character. The inhabitants of Jerusalem do not give the name wadi to the cross valley, and many deny in explicit terms, Tobler[1] among the more recent, that any depression is to be traced there. Gadow, whose observations had been exceedingly close, in detecting the errors connected with the location of the northern wall as laid down by his predecessors, confirms the view taken in opposition to Robinson's, and traces the Tyropœon northward to the Damascus gate, indicating its origin by the expressive term "valley basin" (Thalbecken[2]).

The tongue of high land on which the city stands extends, according to Gadow, from the north-west corner past the Jeremiah grotto northward about 1200 paces, to the western bend of the Valley of Jehoshaphat. Its level is almost identical with that in the north-western portion of the city proper; only the rocky hill out of which the grotto of Jeremiah is hewn rises some sixty feet above this level. Where it forms the western wall of the Valley of Jehoshaphat, it is steep and bold; scarcely less so on the southern side of a valley running westward across the northern part of the great plateau. Here may be seen some large caves excavated in the rock, one of which is known as the Potters' Grave; on which account the Potters' Field has been located there, although there is no resemblance between the rock tombs found there and the other important ancient ones discovered

[1] Tobler, in *Ausland*, 1848, No. 18, p. 70.
[2] Gadow, in *Zeitsch. d. deutsch. Morgenl. Ges.* vol. iii. pp. 38–40.

in the neighbourhood of Jerusalem. Following the southern border of the valley indicated just above, a short distance brings one to the so-called tombs of the kings, known otherwise as Helena's grave. The spot lies forty paces east of the Damascus road, and eleven hundred paces north of the Damascus gate. This whole space would be ascribed to the former northern portion of the city, of which we have already spoken, if the course of Agrippa's wall, as traced by Schultz, coincided with the natural boundaries.

From the tombs north of the city to the city itself the Damascus road runs, while north of the tombs there may be seen traces of an ancient highway passing along the line of the watershed. A gentle hollow or furrow may be traced as far as to the eminence on which stands the north-western portion of the city, with the remains of the third or Agrippa wall (the Psephinos near the Goliath citadel). On the west of the Damascus gate this gentle depression deepens into a perceptible basin or valley, which passes southward through the city from that gate, and issues at the Ain Silwan as the indisputable Tyropœon of Josephus. Between the northwest corner of the wall, near the Latin Convent and the Damascus gate, the watchful Gadow discovered some ruined cisterns buried among the tangled roots of a very high and large tree (the great terebinth mentioned by Robinson), which threw its shade gratefully around, and afforded him a fine resting-place, whence, as the sun was gilding the west with his declining rays, he could look out and enjoy the prospect. He also mentioned the line of old walls, which lies about ten paces outside of the new one, and forms the border of a low terrace mentioned some pages further back.[1]

Wilson, who also noticed carefully the same part of the city which seemed to Bartlett, generally so prompt to see what was noteworthy, not particularly interesting, says of the wall which bounds the city on the north-west, and which had been examined with very little care, that he thought it well worthy of careful study, with a view to determine the line of ancient circumvallation. About three hundred feet south-west of

[1] Gadow, *i.a.l.* iii. p. 41.

the Damascus gate, he noticed that the stones bore in many places marks of great antiquity.[1] He was led to the conjecture, that the Saracens merely made their notches in the old wall, in order to make it resemble the new parts which they were building; and that from this cause the great antiquity of this part of the wall has been overlooked. He thinks it evidently a portion of the second wall, described by Josephus.

This Damascus gate, according to Bartlett, is a fine piece of Saracenic architecture; but Robinson[2] has the credit of first calling attention to the fact that it dates back to a still more remote antiquity than Bartlett supposes. Every one had noticed the large old-hewn stones which lay within the gate, on the east side; but on going round them and exploring them, he discovered a square dark room, evidently in close connection with the wall. Its sides were of the same character with the corners of the temple enclosure: the stones of colossal size, bevelled, and nicely finished, clearly resembling the carefully executed work which he had noticed in the Hippicus tower, but evidently of an older date. A staircase close by leads to the top of the wall, and is of the same architectural appearance. On the west side of the gate there is a second chamber, similar in appearance to the first, but much changed by modern alterations. The old stones of six, seven, and eight feet in length are undisturbed, however: they seem to be the relics of towers erected before the time of Herod, or, it may be, the remains of the watch-towers of a former gate, which, according to Robinson's theory, could only have belonged to the second wall, of which, however, he failed to detect the least trace in any other part of the city. It is plain, therefore, that in lack of any adequate data, all efforts to reconstruct that wall on a chart must be hypothetical; although some of the later antiquarians believe that they have here and there discovered traces of it.

According to the view of one of the latest observers—Krafft—who leans to the theory of the extension of the city northward to Agrippa's wall as located by Schultz, the third

---

[1] Wilson, *Lands of the Bible*, i. p. 421; Bartlett, *Walks, etc.*, p. 131.
[2] Robinson, *Bib. Research.* i. p. 313.

wall and not the second coincides with the present one, the one built by the Turks under Suleiman. His main reason for forming this conjecture is, that the Turks could find no better line of fortification than the ancient Romans, or than the Jewish defenders of the city. There is no part of the whole city and the district adjacent more favourably situated than those chosen by the Turks; and as a strong natural position was lacking at the north-west corner, the deficiency must always have been made good by such strongholds as the Psephinos and the Goliath citadel.

The weakest spot of all was the one where the wall crossed the shallow wadi of the Tyropœon in the neighbourhood of the Damascus gate. Krafft[1] conjectures that that is the reason for the occurrence there of the colossal stones on which the Damascus gate now rests, and which once played a prominent part, it would seem, in the defence of that part of the city.

This Bab Amud el Ghurab of the Arabic authors (Mejer ed Din), the present Bab el Amud, *i.e.* Pillared Gate, derives its name from the decoration of its pinnacles. Two towers, says Krafft, stand fifty feet apart, and on both sides, having very strong foundations, and connected with the side chambers, in which Robinson was led to see more plainly than any one had before done, by noticing the exterior, the peculiarities of the architecture. Traces of pipes have been found in the eastern tower, probably used to conduct water to cisterns.

The large quarried stones which are seen in these towers are some four feet high, in part furrowed at the edges, and roughly brought out in relief at the corners, yet on the whole more rudely executed than those at the Haram. Wilson thinks that they bear traces of the chisel, which would make them more analogous to the results of Saracen workmanship. He believes that he sees in this gate the gate of Ephraim of the Old Testament, because, looking northward as it does, it seems to indicate the way to the capital of Israel. Unquestionably it was once the main place of exit for those going northward

[1] Krafft, *Topogr.* pp. 42, 131.

to Nablus, by way of Ramah and Gibeah (now Jeba); and westward, by way of Gibeon and Antipatris, over the same road along which the Apostle Paul was carried as a prisoner to Cæsarea.

Titus, on his approach to Jerusalem, advanced from Gibeath Saul, where his camp was, on a tour of reconnaissance, accompanied by a guard of chosen men. So long as he held his course over the main road, no one appeared before the gate. But as he turned his course westward towards the Psephinos Tower, there suddenly issued from the gate opposite to Helena's tomb, and between the "Women's Towers," according to Josephus, a strong body of Jewish warriors, who divided his band, cutting off those in advance from those in the rear, and putting the emperor in the greatest peril. He succeeded, however, by dint of great personal bravery, in breaking through their ranks and effecting his escape. It is only this Damascus gate, whose two side towers closely correspond with the two mentioned by Josephus, the γυναικεῖοι πύργοι, which can be thought to have been the place of exit for the Jews engaged in this *sortie*.

East of the gate there is a great cistern, the remains, it may be, of a ditch or fosse once excavated from the rock, which made the towers, situated in a natural depression, seem still higher, and gave them increased strength for military purposes. Gadow did not fail to observe, that the rooms which are contained in the two sides of the gate are lower than the level of the street. In his map he has drawn this cistern or basin, whose original purpose is somewhat obscure, as a broad, deep excavation; and remarks that its southern side is the solid rock on which the city wall stands. This wide, deep excavation forms a part of the ditch which extends along the city wall; yet its south-eastern corner sweeps around in the form of a bow, and displays lines chiselled in the stone, which lead to the suspicion that they were once used to support some kind of wood-work frame. The northern border of the deep excavation is at present partially buried beneath the rubbish which has been heaped up there; but it shows that its height is just about the same as that of the two chambers which have been brought to light within the two towers. A flight of steps leads down to the great cistern, and to an arch which is in it, and which is supported by a single pillar.

Not far from this point northward lies the so-called grotto or cave of Jeremiah, whose entrance is hewn through a steep wall of rock, which makes an impression as if it were once a quarry, since the old quarries in this neighbourhood are very much used for tombs and caves; while, on the contrary, the old tombs and caves have perhaps ante-dated the existence of quarries opened where they once had been.[1] This grotto is at present in the possession of the Mohammedans, and has been used as the burial-place of some of their saints. On the eastern side there is an old broad cistern, whose arch is supported by a single pillar. In the stillness of night, when one lays his ear to the ground outside of the Damascus gate, a sound of running water is heard. There is an old story related by Antoninus Martyr, in the year 600, regarding a cave which he says lies by the side of a rock "near the Altar of Abraham," a location hard to make out: "Juxta ipsum altare est crypta, ubi, si pones aurem, audies flumina aquarum, et si jactas intus pomum aut quod natare potest, vade ad Siloam fontem et ibi illud suscipies." In another passage Antoninus repeats his story respecting the water heard running beneath the street which passes the ruins of Solomon's temple, but the spot is just as difficult to localize: "Ante ruinas vero templi Salomonis sub platea aqua decurrit ad fontem Siloam secus porticum Salomonis."[2]

Statements in confirmation of these singular facts have been made by Robinson and Wolcott, subsequently by Gadow, and still more recently by other travellers, which, when combined, give much interest to the position of the Damascus gate and its immediate neighbourhood, since these seem to give the key to the account of the water-works constructed by Hezekiah, and at the same time appear to show us the true commencement of the Tyropœon.

[1] Schultz, *Jerusalem*, p. 35.
[2] Antoninus Martyr, *Itin.* pp. 15, 18.

The first important discoveries of a subterranean canal were made by Robinson and Eli Smith[1] outside of the southern part of the city. It conducts the waters of the Virgin's Spring to Siloah. Its discovery led the way to further inquiries where the first source is to be found. It was traced farther back, and a body of water was discovered deep below the rocks of the Mosque of Omar, on the city side of the temple enclosure; the medicinal springs are fed by it at a distance of only a hundred and twenty-five feet from the western side of this enclosure.

In January 1842, Mr Wolcott undertook an exploration of these watercourses, and carried out the bold scheme, though not without danger.[2] He was let down into a well eighty feet deep, situated in the heart of the Turkish quarter; he then followed the watercourse horizontally for a distance of a hundred and twenty-four feet eastward, until he came to other borings leading to the west wall of the Haram; yet he had to turn and retrace his steps at a distance of forty feet from that wall, and failed in effecting a passage as far as to the reservoirs which are supposed to exist below the Mosque of Omar.

Robinson had already known of this deep and well-supplied well, from which the neighbouring baths are daily furnished with water brought by peasants, but he did not succeed in obtaining more particular information. These baths are the so-called mineral ones of Hammam es Shefat. The way thither is through the now ruined Cotton Bazaar, which is located near the Cotton Gate (Bab el Katanin). The building of this well so near to the Haram, together with two other marble wells, is ascribed by Mejr ed Din to the Sultan Selim I.; and this is confirmed by the Jewish pilgrim Jichus ha-Abot,[3] who wrote in 1537. The Bab el Katanin had, however, been restored in 1537 by a Sultan Mohammed, a son of Kelavun, in 1387. What Wolcott was obliged to leave incomplete, Tobler completed[4] under more favourable circum-

---

[1] Robinson, *Bib. Research.* i. pp. 343–349.
[2] Wolcott, *Bib. Sacra*, 1843, Feb. pp. 24–28.
[3] Carmoly, *Itin. Jichus ha-Abot*, pp. 436, 437, Note 43, p. 468.
[4] Tobler, in *Ausland*, 1848, No. 19, p. 73.

stances in March 1846. He names the medical baths of Ain es Shefa "the Hygæan Springs." He descended into the deep well in order to trace it to the "water chamber of the rocks." Without difficulty, says his brief account, he passed through the watercourse, lying far below the surface, following no straight line, but turning a sharp angle, till at length he reached a cistern in the rocks. The measurements which he made, and the direction of the needle, showed him that it lay west of the temple area, and outside of it. He was unable to learn whether in the dry season the water sinks very low in this cistern, and after heavy rains rises so much as to throw its waters far up beneath the site of the ancient temple. A more detailed account was promised by Tobler in his full description of Jerusalem.

The lack in local springs which characterizes the soil of Jerusalem was made good by cisterns, reservoirs, and aqueducts, which are among the greatest local peculiarities of the city, and which have had the most momentous influence on its history. The want of brooks in the neighbourhood; the great differences in the amount of rain that falls, and the very small aggregate which characterizes some years; the deficiency in fountains; the number of cisterns, with one or more of which almost every house in the city is supplied; the size of some of the subterranean reservoirs—the one, for example, near the so-called Treasure House of Helena, to which fifty-two steps [1] descend before the surface of the water is reached; and, in addition to this, the immense supplies of water which we know from the records of Jewish antiquity and the middle ages were stored beneath the Haram and the temple terrace; [2] the great pools still visible, from the colossal ones known as Solomon's near Bethlehem, to those of various size in and around Jerusalem;—all these things, when combined, and studied in their relation to the history of the city, solve many an obscure question, and enable us to understand how it was that armed foes sometimes surrounded the city, and were compelled to withdraw from a lack of water, while those

[1] Krafft, *Topog.* p. 183.
[2] Williams, *Holy City*, ii. p. 462.

within were amply supplied. It cannot be denied that a natural valley like the Tyropœon, which crossed the city from north to south, was especially well adapted to supply Jerusalem with water, if the art of man was advanced enough to ensure the perpetuity of the supply.

This shallow depression, which extends from the Damascus gate southward, and which seems admirably adapted by nature to gather into itself an ample supply of rain-water, would be still more adapted for its purpose by deepening the channel. Such a hypothesis is fully supported by statements made in the Old Testament; for if Solomon supplied the city with water at the period of its greatest splendour, Hezekiah thought upon the problem in the time of distress, how to cut off the water which was brought into Jerusalem on the northern side from the advancing enemy, and to accumulate it within the city for the use of the people. It is on this undertaking that his fame mainly rests,[1] and we find marked allusions to it in the books of Kings, Chronicles, and Ecclesiasticus.

Under Ahaz (B.C. 741–726), the father of Hezekiah, Jerusalem was threatened with an assault, evidently on the north side, by the combined forces of Rezin king of Syria, and Pekah king of Israel. The prophet Isaiah received command to go out to Ahaz "at the end of the conduit of the upper pool in the highway of the fullers' field," and to bid him have no fear, but to take courage, for the victory would be in his hands. Ahaz had unquestionably gone out at the northern gate to the place where he must meet the enemy on their approach, and where lay the upper pool and its pipes, in order to see how he might cut off the supply of water, and strengthen the city more perfectly. But Ahaz,[2] not daring to trust himself to the promises of Isaiah, made a treaty with Tiglath-pileser, the king of Assyria, purchasing his good services with treasures from the palace and temple. The Assyrian monarch, in fulfilment of his share of the agreement, attacked the enemy, and compelled Rezin and Pekah to withdraw to their own country, in order to defend

---

[1] Gesenius, *Comment. zu Jesaias*, i. pp. 691, 692.

[2] Krafft, *Topogr.* p. 114.

it against him. By this step Ahaz became a vassal of Tiglath-pileser, and paid him homage at Damascus; and in order to remain more sure of a continuance of his protection, he introduced the heathen worship, and brought disgrace upon the temple of God (2 Kings xvi. 12).

After the death of Ahaz, Hezekiah his son purified the worship of Jehovah from the idolatries which his father had introduced, and threw off the yoke of vassalage to the Assyrian monarch, after Shalmaneser had utterly destroyed the rival kingdom of Israel. The result was, that in the fourteenth year of Hezekiah's reign, Sennacherib attempted the subjugation of Judah and the capture of Jerusalem; but before attacking the Jewish capital, he passed onward to Egypt, with the hope of conquering Tirhakah, the monarch of that country. Anticipating a future attack from the Assyrians, Hezekiah took counsel with his chiefs (2 Chron. xxxii. 2-6), and decided " to stop the waters of the fountains which were without the city; and they did help him. So there was gathered much people together, who stopped all the fountains, and the brook which ran through the midst of the land [the Kedron, or the waters of the Siloah spring, which may have been uncovered at that time, and been in full sight], saying, Why should the kings of Assyria come, and find much water? And he strengthened himself, and built up the wall that was broken, and raised it up to the towers, and another wall without [this Schultz considers the second wall of Josephus],[1] and repaired Millo in the city of David, and made darts and shields in abundance."

The carrying out of this plan, whose object was to cut the enemies outside the city off from supplies of water, while the people within should have an abundance, was what ensured to Hezekiah his greatest fame, as we learn from 2 Chron. xxxii. 30, 2 Kings xx. 20. This is confirmed by the language of Ecclus. xlviii. 17: " He fortified the city, and brought in water into the midst thereof: he digged the hard rock with iron, and made wells for waters." In the statement made in Chronicles, we are expressly told that the

[1] Schultz, *Jerusalem*, p. 89.

CIRCUIT OF THE PRESENT WALLS. 91

course of the channels which he opened was from north to south. The words are: he "stopped the upper watercourse of Gihon, and brought it straight down to the west side of the city of David." When the Assyrian king had passed victoriously over southern Judah, taking many of the fortified cities, instead of turning directly up towards Jerusalem, he went on as far as Lachish, in order to attack Tirhakah the Egyptian monarch. Here, however, he was brought to a halt. He then sent his ministers Rabshakeh and Tartan to Jerusalem, to have a conference with king Hezekiah, and to demand tribute and auxiliaries. The interview took place at "the conduit of the upper pool, in the highway of the fullers' field" (Isa. xxxvi. 2). But in spite of all the threatenings of the Assyrian princes and their master, and in spite of the formidable display of their immense army of 185,000 men encamped on the plain before Jerusalem, the city was spared; for we read in Isa. xxxvii. 36, that the angel of the Lord passed through the enemy's camp in the night, and smote the whole army. The Assyrian king escaped, and fled to Nineveh. This rescuing from impending destruction is referred to in the reproaches poured out in Isa. xxii. 9–11 upon those timid souls who lost all confidence in God, and all remembrance of Him. But, in the words of the verses cited, there are some marked indications of the situation of the upper pool, and of the collecting of water, after the closing of the "old pool," in a new reservoir within the city, and between the two walls. The locality here indicated appears from historic reasons, and from the apparent connection of the place with the Potters' Field, to have been beyond all question north of the city, and before the Damascus gate. The fact that the Assyrians encamped there is confirmed by Josephus,[1] who states that Titus took up his position in the same place which had once been held by the Assyrians for that purpose. This may, with Schultz,[2] be located more on the west side of the gate, and in the neighbourhood of the present Latin Convent, or with Krafft, on the east side near the

[1] Krafft, *Topogr.* p. 47.
[2] Schultz, *Jerusalem*, pp. 72–74; Krafft, *Topogr.* pp. 81–83.

Kedron valley; yet the main fact is unchanged, only so far as it may be affected by the allusions to the Potters' Field, the Fullers' Field, the Dyers' Field, and the local monuments of the place. The same doubt rests upon the locality of the royal caves mentioned by Josephus, and which he once calls the Tombs of Herod, but which, in his description of the beginning and the end of the wall which Titus erected to starve the city into surrender, he tells us lay at the extremity of the military lines. Schultz looked for these on the west side of the Mamilla Pool, but Krafft identified them with the grotto of Jeremiah.[1] They have only borne this name, according to him, since the fourteenth century, it having supplanted the more ancient one, the "Royal Caverns." Their place is, however, clearly designated by Josephus, as, in connection with the "Tombs of Herod," he speaks of the "Serpents' Pool," which indicates the great reservoir in front of the Damascus gate, and the grotto of Jeremiah, and the situation of the old pool of Hezekiah's times, even though under a changed name.[2]

This great cistern is a mere temporary and casual receptacle for water: it is always full, and the taste of the water in it is said by all to correspond with that of the Virgin's Fountain and that of Siloah. This taste is a very peculiar one;[3] and a chemical examination might show that the components are identical, and that they must have a common source. The current story of the Jerusalemites, reported by Gadow, Krafft found to be true, that if one puts his ear to the ground on going out from the Damascus gate, and bearing a little to the right, he hears the sound of running water, which may be traced through the middle of the city as far as below es-Sakrah, *i.e.* the rock where the springs are found, below the Mosque of Omar. This story, which I have already cited from Antoninus Martyr, is repeated in the Arab Mejr ed Din's description of Jerusalem, written in 1495, who, in his account of the Maghar Katanin, *i.e.* Grotte de Cotton, by which he seems to mean the neighbourhood of Jeremiah's

---

[1] Krafft, *Topogr.* p. 220.   [2] *Ibid.* p. 220.
[3] Williams, *Holy City*, ii. p. 455.

cave, writes these remarkable words: " Opposite es-Sakrah, above the north side of the city wall, there is a large and long cave, called the Katanin, or Cotton Grotto, of which some assert that it follows a subterranean course till it reaches the rock es-Sakrah below. It has unquestionably received its name from the circumstance that its direction is towards the Mosque Kubbet es Sukhra, since the same authority speaks of a Cotton Gate on the west side of the Haram."[1] Of this Cotton Gate, Jichus ha-Abot[2] says that it is so named from the cotton bazaar close by, to which it leads, where was one of the three most copious marble fountains erected by the Sultan Selim I.: unquestionably the same with the one discovered by Wolcott, and which, according to Williams, is used at the present time to supply the Healing Baths (Hammam es Shefa). These three wells in the western neighbourhood of the temple enclosure pour their waters, according to Mejr ed Din, into basins of white marble, and supply the wants of Jews, Arabs, and Christians. He adds, that in the mosque there were thirty-four cisterns in which water was collected. Robinson,[3] to whom the investigation of this subject is indebted for an entirely new impulse, made preparations for extensive researches regarding the supply of water for Jerusalem, but he was hindered by external circumstances. He called particular attention to a remarkable passage in Tacitus, *Hist.* v. 12—" Templum in modum arcis. . . . Fons perennis aquæ, cavati sub terra montes, et piscinæ cisternæque servandis imbribus," etc.—which probably shows that he was acquainted with a fact which was entrusted as a secret to the high priest, and of which Josephus does not seem to have ventured to speak. Robinson also pointed out the statement made by Aristeas, a priest of the time of Ptolemy Philadelphus (285 B.C.), who visited[4] Jerusalem, and who, although with some

[1] Schultz, *Jerus.* pp. 35-37; Williams, *Holy City*, i. App. ii.; Mejr ed Din, pp. 150, 163.
[2] Carmoly, *Itin.* cited as above.
[3] Robinson, *Bib. Research.* ii. p. 163.
[4] Aristeas, *De Legis divinæ translatione in Josephi Opp.* ii. p. 112; Lightfoot, *Opp.* i. 612.

exaggeration it may be, yet remains an authentic authority regarding the position of the city, the temple, and Akra. Yet Robinson grants that all the circumstances, taken together, make it not improbable that there was once, and still is, a hidden channel through which the water flows from the deep reservoirs under the mosque to the Tyropœon valley. But from what region they are conducted to these cavernous reservoirs, is a question still beset with the greatest difficulties. That the whole was artificial, he adds, there is little reason to doubt; and it could perhaps be suspected with good reason that these water-passages were connected with the old Gihon spring on the high ground west of the city. To me, however, it seems much more probable, much more conformable to all the conditions of the case, that they were connected with the north side of Jerusalem.

Aristeas relates,[1] among other things, that beneath the temple there is a constant supply of water, as if a profuse natural spring were throwing up its waters there. His words may be taken, however, to mean, that instead of there being such a spring, there is the free outflowing of water from a conduit. He remarks that there are wonderful and indeed indescribable reservoirs under the earth, found within a circle of five stadia from the foundation of the temple. These reservoirs are connected together by means of pipes; their bottoms and walls are covered with lead, while they themselves lie beneath deep accumulations of earth. Numerous openings lead to them, but these openings are all out of sight, and are only known to those whose duty it is to attend to them. Aristeas goes on to state what he has learned regarding the aqueducts employed to conduct the water: "More than four stadia from the temple, and outside of the wall, I was bidden to kneel and listen. I heard the sound of running water, and I understood that I must have been right regarding the size of the main canal as I have already given it."

The cistern at the Damascus gate, says Krafft, is about four stadia from the temple; and he does not doubt that

---

[1] Robinson, *Bib. Researches*, i. p. 345, Note 2; Krafft, *Topogr.* p. 131; Williams, *Holy City*, ii. p. 462.

Aristeas was conducted to this point, and perhaps a little way farther north, and that he heard that rushing sound which is heard even now at certain times. This is probably the same channel which was stopped by Hezekiah—the upper outlet of Gihon; the brook which "flows through the land" which Hezekiah covered (2 Chron. xxxii. 4); the ditch which Isaiah mentions (xxii. 11), which was made between the two walls for the water of the "old pool." How the stopping and covering of such watercourses from the upper to the lower pool was effected by means of subterranean chambers,[1] is shown by the example already cited of the cisterns of Solomon at Etham. From it we can study with an almost assured feeling of certainty the manner in which Hezekiah accomplished the great operations which made his reign so conspicuous. That the pipes which convey water to all sides of the Haram run more naturally down the Tyropœon valley from the Damascus gate than from the Zion side, is now, I think, evident. The writings of the Talmudists are full of allusions to a vast system of water conduits under the temple, which indeed the great number of sacrifices and the later uses of the Haram rendered necessary; and in all the history of the place, while there have been during times of siege many instances of suffering from a lack of food, there is not one of a scanty supply of water. And even now it is a noticeable fact, that the three most profuse sources around the temple enclosure yield a water with the peculiar, insipid, and salt taste of Siloah. This Robinson noticed at the well known as Hammam es Shefa,[2] and Krafft at the cistern in the neighbourhood of the old pool, between the Damascus gate and the cave of Jeremiah. The natural inference from this is, that the water in all comes from a common source. On the north side of the temple enclosure and the Serai, *i.e.* the dwelling of the Turkish governors (the Pretorium of the Roman epoch, where Pilate's house is located), there is a very deep and profusely supplied well, of great width from north to south, and cut in the solid rock. It belongs to the partially ruined Franciscan church, the

[1] Robinson, *Bib. Research.* i. p. 346.   [2] *Ibid.* i. p. 343.

Chapel of Flagellation,[1] on the Via Dolorosa, and probably receives its water from an eastern arm of the main channel which runs along the west side of the Haram. The taste is exactly like that of the Siloah spring, according to Williams;[2] so peculiar, indeed, that it can be confounded with no other. The same is true of the well at the so-called Bab el Katanin, which supplies water to the Healing Baths (Hammam es Shefa).

But whether there is a systematic connection of subterranean water conduits permeating the entire city, and communicating between the Haram well of es-Sakhrah and the Siloah spring at the southern part of the Tyropœon, can only be determined in the future. The prophets Ezekiel (xlvii. 1-12), Zechariah (xiv. 8), and others, draw impressive pictures in their visions of the waters of true life which stream out from the temple; and in the Apocalypse (xxii. 1) we have the conception of a stream of pure water flowing forth clear as crystal from beneath the throne of God and the Lamb.

We now turn once more to the northern wall of the city, on the east side of the Damascus gate. There is but a short distance to be traversed before we come to the north-east corner, and then rounding this to St Stephen's Gate, which completes our circuit of the city. East of the Damascus gate and the so-called cave of Jeremiah,[3] which, although so called, has no historical connection with the prophet, the wall makes a well-defined bend, and runs above a perpendicular wall of rock hewn by hand, and exhibiting the same stratification as that of the opposite wall of the knoll in which the grotto of Jeremiah is seen. The height to which the rock formation is carried is the same in both, and they were unquestionably both connected: the passage which now divides them is artificial, and dates back to the time of Herod Agrippa, who here erected his third wall, upon whose site the present Turkish circumvallation was built. In the neighbourhood of the Maulawiyyeh Mosque, at an early period one of the most

---

[1] Robinson, *Bib. Research.* i. p. 252.
[2] Williams, *Holy City*, ii. p. 461.
[3] Gadow, in *Zeitsch. d. deutsch. Morgenl. Ges.* iii. p. 39.

important Christian churches, the wall again runs in toward the city; yet the natural rock foundation on which it stands, on which towers stand which rise eighty or a hundred feet above the fosse, extends as far as to the little Bab el Zahari[1] or Flower Gate, which takes its name from its decorations. It is also sometimes called the Herod Gate, probably from the palace of Herod Agrippa, which is reputed to have stood there. It has been walled up in our time by Ibrahim Pasha. East of it lie the small pool known as Birket el Hijeh, or Haj, and the great Corner Tower of Josephus, whose primitive foundations are perhaps still recognisable. I have already spoken of it, and will therefore omit any further description at this time. This end of the wall runs parallel with piles of rock fragments and rubbish, partly filling the fosse, which is hewn out of the solid rock along the whole length of this northern portion of the wall. Upon these heaps there are some olive trees growing; and in the rainy season there are patches of barley and wheat extending to the north-east corner, where we come again to the sterile primitive rock. The Herod Gate is mentioned by the pilgrims of the middle ages, Arculfus for example, as the Porta Villæ Fullonis, because in a side recess there is a fuller's monument which is somewhat visited. It may have some connection with the passages already cited in Isaiah relating to the Fullers' Field, and the site of the Assyrian camp.[2] It is a simple, well-preserved, and large rock-tomb, beautifully environed by olive trees and vines. East of it are many other similar remains of graves,—objects of less interest, however. This whole northern side of the city is dotted with groups of olive trees, and traversed by beautiful paths as far as to the north side of the cave of Jeremiah and the burying-ground of the Mohammedans. It is a favourite evening resort of the leading Moslems of the city. In this same locality, that of the ancient Fullers' Field, may be seen the Pool of the Soap Plant,[3] mentioned by Josephus, near which, during the Roman

---

[1] Schultz, *Jerusalem*, p. 37; Krafft, *Topog.* pp. 44–47.
[2] Krafft, *i.a.l.* pp. 118, 121, 138, etc.
[3] Tobler, in *Ausland*, 1848, No. 20, p. 78.

siege, some walls were thrown up. The word στρούθιον is said to be the Greek name of the *Saponaria*, which corresponds to the *herba fullonum* alluded to in Isaiah's account of the Fullers' Field. Mejr ed Din[1] says that in the neighbourhood of Bab el Zahari there were soap factories in his time.

DISCURSION IV.

THE INTERIOR OF THE CITY OF JERUSALEM: ITS PRESENT PHYSICAL CHARACTER, AND DIVISION INTO STREETS: EL-WADI, OR THE STREET OF MILLS; THE TYROPŒON; THE SITUATION OF THE BATHS AND WELLS ON THE WEST SIDE OF THE HARAM. THE DOUBTFUL SITUATION OF AKRA; THE ANTONIA; THE SERAI; THE TEMPLE ENCLOSURE ON MOUNT MORIAH; AND THE MOSQUE OF OMAR—KUBBET ES SUKHRAH.

Coming from a survey of the outer wall of the city, which is so extensively planned and executed as to impress the observer with the fact, confirmed by the character of the country immediately adjacent, that the place is by nature a true royal residence; and also from the preliminary survey from the Mount of Olives of the region within the range of the eye, which shows the remarkable beauty of the situation,—entering the city, the piles of rubbish and the narrow streets compel us to recognise the fact that it is no longer a royal capital, princely in its magnificence, but a squalid town, which shows only too plainly its humiliation and poverty. As a recent traveller has truly and beautifully said, To him who does not see this city with the eye of faith, and who, amid all the strife which now divides the church, does not look forward to the glorious triumph which awaits it, Jerusalem is only a little eastern city covered with the wrecks of past desolation, suffering under want and oppression, and from which the casual traveller hastens as rapidly as possible. But the classic ground, with its history extending over thousands of years, remains, under all its rubbish and ruins, still classic; and so it will remain, rewarding, as does the equally imperishable Rome, the patient

[1] Mejr ed Din, in Williams, *Holy City*, i. app. ii. p. 159.

THE INTERIOR OF THE CITY.

inquirer who penetrates into its deposits with discoveries of great value.

There are a number of circumstances which combine to make it extremely difficult to overleap the present condition of the city, and to come to results regarding its former appearance which shall be accepted by all as true. The task which I have prescribed to myself is by no means to enter into all questions, and to discuss the grounds on which they rest; for the ultimate decision regarding them is to be made by those who shall visit the place itself, and enter into a thorough examination of all the topics involved. As, however, there are certain names given to the various parts of the city, and a certain amount of knowledge presupposed regarding it, I cannot avoid, in order to afford the reader the opportunity of judging of the disputed questions regarding the topography of the city, giving a brief *resumé* of the subject, drawn in great part from the work of one of the most recent scientific travellers, Dr Philip Wolff, who in his seventeen brief papers has condensed a great deal of valuable matter, and has hinted at most of the points which are held by some, and the grounds on which they are disputed and denied.[1]

Fortunately there are some leading points, such as Mount Zion, Mount Moriah, the Valley of Jehoshaphat, the Valley of Hinnom, the Mount of Olives, the Tower of Hippicus, the Fountain of Siloam, etc., regarding whose situation there is no question; on the other hand, as we go into the city, we are at once confronted with difficulties in making out the various details of the topography. In the first place,[2] within the city there is a lack of established names for the streets,— a want which we feel at once when we wish to define in brief terms the main places of interest. The map of the city does not by any means remove this difficulty; for even in the best delineation, the countless little corners and inaccessible quarters of the city are left in a very imperfect manner, and throw little light upon its topography. Sometimes there is an unexpected settling of the great accumulation of rubbish,

---

[1] Dr Philip Wolff, *Reise in das Gelobte Land*, pp. 75–89.
[2] Gadow, *Mitth.* in *Zeitsch. etc.* iii. p. 42.

which introduces great confusion; sometimes whole quarters of the city are forbidden ground to the stranger, and can only be entered by the natives. Meanwhile there is one marked physical feature in the city which can be traced from one side to another: this is the depression which extends from the Damascus gate south-south-eastward, and which in its lowest portion forms unquestionably the ancient Tyropœon. This divides this city of hills into two distinct parts: Mount Zion, with the high dome of the Church of the Holy Sepulchre and Golgotha, on the one side; and on the other Mount Moriah, with the Mosque of Omar, with Ophel on the south and the Turkish government buildings, the Antonia fortress, and the Bezetha quarter on the north. Akra, whose situation Josephus has so indefinitely located, is supposed by Robinson and Raumer to have been on the west side; by Williams, Schultz, and Krafft, to have been on the east.

As one leaves the Damascus gate and enters the city, the street runs for fifty or sixty paces down a gentle slope (according to my view, down the upper course of the Tyropœon) to a small open square, from which five streets diverge towards the south-east, two of them being partially hidden by fragments of old buildings, lying nearly as low as the two side chambers of the Damascus gate.[1]

The street which runs southward from the little square just mentioned, I will speak of first. It is generally called the Damascus Street, and passes through the entire city as the main thoroughfare, bending a little, and then crossing Mount Zion, from which cause it is sometimes called Zion Street. It terminates a little distance east of the Zion gate, and near the houses of the lepers. In its northern half it forms the eastern boundary of the Christian quarter, and the western boundary of the Mohammedan. It passes the so-called Corner Gate (Porta Judicialis), leaves on the west the Church of the Holy Sepulchre and the terrace of St John's

[1] The following general description of the city can only be understood by comparing the revised plan of Gadow with the survey of Aldrich and Symonds, which differs from it as well as from that of Tobler.

Convent, with its gate on which are still to be seen the traces of representations of the animal world, and passes through the long bazaar in the heart of the city, extending up through many vaulted lanes to Mount Zion. Up to this point the course of the second wall, as laid down on Schultz's map, coincides with this. East of this Zion Street, and hard by it, is a parallel one, which Tobler calls Haret el Jehud, the Jews' Market Street,[1] or Street of the Arch of Jehuda on Symonds' map. It is not merely a thoroughfare, but a real though slight valley, which divides Mount Zion into an eastern and western half, which long passed unobserved, unquestionably because it was seldom traversed. A second main street, which comes from the Jaffa gate, crosses the street just mentioned at right angles, and after making several bends at the southern part of the bazaar, pursues a generally straight course towards the west wall of the Haram, and the Mekhemeh or Council House. It separates the Jewish quarter in the south from the Mohammedan quarter in the north. As it approaches the Haram it divides into a number of smaller streets, which form the Haret el Mugharibeh. This street, which passes the northern base of Mount Zion, and which, according to Robinson, is the modern representative of the ancient Tyropœon, is commonly called the Street of David, excepting in its more eastern portion near the Haram, which bears the name of the Temple Street. Tobler insists that to the eye, as it looks down this street, there is no appearance whatever, at the present time, that it was ever a valley. Neither can the slight hollow directly north of the castle at the Jaffa gate, between the Square or Place of the Citadel and the Latin Convent, be designated as a valley. From the Damascus Street—which, when carefully observed, slopes gradually towards the cross David Street—there run eastward several lanes towards the Mohammedan quarter, which have this one feature in common, that their course is towards the Tyropœon or depression from the Damascus gate southward, as Schultz and Krafft understand it, in contradistinction to Robinson. All these streets, therefore,

[1] Tobler, in *Ausland*, 1848, No. 18, p. 70.

have the same slope. The hollow which passes through the city is designated upon the map of Aldrich and Symonds as the Street of the Mill Valley. On the earlier one of Catherwood it is entirely wanting, on Gadow's it has no name. Although Mejr ed Din[1] described it fully, yet, as it lies in the heart of the Mohammedan quarter, it has not been considered advisable for Christians to walk through it, and so has been little known. Only on Tobler's map of the city, which contains many new details that supplement that of Symonds and Gadow, is this hollow designated by the name given it by the inhabitants—el-Wadi. They apply the term to its whole extent, from el-Mugharibeh as far as to the Damascus gate. In his text he remarks that this designation is in common use from the Sultan's Baths (Hammam es Sultan) in the north, as far as to the Suk Bab es Sinesleh (*i.e.* to the so-called earth wall of David Street), near the Haret el Mugharibeh.[2]

Gadow, who earliest called attention to the sloping direction noticeable in some of the minor avenues, says that the first cross street running from the small square already referred to near the Damascus gate, runs across the Damascus Street in a north-easterly direction.[3] The next prominent thoroughfare running in a similar course is the Via Dolorosa, or Tharik el Alam, which crosses the Damascus Street at nearly right angles near Porta Judicialis, and runs eastwardly towards the Street of the Mill Valley: it then runs northward for a short distance following this Valley Street, until it comes to the spot where Simon of Cyrene is represented as taking up the cross, where its eastern course begins, and gradually ascends to the house of the Turkish governors and the Chapel of Scourging. This Via Dolorosa has only been so called since the time of the crusaders: before Marin Sanutus no mention is made of it.

All the streets running from the main thoroughfare, or Damascus Street, as well as those which run from the Church of the Holy Sepulchre, the Prussian Consulate, and which

---

[1] Mejr ed Din, in Williams, *Holy City*, i. app. ii. p. 158.
[2] Tobler, *i.a.l.* p. 70.   [3] Gadow, as above quoted.

pass the Tekiyeh or Hospital of Helena (Akbet el Tekiyyeh el Sahahira, an institution for the poor, now in ruins, but a monument of Moorish architectural skill and of Mohammedan beneficence[1]), and run eastward through the Moslem quarter, display a marked slope towards the deepest part of the valley.

The street, says Gadow, which runs eastward from the square already mentioned, near the Damascus gate, and which crosses the Via Dolorosa [he refers, without naming it, to the Street of the Mill Valley, or el-Wadi], forms, up to the place of its junction with the David or Temple Street, the border of the slope[2] which declines towards it from the west. As the inclination towards it of all the streets of the eastern side shows, it follows a deep valley which separates the north-eastern portion of the city from the north-western.

The Via Dolorosa—a street visited by all pilgrims on account of its reputed connection with the toilsome walk of our Lord to the scene of His crucifixion, and now dotted all along with the "stations" which the monks have set up—leads from the above-mentioned bend at the station named after Simon of Cyrene, passes a rubbish-heap surrounded by a loosely constructed wall, then the Ecce Homo arch (the reputed place where the Saviour was crowned with thorns), the Scala Santa, and the Chapel of Scourging, imperceptibly ascending to the official residence of the Turkish governors. The cross alleys running northward from it run steeply up to Bezetha, passing the Chapel of Scourging.

About sixteen paces east of this chapel, the Via Dolorosa slopes perceptibly towards a hollow course running thence northward; the street follows this, passing the ruined convent Deir el Addas (place of Mary Magdalene's repentance), as far as to the now closed gate of Herod, or Bab el Zahari. In this way, the north-easterly portion of the city is subdivided into its eastern and western hills, the latter one of which extends beyond the northern wall as solid rock, there to be divided artificially, as I have already mentioned, by

---

[1] Schultz, *Jerusalem*, p. 32.
[2] This place is omitted in Gadow's text.

the passage which separates the cave of Jeremiah from the city.

We now leave the rising land on the north and north-west sides of the Haram, which, according to the view of Schultz and Krafft, are interesting in connection with the ancient Acra: we pass to the west side of the city, the Lower Town, or Acra of Josephus, according to Robinson and Raumer. Despite all the efforts which have thus far been made, the question is not yet freed from many great difficulties.[1] This north-western portion of the city, the Christian quarter, lies upon a slope, says Gadow,[2] which descends very uniformly and gradually on the street leading from the Jaffa gate to the Bazaar, but far more steeply on that which runs along the north side of the Patriarchs' or Hezekiah's Pool towards the Church of the Holy Sepulchre. The slope is less steep, however, farther north, from the Latin Convent past the Greek Patriarch's Court to the Porta Judicialis on the Via Dolorosa.

The part of the city which the ruins of the St John's Convent and the Church of the Holy Sepulchre, with the extensive buildings which were connected with them, occupy, extending as far north as to the former palace of the Latin Patriarch, lies upon a uniformly level tract, forming a bow-shaped section taken out of the whole slope which runs westward from the Patriarchs' or Hezekiah's Pool, and from the present residence of the German Protestant bishop, and which is connected by a kind of isthmus with Mount Zion, and more directly with the citadel at the Jaffa gate. Here there must be a natural hollow, a shallow wadi, as at the Damascus gate, if the real Tyropœon valley connected the waters which issued from the upper pool, and which were stopped by Hezekiah, in the upper Gihon valley, and brought them into the interior of the city. In order to show the possibility of such an enterprise, Robinson calls attention to the vast collection of rubbish which has collected north of Mount Zion, extending in some cases to a depth of twenty or thirty feet: he adduces the discovery of subterranean

---

[1] Tobler, quoted as above.   [2] Gadow, as above.

canals and vaulted passage-ways, whose original purpose seems exceedingly uncertain, and which are conjectured to have had some connection with the works which underlay Herod's palace and gardens, regarding the means of watering which, as well as regarding a conduit which led to the Hippicus, we have allusions in Josephus.[1] Tobler, as I have already mentioned, does not admit the existence of any valley there. The old residence of the patriarchs, which gives the name to the el-Batrak Pool, lies not close by the pool, but some distance farther north, on the street leading from the Holy Sepulchre to the Latin Convent, and to the present residence of Nakib el-Ashrâf.

The part of Zion within the city attains its greatest height at a line extending from the Armenian Convent to the great synagogue of the Sephardim. It sinks gradually northward towards the cross street which runs from the Jaffa gate; and in former times the descent must have been even greater than now on that side, if we can judge from the depth to which Mr Whiting has gone[2] in the course of his excavations. The Hippicus served mainly to defend the narrow band or isthmus which connected Mount Zion with the north-western portion of the city. On the eastern side, directly opposite Moriah, it falls away very steeply, and houses press close to the very margin of this abrupt descent. There may be seen in one place a path three or four feet wide, hewn in the rock, running from south-west to north-east for a little way, till its course is stopped by the accumulations of rubbish; but there are no other traces of the existence of a bridge once spanning the valley, and leading to the side of Moriah.

From the place where the Aqueduct of Pilate enters the city to the neighbourhood of the synagogue of the Sephardim, there are rubbish piles of such size as to tower far above the city wall, on which account the land on the inside seems to rise much above that which lies along the western border of the old Tyropœon. I have already alluded sufficiently fully to the small quarter, the Haret el Mugharibeh, situated

---

[1] Robinson, *Bib. Research.* i. p. 346.
[2] *Zeitsch. d. deutsch. Morgenl. Ges.* ii. p. 231.

in the hollow between Zion and Moriah, the Jews' Wailing Place, and the traces of the arch discovered by Robinson on the side of Moriah.

It will be seen by one who has gone with me thus far, that the main facts regarding the physical character of Jerusalem have been ascertained as well as is possible without a thorough survey; but how to apply Josephus' imperfect description to it, will remain a matter of difficulty so long as we are unacquainted with the exact nature and situation of the primitive foundation on which the structures of three thousand years have been placed. The statement made in a general form, and occurring in the *Bell. Jud.* v. 4, is as follows: "The city was fortified by three walls where it was not made inaccessible by the valleys which are on some sides of it: where this was the case, there was but a single wall. It was built in two parts confronting each other, and upon two hills parted by an intervening valley, to the very outer edges of which the houses of the city crowded down on both sides. The one of the two hills on which the upper town lay was much the higher. This was formerly named the City of David; but in Josephus' time it was usually known as the Upper Market. The other hill, called Akra, on which the lower town lay, was rounded [shelving] on both sides [ἀμφίκυρτος]. Over against this lay a third hill, by nature lower than Akra, and at an earlier period separated from it by a broad valley; but during the reign of the Maccabees [Asmonæans] this valley was filled up in order to connect the city with the temple. When the top of Akra was removed, the other hill was lowered also, so that the temple still towered above it. The valley known as the Tyropœon separated the upper and the lower hills of the city, and extended as far southward as to the Fountain of Siloah, whose waters are sweet and abundant. On the outside both of the hills were hemmed in by deep valleys, and on account of their steep slopes there was no approach afforded." This is the account of Josephus which has offered such trouble to travellers and to commentators.

Looking only at the first and the last changes of this

remarkable passage in Josephus, confirmed by the Maccabees, relating to the lowering of the mountain and the razing of the fortress overlooking the temple, clearly indicate the completely changed character of the surface of the land. For the fortress built in the lower city on Akra by the Syrians, named Antiochia from the builder Antiochus Epiphanes, and held by this inveterate enemy of the Jewish worship, was at last wrested from the Syrian forces, after a possession of twenty-six years, by Simon the son of Matthias the Maccabæan. In order to ensure the exemption of the temple from all danger in the future, always a thing to be apprehended from the great fortress on Akra, Simon made a proposition to the Jewish leaders to raze the stronghold, and to reduce the height of the mountain itself. The project was received with favour, and the people worked night and day for three years to accomplish the work. The result was the so complete change in the relative heights of the two parts of the eminence, that the part on which the temple stood was the highest: the portion on which Akra (or Antonia) stood was comparatively insignificant in height, while the valley which had heretofore divided them was entirely filled up. When the great danger had been forestalled which had seemed to be impending in consequence of the commanding position of the Antiochia fortress in relation to the temple, Simon the Maccabæan had the wisdom to fortify anew the strong strategic position north of the temple, known by the Persian name of Baris. His successors took the place as their residence. Herod the Great strengthened it still more, making the fortress at the north-west corner of the temple enclosure (the ἀκρόπολις ἐγγώνιος of Josephus) the acropolis of the lower city and of the temple, and naming it the Antonia in honour of his friend Antonius.[1] These were the defences which at a later period proved so prominent in warding off the attacks of four different assailants: Pompey, Herod, Cestius the Roman prefect of Syria, and lastly of Titus.[2] They confirm the view that the whole width of the temple enclosure was not covered

[1] Robinson, *Bib. Research.* i. pp. 292-295.
[2] Krafft, *Topogr.* p. 74.

with these works, but that from the Antonia eastward a deep ditch or fosse was cut in the rock, sixty feet in depth, and two hundred and fifty in breadth. This had to be filled up by the beleaguering forces in order to enable them to reach the temple. The last traces of this fosse are apparently to be seen in the so-called Pool of Bethesda, more correctly called the Birket Israin.[1] It apparently served the double purpose of protecting the city, and also of acting as a reservoir of water. Towards the north the Antonia could not have extended farther than to the hollow where runs the Via Dolorosa, which, as I have already shown, passes the north side of the Turkish governor's palace, to which as well as to the former Roman acropolis, and castle and residence of Pilate, the legend transfers the Scala Santa and the scene of the scourging, since north of the Via Dolorosa the most northern eminence begins to ascend, on which stood Bezetha, the most recent quarter of the city, which must be passed through before Antonia could be reached. It was to this castle, according to Robinson, which lay close to the temple, that the Apostle Paul was brought bound through the excited ranks of the people, and it was from its steps that he addressed them in the court below. This is the $\pi\alpha\rho\epsilon\mu\beta o\lambda\acute{\eta}$ of Acts xxi. 34–37.

The extent of territory covered by the Antonia would appear too limited upon the present plan of the city, and would excite doubts regarding its former situation there, if it were not made certain by various passages that the fortress once pushed itself southward with a marked angle into the temple court, and thus gained room for itself. We are equally certain, too, that the Maccabees, or more correctly the Asmonæans, by their reduction of the rock which towered above the temple, gained a considerable amount of space. This is manifested by the clear horizontal face of rock still to be seen in the regular quadrangular form of the Haram area,[2] as well as from the character of the southern side of the old temple

---

[1] Robinson, *Bib. Research.* i. pp. 330, 331.

[2] Comp. the sketch in Krafft, pp. 12, 13, 76–79; and Bartlett, *Walks about Jerusalem,* p. 161.

mountain, at the spot where the Turkish governor's palace now stands. From this place the general character of the locality can be the best observed, as it cannot be entered by Christians. The place last mentioned, Krafft was obliged to inspect under the heat of a most oppressive sun, when not a creature was to be seen stirring in the streets. He ventured to go so far as to pass through the north-west gate—probably the Bab el Ghowarneh of Tobler's Plan No. 41, more correctly, according to Tuch, the Bab el Ghawarimch—and to advance as far as to the inner court of the Haram. He did not venture to remain long, however, for he was in peril of his life every moment. Josephus tells us himself, that the temple enclosure was irregular in shape on the north side: he remarks also, that after the destruction of the Antonia the Jews converted the sacred precincts into a square, although they had a tradition that the city and the temple should perish, if the court should ever assume a quadrangular shape. Not only the ravages of Titus, but also the later erections of Hadrian—his temple of Jupiter with its adjacent halls and other structures, including the rebuilt Antonia [1]— must have materially changed the form of the temple court. This must have been the case also with the magnificent works which owe their existence to Herod, with those carried out by Justinian, and the still more recent Mosque el Aksa. The great changes effected by the erection of so many successive edifices upon Moriah, and immediately around the temple court, would make it almost impossible, even with the most careful use of all the measurements of past as well as present times, to ascertain with any degree of correctness the precise character and extent of the structures which covered it. All that remains is to leave the matter to the hands of the most persevering and zealous antiquarians to explore and settle so far as it may be possible.

The spirit of inquiry has not been lessened in its desire to trace the early character[2] of the temple by the complete, the only too literal fulfilment, in fact, of the prophecies regard-

[1] Krafft, *Topogr.* p. 228.
[2] See Ewald, *Gesch. des Volks Israel*, iii. pp. 35-38.

ing it, one of which is contained in Matt. xxiv. 1, 2: "And Jesus went out, and departed from the temple: and His disciples came to Him, for to show Him the buildings of the temple. And Jesus said unto them, See ye not all these things? Verily I say unto you, There shall not be left here one stone upon another, that shall not be thrown down." Another is contained in Mark xiii. 1, 2: "And as He went out of the temple, one of His disciples saith unto Him, Master, see what manner of stones and what buildings are here! And Jesus answering, said unto him, Seest thou these great buildings? there shall not be left one stone upon another, that shall not be thrown down." Every reader knows how thoroughly the prediction has been fulfilled. Long after these words were spoken, all the sacredness which had rested upon Moriah was swept away: of the outermost walls of the hallowed places only a few fragments remained; for at the time of the destruction by the Romans all the upper storeys were hurled down, and nothing was left but the lowest layer, to the amazement of centuries to come at this complete confirmation of the word of prophecy.

Dio Cassius has given us no full and reliable statements regarding the erection of a temple of Jupiter upon the same area by the Emperor Hadrian. The attempt of the Jews during the reign of Julian the apostate, in A.D. 363, to rebuild their temple, failed in consequence of the outbreaking of flames, and the terror which they are said to have inspired. The erection of the church of the Emperor Justinian in honour of Mary, and in the place of the former temple, is veiled in darkness: it seems, however, not to have been on the precise site of the temple of Jehovah, but, so far as we can gather from Procopius, towards the southern extremity of the temple enclosure.

With regard also to the erection of the mosque upon the temple court, we have no authentic account; nothing, in fact, more valuable than the statements of those who lived some centuries later.[1] Omar captured the city in the year 636, and determined to build a great mosque on the site of

[1] Robinson, *Bib. Research.* i. p. 298 et sq.

the ancient Jewish temple. Upon his asking the patriarch Sophronius for the place, he was directed, after some evasions, according to the crusaders, to a great church, to whose court a staircase ascended. Not far from it were the traces of former structures. According to the Arabian authorities, the place to which he was directed was the celebrated rock es-Sukrah, but they add that the place was then covered with all kinds of filth in order to show the scorn felt for the Jews. This, as the story runs, he had cleared away, and subsequently erected the mosque, which is generally supposed to be the one bearing his name; yet it is more probable that the Church of Justinian was converted into the el-Aksa, the old foundations being allowed to remain. The Arabian authors assert expressly that it was not till 686 that the Caliph Abd al Melek, whose capital was at Kufa on the Euphrates, and who forbade the pilgrimage to Mecca[1] to the Syrians, erected the magnificent mosque Kubbet es Sukrah, *i.e.* Dome of the Rock, which was finished in seven years, and which was especially intended to make a pilgrimage to Mecca unnecessary. In this way he hoped to weaken the opposite party. This magnificent work, with its surroundings, the last resort of the Mohammedans at the time of the conquest by the crusaders in 1099, became the deathbed of many thousand slaughtered followers of Mohammed, who defended themselves here to the last. Subsequently cleansed, re-dedicated, and changed by the erection of a choir and altar into a Templum Domini, Godfrey, the king of Jerusalem, transformed the old temple of Jehovah into a cathedral, having all the character of a church of the West, and as freely opened. The buildings in immediate connection with it he gave to the various orders of ministers to live in, and made over to the Knights Templar the guards' quarters in the old palace at the southern end of the enclosure.

When Sultan Salad ed Din came in 1187 into the possession of Jerusalem, the sacred spot underwent another transformation. The golden cross upon the lofty dome was torn[2]

[1] G. Weil, *Gesch. der Chalifen*, i. p. 414, Note 1.
[2] Wilken, *Gesch. der Kreuz.* iii. p. 312.

from its place and dragged in the earth, while the crescent became its substitute. All the Christian edifices and decorations were removed; a pulpit was erected for the praise of Allah, in the place of an altar to the glory of God; the whole precincts were sprinkled with rose-water brought by five camels from Damascus; and instead of the Christian songs which had been heard there, but not during a whole century in all, the wild cry of the Moslems was heard, as they intoned the verses of the Koran. As it was left then, it has remained to the present day. The rock es-Sukhrah under the great dome, with the apartments hollowed out beneath it, is the most sacred shrine of the Moslems: it is claimed for it that it was the rock on which Jacob slept when the angels appeared to him in a dream, and Mohammed called it one of the rocks of Paradise. It hence took the name Haram esh Sherif, or the Most Holy Place, and has been forbidden ground to all unbelievers.

The history and description of the ancient Jewish temple and its enclosure[1] or court have been given with great fidelity by Winer[2] in his *Biblisches Realwörterbuch*: the unfounded and whimsical hypotheses of Ferguson, who seeks to identify the Mosque of Omar with the ancient Christian Church of the Holy Sepulchre, have been satisfactorily confuted by Williams.[3] The most accurate description which we now have of the temple enclosure and the mosque, is that given by Catherwood, the result of his six weeks' measurements and observations. These Bartlett[4] has accompanied with exceedingly beautiful drawings, taken from the summit of the Mount of Olives, and from the top of the Government building, from the east and the north sides of the city. These give us the best view[5] of the whole of Moriah, whose top, so

[1] Robinson, *Bib. Research.* i. p. 281 et sq.
[2] Winer, *Bib. Realw.* ii. pp. 569-591.
[3] G. Williams, *Holy City*, i. p. 300, ii. pp. 100-110.
[4] Catherwood's adventure with Bartlett, *Walks, etc.*, pp. 148-165, Tab. x. p. 100, and Tab. xx. p. 143. See *Christian in Palestine*, p. 154, Tab. liii.
[5] Krafft, *Topog.* p. 68; Strauss, *Sinai und Golgotha*, pp. 258, 259.

far as it is occupied by the Haram and its wall, appears to be levelled uniformly off and to be covered with buildings. I must content myself with the most general sketch of the whole field.

The temple square appears at the present day in the form of an open terrace, covered with the mosque, and with gardens where jets of water leap from marble fountains, and where noble cypresses and other trees cast a refreshing shade. The place is a Moslem paradise. Near the middle of the whole enclosure there appears an immense platform, fifteen feet high, five hundred and fifty long, and four hundred and fifty feet wide. This is covered with slabs of bluish-white marble, and was the site of the ancient temple: steps ascend to it on the north, east, south, and west. In the Jewish times the temple probably stood in the middle of this elevated place; while at a lower elevation, and falling away from it as in a series of terraces, were the successive courts of the Jews and the Gentiles. The arrangement seems to have been not unlike that now seen in the gardens which surround the most conspicuous object of all, the Mosque of Omar, which stands at the centre. South of the platform of which I just spoke, and from which a marble staircase descends, there is a great marble basin, surrounded by green and fresh grass-plots, and olive, lemon, and cypress trees. This part of the area is filled with stations which the Moslem visits with the same fidelity, on account of the legends connected with them, with which Christian pilgrims visit those in other parts of the city which have their own special traditions. It is closed on the south by the Mosque of el-Aksa, and by the beautiful Basilica of Justinian, built in honour of the Virgin Mary. The fine architecture of the two ennobles the whole scene, from whatever point it is observed. If the eye wanders away past the walls which encompass the enclosure, it falls upon striking objects on every side: the threefold summit of the green Mount of Olives on the east; the massive and lofty pile of houses on Mount Zion on the south; and towards the north-west the group of buildings with the Church of the Holy Sepulchre in the middle, and Golgotha close by on the slope of the ridge, and at the north-west corner of the present city, and probably entirely beyond the limits of the ancient Jerusalem.

The entire area enclosed within the massive walls of the present Haram forms an imperfect parallelogram, whose northern side, according to Robinson's measurements, is about thirty paces longer than the south side. The entire distance from north to south is greater than that from east to west. Robinson took the dimensions of the south and east sides. According to the results gained by him, the south side is nine hundred and fifty-five feet in length, or about double the width of the above-mentioned platform,—the east side fifteen hundred and twenty-eight feet, the west side a thousand and sixty feet. The entire extent surpasses, therefore, that which Josephus ascribes to the ancient temple; but it corresponds well with the dimensions which the whole area attained after the accessions of territory on the north and south sides. The entrance on the north is through the house of the Turkish governors and the barracks of the garrison stationed there, which occupy the place, I may say, speaking generally, which was covered by the ancient Antonia. On the west side there are five entrances into the square from the various streets of the Turkish quarter, one of which, the fourth, leads to it over the deserted bazaar called the Cotton Market; another, the fifth, runs farther south, by the Mekhemeh, *i.e.* the City Hall and Court House, which lie close to the Haram. The eastern wall has no other entrance than the now closed Golden Gate; the southern one is equally shut in by the Mosque el Aksa and the adjacent edifices.

On the western side of the temple enclosure—namely, the one towards the city—the space is occupied by long rows of buildings of fine Saracenic architecture, and used as the colleges of the dervishes, as Turkish schools, and for other purposes of the mosque, as well as for the reception of pilgrims. On the west side of these structures, directly outside the elevated area, there are several baths of the Mohammedans, among them the Hammam es Sultan; and farther south, at the Suk el Katanin, or Cotton Bazaar, the Baths of Healing, as they are usually called. Regarding these,

both Tobler and Wolcott have learned that they are fed from
the deep and large water-chamber cut out of the rock under
the mosque. Farther south, on the same west side, there are
the Mekhemeh, or City Hall and Court House, where the
cadi lives, and the earth wall where the situation of the
γέφυρα to the Xystus was supposed to be. Following on in
the same line, there are the localities alluded to in a previous
page—the Wailing Place of the Jews, the Mosque of the
Africans, and the broken arch discovered by Robinson—all
of which are close to the Haret el Mugharibeh. With regard
to all the edifices on this south side of the Haram, Wolcott's [1]
accounts, though brief, fragmentary, and difficult to follow
by one who has not been an eye-witness, are to be consulted.
With regard to this part of the city, which was formerly so
much neglected, Tobler [2] has communicated the result of his
discoveries made in 1846, unfortunately in language so brief
that it is difficult to draw from them their full meaning.
I quote his own words: "At the time of my visit, all my
efforts were in vain to penetrate the arches which sustain the
southern side of the Haram. Mr Nathan and I were, however, more fortunate at another place directly west of the
arches which lie at the south-east corner of the temple area.
An inhabitant of the Haret el Mogharibeh (African quarter)
allowed himself to be induced by the sight of money to
break a hole in the wall, through which we passed the very
fine, ancient double gate very seldom visited by Christians;
thence we went to the long arch running northward directly
beneath the Mosque el Aksa, certainly at the peril of our
lives. There is still an unexplored [3] tract between this arch
and the western wall of the temple enclosure. Yet we
pressed on into other arches, not indeed under the temple
enclosure, but close by its wall, and northward of the Jews'
Wailing Place. These arches sustain the house [4] of the cadi,

[1] Wolcott, in *Bib. Sacra*, 1843, pp. 19-24.
[2] Tobler, in *Ausland*, 1848, No. 19, p. 73.
[3] Barclay's interesting account of his personal investigations beneath the temple must not be overlooked by the reader.
[4] *Zeitsch. d. deutsch. Morgenl. Gesell.* vol. v. p. 376.

the Suk Bab es Sinesleh, and the aqueduct which comes from Etham. This Suk, which is now known to have been a kind of causeway, has only been made the object of investigation within modern times. My own researches—the first, so far as I am aware, which have been made in these vaults on the part of Franks since 1187—indicate that they form a part of a connection between Mount Moriah and Mount Zion; and if my interpretation of some difficult passages is correct, at the time of the Frankish possession of Jerusalem the arches were broken, or at any rate so far traversable, that one could pass from what is now called the Damascus gate directly to the Dung gate, without passing the Suk Bab es Sinesleh and the African quarter. In the course of our investigations, we discovered a curiosity which I cannot pass over in silence. It was a tolerably large pool, called by the Arabs the Birket el Obrat: it was probably known to the crusaders, but subsequently was forgotten." The only observations which preceded those of Tobler in this quarter are, as far as I know, those of Wolcott[1] and Tipping, who, after various attempts to follow the Etham aqueduct after it passes beneath the surface of the ground in the lower Tyropœon, succeeded in detecting its course by means of occasionally touching pipes, tunnels, and arches, for a distance of four or five hundred feet within the city. This led them along the wall of the Haram, and to the neighbourhood of the old arch discovered by Robinson. They failed, however, in discovering a connection between it and the Haram basin. If I understand the fragment aright, Tobler has been able to follow this aqueduct still farther north.

Regarding the temple terrace as it appears at the present time, Catherwood gives us the most authentic account.[2] The main entrance to the enclosure is from the west side through the now deserted Cotton Bazaar; although there are two others of less importance farther north. From this bazaar the way runs due east a hundred and fifty feet to the mosque, passing several Turkish places for prayer, and two elegant

[1] Wolcott, in *Bib. Sacra*, 1843, pp. 31-33.
[2] Catherwood, in Bartlett, *Walks*, p. 152; Ferguson, Tab. iv.

fountains, overarched with beautiful cupolas, and shaded by fine cypress and plane trees. The great platform which surrounds the mosque is, as has been already said, about fifteen or sixteen feet above the level of the general area; on the west side three staircases ascend to it, surmounted by elegant arches, apparently of the same age as the mosque itself. On the south and the north sides of the platform there are two of these staircases, on the east there is but one. There are also scattered here and there, among the buildings which surround the princely pile at the centre, places of entertainment for the poorest pilgrims: here they are cared for at the charge of the mosque. One of these hospices is set apart for the express use of the African pilgrims.

This great platform has an extent from east to west of four hundred and fifty feet, and five hundred and fifty from north to south. In these measurements Catherwood confirms Robinson. It is partly laid with marble, and upon it there are several elegant stands for prayer. One of these is particularly noticeable, and bears the name of Fatima, after the daughter of Mohammed. On the south side of the external protection there are extremely rich pulpits, made of the most costly materials. On the east side, and not many steps from the mosque, are a fountain and a praying place, the latter of which is so arranged as to look towards Mecca, and is called the Judgment Throne of King David.

The great Mosque of Omar, which stands upon the platform, is eight-sided in form, every one of its sides measuring sixty-seven feet. The lower portion of the wall consists of parti-coloured marble slabs, arranged in elegant and artistic patterns. The upper portion displays fifty-six arched windows, whose glass is of the most beautiful colours, surpassing in richness those of many a church of the West. The pilasters between the windows, on the outside of the mosque, are composed of glass tiles, laid in very attractive patterns; and no less striking is the upper part of the wall which supports the dome, which is composed entirely of fine lattice-work in wood, swelling upward with great regularity and beauty, covered with tin, and supporting the immense crescent. Four gates beneath marble arches lead from the four cardinal points to the interior: at the western one of these there is a fountain, which is perhaps connected with the subterranean basin discovered by Wolcott and Tobler. The southern portal is sustained by marble columns. A corridor of about thirteen feet in breadth passes around the whole interior[1] of the mosque, bordered by eight pillars and sixteen marble shafts, which appear to have been taken from some ancient Roman building. They are spanned by arches which sustain the circular wall, which in its turn bears up the dome. The inner walls and the dome are stuccoed with gold, in the arabesque style, as in the Alhambra. The dome, which is of great antiquity, consists wholly of wooden rafters, skilfully carved, but altogether out of sight. It is sixty-six feet in diameter, and is supported by four massive stone pillars and twelve antique Corinthian columns, which seem to have belonged to some Jewish or heathen temple occupying the place. There is a second corridor found farther towards the centre of the building than the first. This one is thirty feet broad, and yet there remains within it a circle of still ninety-eight in diameter. Beneath the dome lies the singular limestone rock of irregular shape, from which the whole building derives its name, Kubbet es Sukhrah, or Dome of Rock. According to Ferguson's new theory,[2] this is the Holy Sepulchre; and the Corinthian columns are considered, though without any foundation, to have belonged to the Byzantine church which Justinian erected, not on the site of el-Aksa, but here.

The greater part of the rock lies out of sight, beneath the entire extent of the mosque; but a gilded iron railing surrounds the part which can be seen, and is intended to prevent the touch of the countless pilgrims who visit it. It appears to be a relic of the natural and primitive summit of Moriah: it is only in scattered places that chisel-marks can be seen. It is covered now with a purple canopy. In the south-east corner of the rock there is a chamber hollowed out, the

[1] Bartlett, *Walks*, Note, p. 164; Williams, *Holy City*, ii. p. 114.
[2] Ferguson, *Essay*, Frontispiece and Plate i.

sacred cavern of the Moslems: a flight of steps leads down to it. This chamber is of irregular form, seven feet high, and about six hundred feet in length. The legend of the place claims for it, that it was the resort for prayer of Abraham, David, Solomon, and Jesus Christ; and in confirmation of the story, the places are shown where altars are said to have stood. In the rocky floor there is a circular marble slab, which, if struck, yields a hollow sound, proceeding from a cavern beneath, known as Bir Arruah, the fountain of [wicked] souls, and held by the Moslems to be the entrance to hell. It is said to have been open about sixty years ago, and to have been accessible to those who wished to hold converse with the dead; but since then, owing to the dangers which beset those who entered it, it was closed. The corridors in the mosque, so brilliantly lighted by the number of the windows and the reflection of the sun from the marble, are in strong contrast with the darkness in the vault beneath, where one strives in vain to read the quotations from the Koran which cover the walls. The crowd of wandering pilgrims from all parts of the earth, of all races, in all costumes, mingled with the dervishes in their green talars, who act as their guides, and who prostrate themselves with them in a common devotion, gives rise to the strangest scenes, says Catherwood. They meet at a spot which some of them have spent years in reaching, coming from Calcutta, Morocco, Central Africa, and the farthest parts of Moslem soil; and now, in ecstasy at having attained the end of their journey, and at seeing a free access to paradise ensured to them, they are ready to return, wearing the title of Haji, having visited a spot only second in holiness to Mecca itself.

South of the Mosque of Omar there is a space three hundred and fifty feet in extent, filled with leafy cypresses and other trees, among which are seen many stations hallowed to Moslem eyes, and erected in honour of Mohammed, Ali, Omar, Fatima, and other saints. Through these the way leads to the beautiful architectural remains of the Church of Justinian and the Mosque el Aksa, with its adjacent buildings, of whose position I have spoken on a preceding page.

DISCURSION V.

THE CHRISTIAN QUARTER OF JERUSALEM, WITH GOLGOTHA AND THE CHURCH OF THE HOLY SEPULCHRE.

It only remains that I should speak of the north-west portion of the city, the real centre of the Christian life of Jerusalem, and the point of attraction for pilgrims from every part of Christendom, because it contains the reputed sepulchre of Christ, and Golgotha. This place has been held in reverence for more than fifteen hundred years; and whatever may be the influences from it, there is no doubt about the fact. Although within late years there have been criticisms raised adverse to the arguments derived from tradition, architectural proofs, and various authorities, yet it must in fairness be confessed, that in the present stage of the controversy we are far removed from being able to attain to any settled conclusions regarding the identity of the ancient Jewish Golgotha, and the place which bears the name at the present day. I am compelled to pass over these various discussions, which have been settled in a different way in almost every work which has appeared on Jerusalem, and to refer the reader to Tobler's[1] works on the Holy City, in which he has worked out the whole subject even in its minutest details, and cited all the authors who have brought out their various and conflicting theories. I can only say that the questions raised cannot be settled till there be a more thorough search after existing historical monuments; and until that is done there will be little opportunity for fresh results, the field of hypothesis being nearly exhausted. The manner in which these questions are settled, depends confessedly upon the course of the second wall of the city, which ran between the first and the third, that of Agrippa, which Josephus has described, and whose course I have on a preceding page indicated in general terms as extending northward through the heart of the present city. For only on its outside could the Place of Skulls lie at the time of Christ (the third wall of Agrippa not being built

[1] Tobler, *Golgotha, seine Kirchen, und Klöster nach Quellen und Anschau.*

till ten or twelve years after the Saviour's death) : it could not be in the interior of the city, since the Jews would suffer no graves but those of their kings to be within the city walls But since this second wall as well as every part of Jerusalem was completely destroyed by Titus, and since, from the account in Josephus, the course which it followed can be gathered only in the most general way, to try to follow it after its destruction must be a matter of the greatest difficulty. There was left, therefore, a wide field for the play of the fancy, and one could delineate the course of the second wall from its beginning to its end in this way, another in that, according as each applied the measurements given by Josephus to the city walls, and to those architectural remains which are supposed to be the relics of ancient walls, and which are now to be traced incorporated in the fortifications as they exist at present. Others have taken the physical character, and especially the heights and slopes of the various parts of the city, most into account; others the old legends and traditions, deciding in conformity to them whether the reputed site of the Holy Sepulchre and the Place of Skulls lay outside or inside of that second wall. The second wall, says Josephus, began at a gate called Gennath in the first wall, and passing only the northern part of the city, ran up as far as the Antonia. He gives us nothing but this; and the position even of the gate Gennath is unknown to us, unless occasional gate-pillars and arches found in the neighbourhood of the Hippicus or in the Street of David, do not hint with more or less probability at the existence of such a gate. But with regard to the course of the second wall northward to the Antonia, there are nothing but conjectures, possibilities, perhaps here and there probabilities, but nowhere certainty. For the same ancient fragments which, for example, Schultz[1] and Williams consider to be architectural traces of former walls, proving that the Church of the Holy Sepulchre was outside of the lower town and beyond the second wall, appear to Krafft[2] and Tobler to be no proofs of a line of

[1] Schultz, *Jerusalem*, pp. 59-62; as well as Schultz and Gadow's plans.
[2] Krafft, *Topogr.* pp. 25-34.

circumvallation once existing there. Although the former of these two, following the indications of an eastern position of the old gate Gennath and the slopes of the district, locates this second wall, like his predecessors, on the east side of the Church of the Holy Sepulchre, yet he does not trace the wall to the Damascus gate, but to the Porta Judicaria, and thence, after encountering a sharp angle, eastward to the Antonia. Wolff,[1] too, doubts whether the relics referred to can be considered the remains of a connected wall, although there appear to him to be good grounds for the possibility, and even the probability, of the genuineness of the Holy Sepulchre. Tobler[2] holds decisively to the theory that the so-called Church of the Holy Sepulchre occupies a site within the second wall; but he bases his inference upon measurements drawn from Josephus, which, however, are held by others as by far too unreliable to be accepted. Tobler holds them so weighty, however, that in view of them he does not hesitate to reject the story of the grave and the discovery of the cross on the site of the present Church of the Holy Sepulchre. Others before him have not laid much stress upon the history as it has come down, but at the same time have not been able to confess that it is possible for tradition to mislead entirely regarding a locality which is so fraught with interest. According to the opinion of my friend, Finlay,[3] the author of a small work, *On the Site of the Holy Sepulchre*, the exactness of the Roman municipal arrangements, the detailed lists of names of places which they made out for the collecting of customs as well as the exacting of taxes, the strict care which they took of their archives, corroborated by the fact that the Pandects speak of the colony of Ælia Capitolina, and the regular census taken by the Roman authorities in Palestine, afford sufficient proof that it was impossible for so important a place as Golgotha to fall into such entire oblivion, that at the time of Constantine and Helena a wrong location should have been assigned to it. Williams,[4] on the contrary,

---

[1] Wolff, *Reise*, p. 81.   [2] Tobler, *Golgotha*, p. 160.
[3] Finlay, *On the Site of the Holy Sepulchre*, p. 40.
[4] Williams, *Holy City*, ii. p. 68.

glories in following tradition with perfect confidence, notwithstanding that its voice is so uncertain after the long lapse of years, and goes so far as to say that he should have no hesitation in adopting its decisions, even if all topographical arguments were at variance with it. Robinson,[1] adopting an exactly opposite course, had so firmly persuaded himself during his investigations of the entire want of trustworthiness in the traditionary legends current throughout Palestine, that he reposed little confidence in them, and entirely discredited their statements regarding the site of the Holy Sepulchre. Those statements were, that the Emperor Hadrian erected a temple of Venus on the site of the Saviour's burial-place; and that at the time of the Emperor Constantine, and in the presence of the Empress Helena, the true cross was discovered there, the result of which was the building of the Church of the Holy Sepulchre and that of Golgotha upon that site. For the first tradition it does not appear that there is any trustworthy datum existing, since, even if there had been at the time of Constantine a tradition of a former temple of Venus there, the place would never have been a resort for Christian pilgrims, and yet its being so would be the only way that a remembrance of its sacred character could be perpetuated among the heathen. Against the second tradition the silence of Eusebius speaks decisively, who must have known about the existence of such a temple of Hadrian. Later authors, however—Jerome, for example, who wrote seventy years later—give the legend, but in a way which differs from Eusebius' statement regarding the Holy Sepulchre. These objections, resting as they do upon thorough investigations, appear to be incapable of being set aside, even although some may try, as Wilson[2] has done, to correct some minute details in the facts adduced by the opponents of the traditional theory. But when Robinson[3] goes further, and seeks to show that the reputed situation of Golgotha and the Holy Sepulchre cannot be the true one because it lies within the

---

[1] Robinson, *Bib. Research.* i. p. 410 et seq.
[2] Wilson, *Lands of the Bible*, i. pp. 436–438.
[3] Robinson, *Bib. Research.* i. p. 407 et sq.

second wall, which is contradictory to John xix. 17, 20, his argument is just as unsatisfactory as that drawn from tradition, because it rests upon his theory of the course of that second wall, which he, after careful measurements and a minute study of the levels and slopes, puts on the west side of the modern church, but which others, after equally patient researches, have located elsewhere. For the Gennath gate, where the wall began, cannot lie, it is said, so near to the Jaffa gate as Robinson supposed, but must be looked for farther eastward in the Street of David. Another reason which Robinson gives, that the locating of the second wall on the east side of the present Church of the Holy Sepulchre would make the city narrower than we have reason to think that it was, and would have given it a singular and inexplicable form, has been satisfactorily disposed of by Wolff.[1]

If we cannot come to any settled conclusions regarding the location of the Holy Sepulchre, and are in doubt whether we are to accept the site which is now occupied by the church bearing the name, or to reject it, we are equally compelled to doubt whether it is the place which was honoured at the time of Constantine as the scene of the Saviour's sufferings. Yet, although the place is interesting to us as bearing witness to the devout piety of the early Christians, and their affectionate veneration for the scenes which had witnessed such sad scenes, we must be on our guard against assigning any more value to the place than we ought, and of overrating the worth of the localities where Jesus met His death, lest they should lead us to a disguised idolatry, and transform our pilgrimages into pagan devotions. Robinson,[2] in speaking of the opposition of the spirit of the gospel to this, states with great truth and force, that the four Evangelists, although they describe with great minuteness the crucifixion and the resurrection of the Lord, yet touch lightly on his grave, and in general expressions; and although they write scores of years after His death, yet they maintain a perfect silence regarding any special sanctity connected with the place of

[1] Wolff, *Reise*, pp. 81–83.
[2] Robinson, *Bib. Researches*, i. p. 412.

His burial, and do not even allude to it in the subsequent apostolical history, although they mention David's grave in one passage. Even the great Apostle of the Gentiles, whose constant theme is the death and resurrection of the Lord, and the glory of His cross, does not make an allusion in all his writings to the scene of His sufferings, nor to their instrument. On the contrary, the whole purport of the Saviour's teaching, and that of Paul—indeed, of every part of the New Testament—is to draw away the spirits of men from their incessant regard to what is visible, from times and places, and to lead true Christians to worship God not merely in Jerusalem and on Gerizim, but everywhere, in spirit and in truth. But so far as the point is of interest as a geographical question, I am strongly inclined to adopt von Raumer's[1] view, and to regard it as still open, and yet to admit that a pious soul may receive some benefit from standing, not on the unquestioned scene of Jesus' sufferings, but in the immediate neighbourhood, and reviewing them not in the spirit of idolatry, but in that which comes from the recollection, that within the range of sight the Saviour met His death.

The contemporaneous accounts of the Evangelists regarding the place of the crucifixion and the burial of Jesus, yet agree perfectly in this, that they were both outside of the gate of the ancient city, but near to it, and beyond the walls, but not far from them. This is confirmed by the statement in Heb. xiii. 12, "Wherefore Jesus also, that He might sanctify the people with His own blood, suffered without the gate." In Matt. xxvii. 32 we read, "And as they *came out*, they found a man of Cyrene, Simon by name; him they compelled to bear His cross." John xix. 17 states, "And He, bearing His cross, went forth into a place called the place of a skull, which is called in the Hebrew, Golgotha." And in ver. 20, "This title [Jesus of Nazareth, the King of the Jews] read many of the Jews; for the place where Jesus was crucified was nigh to the city." And again, vers. 41, 42, "Now in the place where He was crucified there was

[1] Von Raumer, *Beitr. z. Bibl. Geogr.* pp. 55, 56.

a garden; and in the garden a new sepulchre, wherein was never man yet laid. There laid they Jesus therefore, because of the Jews' preparation-day; for the sepulchre was nigh at hand."

The present position of the reputed Golgotha and the Holy Sepulchre within the walls of the city appears to militate against that statement: yet it is only an appearance of contradiction, if one thinks of the fact that the whole northwest corner of the old city has been destroyed, that all traces of it have been effaced, that the new wall is entirely independent of the ancient one, and that no opinion can be drawn from the situation of any locality in this part of the modern city, as to its relative situation in the ancient one. Yet, notwithstanding the scriptural statement that the place of crucifixion and of burial were very near each other, it is a little surprising to find that the ecclesiastics have covered both spots with a single roof, and that a mere partition separates the portion which commemorates the place of Jesus' death from that which is said to have received His remains. Struck by this, Thenius[1] has endeavoured to show, and has displayed great learning and acuteness in the effort, that the situation of Golgotha was separated some distance from the burial-place, and that it was in front of the Damascus gate, on the skull-shaped hill already alluded to, in which the cave of Jeremiah is found.

The earlier opinions regarding the identity of the place where the Saviour suffered have already been critically considered by Robinson,[2] from Cotovic to Chateaubriand; but I can only follow here the slight traces which have to serve as our guide over the city as it exists to-day, without sufficing to indicate the circumstantial progress of the changes which have occurred. After Titus' destruction of the city, A.D. 70, there remained but three colossal towers as monuments of the victory gained by the Romans: everything else was reduced to fragments. For half a century Jerusalem remained a mere garrison for Roman soldiers, stationed

[1] Comp. Thenius and Rödiger, in *Hall. Allg. Lit. Zeit.* 1843, No. 110.
[2] Robinson, *Bib. Research.* i. p. 407 et sq.

there to guard and hold the land, till the Emperor Ælias Hadrianus began to carry out his plan of restoring the city, for the purpose of making it the seat of a temple of Jupiter. This was in the year 132. He gave to the Roman colony located there the name of Ælia Capitolina, in honour of his new temple. Although at the time of Titus' conquest the Jewish population of the city was almost entirely annihilated, or scattered through the neighbouring lands, yet their number had increased in the other portions of Palestine, so much so as to foment a formidable spirit of discontent, when they were compelled to bring their tribute up to their old capital, and to contribute it to the servitors of the heathen temple erected in honour of Jupiter Capitolinus. At length the long-cherished hatred and murmurings broke out in a terrible rebellion against the sway of Hadrian,[1] beginning at a place called Bether,[2] in the neighbourhood of Jerusalem, whose position, however, fell into oblivion, till it was discovered by Williams. Kirbet el Jehud, *i.e.* the Ruins of the Jews, is the name given by the Arabs to an old fortress surrounded with a fosse excavated from the rock, near to the new village of Beitir, near Jerusalem. This Beitir, which lies in Wadi Bittir, is thought by Williams[3] to commemorate the site of the ancient Bether. He has also discovered a kind of fortification south-east of the spring Ain Yalo, where the local legend locates the Philip's Fountain. The Arabs still preserve the tradition, without knowing whence it is derived. The Jews converted Bether into a fortification, and some forty cities entered into an alliance. They even succeeded in re-fortifying Jerusalem during the absence of Hadrian. Two years of war followed, full of cruelty and terrible defeat, resulting in the entire restoration of the Roman authority, and the dispersion of the Jews as slaves over the entire earth. All who were not killed in the strife met this ignominious fate, were banished from Palestine, and forbidden under penalty of death ever entering Ælia Capitolina again. The Christians,

---

[1] Robinson, *Bib. Research.* i. p. 366.
[2] Krafft, *Topogr.* p. 224.
[3] Bartlett, *Walks, etc.*, p. 246.

on the contrary, were permitted to have free access. It is probable that the Christian community, which at the time of Titus' siege had withdrawn across the Jordan with its bishops to Pella, at Hadrian's[1] time returned to the city, their former home, though now a heathen capital, whose very name Jerusalem had completely passed away, as we learn both from coins, and also from the works of contemporary Roman writers.

These Christians had suffered not a little from the Palestine Jews during the time of rebellion. The list of their bishops, from the first (Jacobus frater Domini) to the last (Judas, who died at Pella, the fifteenth of the line), were all of the lineage of David, of the countrymen of Jesus; but after their return to Jerusalem, in order to put away every trace of Judaism, they discontinued the custom of selecting their bishops from Jewish proselytes, but chose them from the converted Gentiles, beginning with Marcus the sixteenth, who was contemporary with Hadrian. Thence the number went on till, at the time of Constantine,[2] the twenty-fifth bishop (Macarius) was in power (not the Macarius who was present at the Council of Nice, who was the forty-first of the entire list).

This succession of bishops is the only possible means by which the traditions regarding the location of the Holy Sepulchre could have been preserved up to the time of Constantine. Yet there is no allusion among the church fathers to the site of Golgotha; and it seems as if the sixty-five years' sojourn at Pella so effectually served to efface all recollection of it, that the discovery of it at the time of Constantine and Helena appears more in the light of a miracle than of an ordinary occurrence. From the Jews no recollections of the locality were to be expected; for only after centuries of banishment, and after Constantine had become a Christian, was it permitted to them to enter the city, and drop their tears over the desecrated ruins of their sacred places.

[1] Robinson, *Bib. Research*. i. p. 369 et sq.
[2] Credner, *Nicephori Chronographia brevis*, Pt. ii. pp. 35, 36.

It must be confessed, that subsequently to Hadrian's time there was not an absolute lack of Christians in Jerusalem, although those who were there were subjected to incessant persecutions. Nothing, however, is known definitely regarding their bishops and churches, although Eusebius tells us that the bishop Alexander founded a library there, and that Christian pilgrims[1] were casually there, who, however, at that time—the end of the third and the beginning of the fourth centuries— directed their thoughts to the place where Jesus was born, and where He ascended to heaven. The pilgrimage most fruitful in results was that of the Empress Helena, the mother of Constantine, who in the year 326 visited the Holy Land, in order to return her thanks for the conversion of her son, and for the success which had attended his efforts to disseminate Christianity throughout the entire Roman Empire, and also to pray for the conversion of her grandson. All the church fathers of the fifth century agree in the statement, that in the year above mentioned Helena was present at the discovery of the grave of the Saviour, which had previously been covered with rubbish, and with a temple erected in honour of Venus. Eusebius, however, the contemporary and biographer of Constantine the Great, makes an allusion indeed to the discovery of the grave of Christ, but ascribes it to a divine impulse given to the mind of the emperor himself, who determined to make the place a consecrated spot, and to erect there a house of worship. Eusebius makes no allusion to the discovery of the cross on which the Saviour was crucified; yet Cyril, the bishop of Jerusalem in 348, not a quarter of a century later, states that the true cross had been preserved, and was an object of veneration to the church. Jerome makes incidental allusions also to its existence, but neither of these ascribe its discovery to the efforts of Helena. The so-called Bordeaux pilgrim, who visited the Holy Sepulchre in 333, during the erection of the church there, says nothing about the discovery of the cross: he is, however, the first witness to the fact that then the site of the grave had been fixed, and he is the first to speak of the church

[1] Robinson, *Bib. Research.* i. 372 Krafft, *Topogr.* p. 230.

which had been begun on that site : . . . . a sinistra parte est monticulus Golgotha, ubi Dominus crucifixus est. Inde quasi ad lapidis missum est cripta, ubi corpus ejus positum fuit et tertia die resurrexit. Ibidem modo jusso Constantini Imperatoris basilica facta est, id est dominicum, miræ pulchritudinis, habens ad latus excepturia, unde aqua levatur, et balneum a tergo, ubi infantes lavantur, etc.[1]

It is only when we come to the church historians of the fifth century, that the story of the discovery of the cross on the scene of Jesus' sufferings begins to be draped with all the fantastic habiliments of a circumstantial history, and takes its rank as a miracle. Once accepted on this ground, it not only was retained by the priests, but received, besides, fresh additions and decorations with advancing years. Yet it is not to be discredited that a cross was found in 348 in the same church in which Cyril frequently preached, the same cross before which the pious Paula prostrated himself in devotion in 404, even although the miraculous discovery of this so-called true cross be admitted to be a mere idle legend.

The Empress Helena died directly after her return to Constantinople from Palestine, in the year 327 or 328; but the prestige of her name grew from age to age, till in the fourteenth century there were no less than thirty churches in the land ascribed to her as the founder. Eusebius, in the eulogistic life of Constantine the Great, may have gone quite far enough in denying to Helena the honour of laying the foundation of these establishments, in order the more to enhance the glory of the emperor himself; but it is certain that the credit cannot be denied to the mother of founding two churches,—that at Bethlehem, and that of the Resurrection on the Mount of Olives. These the emperor decorated profusely after the death of Helena, but he did not found them. Eusebius grants to the empress a share in the honour of discovering the true cross; but the building of the Church of the Sepulchre he ascribes to Constantine alone. His statement is, that after the Council of Nice[2] the emperor wished

[1] *Itin. Hierosolym.* in *Itin. Anton.* ed. Parthey, pp. 279, 280.
[2] Robinson, *Bib. Research.* i. p. 373.

to hallow the place of the Saviour's resurrection by building some structure which should be an object of universal reverence; for long before the time of Constantine, frivolous-minded men, or rather the whole race of demons acting through them, had sought to consign to oblivion the place which bore the glorious witness to the immortality of the spirit. Now these words of Eusebius can be understood in this sense,[1] that in order to offer a purposed insult to the Christians, the grave of Christ, which had been known before, had been filled with filth and covered with earth, in order to get a paved foundation to erect a temple of Venus upon. This view is also indicated in Jerome's comments upon Eusebius' account, contained in the *Epist. ad Paulin.* 49, where he says, that from the time of Hadrian to that of Constantine the Great, a hundred and eighty years, a statue of Jupiter (of which Eusebius makes no mention) occupied the place of the resurrection, and a marble one of Venus stood upon Golgotha. The latter, he says, was held in veneration by the heathen; and at the time when it was set there, it was thought that it would be a good means of causing the Christians to lose the recollection of the cross and the resurrection, if the places considered sacred were covered with the images of Roman gods and goddesses. But so far from accomplishing this object, it only impressed the Christians more deeply with reverence for the places so desecrated, and caused them to cherish them with the greater care. The empress took an interest in the restoration of the honour due to the sacred localities, and gained from her son permission that the heathen shrines should be destroyed. Constantine thus coming upon the scene as the avenger of the scandal done to the Christians, overthrew the profane temple and images which had been set up, and had the grave of Jesus cleared of its impurities,—thus indirectly bringing it into the notice of men once more. He gave orders to bishop Macarius that a magnificent house of worship should be erected over it, as we learn from a letter which Eusebius has preserved in his life of Constantine. The imperial edict was also

[1] Krafft, *Topogr.* pp. 172, 173, 234.

given to the governors of the eastern provinces, that money should be collected towards the defrayal of the expenses of erecting this church, the emperor himself promising to provide the marble pillars. In this letter he also inquires of the bishop whether he would prefer a panelled roof, or some other kind.

Bishop Macarius died during the first year after the church was begun; but his successor Maximus prosecuted the work, which progressed so that the pilgrim of Bordeaux visited it in 333, when Helena had been dead six years. In the thirtieth year of the emperor's reign it was completed, and a great council was convened at Tyre for the purpose of taking part in its dedication. On the list of this council was Eusebius, who as an eye-witness has left us a description of the building, difficult to be understood indeed, but which Krafft[1] has compared so successfully with the present condition of the place, as to make it evident that, despite the repeated destructions and changes, the original site is to be recognised in the massive and very peculiar building which covers the reputed Golgotha as well as the place of the crucifixion.

Although, in what has been adduced above, there are no positive proofs, yet there are weighty grounds for believing that the site of the Holy Sepulchre[2] is to be looked for beneath the church erected by Constantine; and that, moreover, Golgotha (only a stone's throw away, says the *Itin. Burdig.*) was the place now bearing the name, and receiving the honour paid to it. The "Monticulus Golgotha" of the *Itin. Burdig.*, thirty-five feet, according to Tobler, above the lowest part of the chapel where the cross is said to have been found; the rock called by Jerome *Crucis rupes*, by which the place of execution—Calvaria, the κρανίον of Luke xxiii. 33, and the κρανίου τόπος of the other Evangelists—can be easily recognised at the present time[3] in the tract covered by the pile of ecclesiastical buildings; and Goath, the older name of the

[1] Krafft, *Topogr.* pp. 236–241.
[2] See also the statements in Scholtz, Williams, and Schultz on this subject.
[3] Krafft, *Topogr.* pp. 157–159, 170, 235, Note.

same locality, designated in Jer. xxxi. 38–40 as the place of execution of malefactors, but afterwards, when the Aramæan dialect became predominant, applied to the skull-shaped top of the hill,—certainly are terms applied to the ground immediately contiguous to the church. Jerome tells us that Christ was buried on the north side of the hill Goas; in which the word Goath of the prophet's time and the later Golgotha are unmistakeable. This Goath lay, however, unquestionably outside of the city walls.

The tomb which was brought to light by the removal of dirt and rubbish, as alluded to on a preceding page, may easily be conceived of as having an aspect different from the other vaults, as might be inferred from the fact that it belonged to the wealthy councillor of Arimathæa. At the time that the great work was erected above it by Constantine, it may have received the form in which Bishop Arculfus[1] found it in the year 698, who describes it as a round chapel hewn out of the rock, ornamented externally with beautiful marble decorations, but tasteless in the interior, and still bearing in the red limestone marks of the chisel. He gives it the Latin name *Tegorium*, or a place roofed over. In the year 614— that is, eighty years before his day—a terrible desolation had been brought to the whole neighbourhood, and indeed to the entire city of Jerusalem, by the invasion of the Persians under Chosroes II.; and even the church erected by Constantine had perished in the flames. The most weighty confirmation which has been adduced within our times in support of the genuineness of the Holy Sepulchre, is that given by Schultz,[2] who has himself discovered beneath it the remains of ancient graves. This makes it certain that the place was outside of the city, and beyond the second wall; and though it does not settle the question of the genuineness of the spot, it at least sets aside one class of objections. In the west wall of the rotunda which surrounds and overarches the reputed Holy Sepulchre, there is, according to Schultz, a door which leads to a little chapel belonging to the Jacobite

---

[1] Krafft, *Topogr.* p. 173; Schultz, *Jerusalem*, p. 98.
[2] Schultz, *Jerusalem*, pp. 96, 97.

Syrians; and out of this, again, leads a second door to a narrow space, in which three men can hardly find place, and in which one can scarcely stand erect. The eastern side forms the wall of the rotunda; on the other sides the primitive rock surrounds it, and in this rock are found the graves running horizontally into the wall. On the floor there are also graves, but these are sunk perpendicularly: they all bear the interchangeable name of the graves of Nicodemus and Arimathæa. The discovery of them is not new indeed: they have been long known, but were held to have been made by the crusaders, by those who overlooked one important point in their structure. Those sunk perpendicularly in the rock may very probably belong to the crusaders, and were excavated, it is not unlikely, out of a natural desire to be buried near to the place where the Lord lay. But those which are sunk horizontally in the rock are almost exactly similar to the most ancient niches found in the old necropolis around Jerusalem. Yet Tobler[1] refuses to admit the validity of this reasoning, and assigns these excavations to the labours of monks. Schultz[2] closes his remarks under this head with the observation, that it seems to him unquestionable that here was an ancient burial-place long before the erection of the Church of the Sepulchre; an ancient Jewish rock-tomb, indeed, excavated prior to the destruction of the city by the Romans. Schultz thinks that Robinson has worked out the history of the Holy Sepulchre with great exactness and learning; and I will only refer to his work for more full statements.

Eusebius has communicated the earliest description of the earliest church—the one dating from the time of Constantine. This has been compared by Krafft, notwithstanding the great difficulty of the attempt, with the structure in its present shape,—a task which Schultz, great as was his facility and adequate his preparation, shrank from. In the manuals of the history of art[3] there are no allusions to this earliest of the

---

[1] Tobler, in *Ausland*, 1848, No. 92, pp. 365, 366.
[2] Schultz, *Jerusalem*, pp. 97, 98; Robinson, *Bib. Research.* i. p. 371.
[3] Kugler, *Handbuch der Kunstgeschichte*, p. 362.

line of churches which have crowned the same site from Constantine's day down to our own. The peculiar style as well as the massiveness of the first of these, corresponding well as it did with the deep and earnest piety of its builders, though often spoken of as a mark of their superstition, and the undeniable traces still to be seen in the basis of the present church, impart to this description of Eusebius a high degree of interest, despite the fact that repeated destructions, the erection of contiguous buildings, and the overloaded adornments lavished by monkish taste, have given to the whole a modernized appearance. The first entire destruction of Jerusalem in the Byzantine[1] period—in the year 614—effected by the Persians under Chosroes II., in which the Church of the Holy Sepulchre perished by fire, and all the Christians, priests, and pilgrims were massacred or carried into captivity, was followed in 1010 by a second complete destruction of the restored church, effected by the Egyptian Caliph Hakem,[2] the prophet of the Druses. And although the building was restored in 1048 by the Emperor Romanus, yet before and after the Crusades—from 1048 to 1808—it underwent various injurious[3] changes, so that its Byzantine character gradually disappeared, without being totally lost. Here and there, parts of walls, gates, pillars, and other ornaments, are found still preserved from the older buildings.

The modern descriptions of the exterior and the interior of the pile of buildings which, taken together, bear the name of the Holy Sepulchre, of the convents, pilgrims' stations, altars, ceremonies, and festivities, particularly those at Easter and Whitsuntide,[4] have been so often[5] repeated that they are sufficiently well known; and I need not dwell on them, but

[1] Robinson, *Bib. Research.* i. p. 385.
[2] *Ibid.* i. p. 395.
[3] *Ibid.* i. pp. 396, 403, 407.
[4] J. M. A. Schultz, *Reise,* pp. 225-230.
[5] Bartlett, *Walks,* pp. 168-185; Roberts, *The Holy Land,* Book i.: frontispiece and various views in the same work. See also Krafft for the best outlines, p. 238; Tobler's *Golgotha;* and Williams' *Holy City,* Tab. i. Pl. ii. and iii.

will pass to Eusebius' description of the first Church of the Holy Sepulchre, undertaken directly after Constantine had sent the letter to Bishop Macarius ordering the edifice to be built.

The work was begun in the new town,[1] opposite to the old,—a fact which gives Eusebius an opportunity to allude piously to the New Jerusalem. At that time it seems to have been well known that this part of the city lay outside the walls at the time of Christ. The grave itself, as the central point of interest, was adorned with carefully selected columns, and with all kinds of decorations. A space paved with particoloured stones surrounded the grave, and was shut in on three sides by rows of pillars (now in the interior of the church): no roof rested upon them, as is the case likewise with Abraham's grave in Hebron, already referred to; but the court with the motley pavement lay under the open sky. Opposite to the fourth side, the entrance to the sepulchre, which faced the rising sun, the Basilica closed the view,—a wonderful work, according to Eusebius, of imposing height, and of great length and breadth. He first describes the central nave, whose inner walls were decorated with tablets of variegated marble, and whose exterior was beautified with polished and well-joined stones. The roof was covered with tin, in order to keep off the rain: the interior was of finely panelled wood-work, richly gilded, and throwing its glancing beams over the whole floor. Then follows the description of the side naves or aisles, which were divided into two parts, one of which was subterranean. Of equal length with the main nave, their roofs were also heavily gilded. The side aisles above the ground were boarded by colossal columns: those below supported their roofs with carved pillars. Three doors at the east end of the Basilica received the throng of pilgrims.

Eusebius returns, in his description, to the spot of main interest, viz. to the Holy Sepulchre itself, and describes its adornments more minutely. It lay at the western end of the Basilica, and was the point where all converging lines natu-

[1] Krafft, *Topog.* pp. 236–239.

rally met. It was shaped like a horse-shoe, there being on the east side three small recesses or chapels, which the Basilica joined still farther east, while the extreme western point or absis was occupied by the sacred grave. This, it will be noticed, took the place of an altar, and was at the w'stern extremity of the whole, instead of being at the eastern. It was surrounded by twelve columns—the number of the disciples; and upon these were twelve caskets, all of them the offerings of piety. Eusebius then passes to a description of the courts east and in front of the Basilica. The one nearest to it had recesses or chapels on both sides, and gates opened from it to a market-place, in whose centre arose other richly ornamented gates; the *propylæa* of the whole hinting to the passer-by what splendour was contained within.

Eusebius makes no allusion to what was found within the ancient sepulchral vault itself; but Antoninus Martyr, writing about the year 600, and before the Persian invasion, says that it was hewn out of the solid rock, after the fashion of a church (the present Chapel of the Resurrection, the Anastasis of the Greeks, and the Kiyameh of the Arabs), and that it was overlaid with silver. According to Arculfus' account, marble took the place of silver at a later period. The half circle, or horse-shoe-shaped space, still exists around the grave, but it is now overarched with a cupola, and a great window is just above the reputed site of the sepulchre. East of this semicircular space we come at once to the central nave of the church. The elevation of the hill Golgotha, which is embraced by the building above the floor at the south-eastern part, is the unquestionable reason why the side naves were constructed both above and below the surface, and why, for the sake of symmetry, the same arrangement was retained in the north side of the main nave of the Basilica. The subterranean Helena Chapel close by the north-east side of the Golgotha Chapel, known as the Martyrion, with the traditional site of the discovery of the cross, which is close by the east side of the main nave of the Church of the Sepulchre, may take the place of one of those side halls or chapels which were east of the Basilica. Eusebius does not speak of the last two of these; but Antoninus Martyr informs us, that in a room connected with the vestibule of the Basilica, the wood of the cross was preserved for the veneration of believers. The remains of a great gate, which dates back to the Byzantine epoch, appear to have served as the *propylœa* of the church erected by Constantine.[1]

The architectural character of the structure of the first Church of the Holy Sepulchre, says Krafft in his instructive delineation, was connected in the closest manner with the situation of Golgotha and of the grave of the Lord. Entirely in antagonism to the usual custom of putting the entrance at the west and the altar at the east, towards the rising sun, in the Church of the Sepulchre the sacred spot which took the place of the altar was at the west. But it was considered necessary to embrace within the compass of the building the hill of Golgotha east of the grave; and at the same time the Holy Sepulchre must remain, as just indicated, the most sacred spot of all,—the place which conferred an especial sanctity upon the whole edifice; or, to use the words of Eusebius, it must take the place of the altar. From this came the idea, that out of the Holy Sepulchre in the west of the Church of the World, a new Sun, namely Christ, had arisen on Easter morning. Further, had the builders crowded the nave into a position west of the grave, through the rapid rise of the land from east to west, the sepulchre would have lain altogether too low in relation to the rest of the building; while, built as it was, it was so high, that, as Antoninus Martyr says, a blessing seemed to flow forth from it upon all the throng of worshippers in the main body of the church. This gives us the key to the remarkable position of the present church; only there has crept in during modern times this change, that a choir has been erected at the eastern end of the nave, and the entrance has been transferred to the southern side. The disagreeable access through the small court, and the unattractive arched gates near it, are so well known from the accounts of travellers, that I need only allude to them. Rev. Professor Willis has given, in Williams'

[1] Krafft, *Topogr.* pp. 30, 239.

Holy City, a circumstantial history of the erection of the Church of the Holy Sepulchre.[1]

It is a very instructive task to trace the influence of this building of Constantine's upon the ecclesiastical style of the West, made very prominent as it was by the influence of two political friends and allies, who rank among the mightiest princes[2] whom the earth has seen—the Caliph Haroun el Raschid and Charlemagne. This influence was not made the less through the East or the West, from the fact that the great Byzantine Empire lay between those monarchs. As the defender of Christianity, Charles sent an ambassador[3] to the Caliph, laden with valuable gifts for the Holy Sepulchre; but the magnanimous eastern ruler not only sent back the envoy with great honour to Aix la Chapelle, but bade him assure his master that, out of favour to the religion of the western monarch, Haroun would entrust him with the entire supervision and control of a place on which the Christians set such value, and which was invested with such an odour of sanctity. Acknowledged now as the possessor of the Holy Sepulchre, Charles sent rich gifts to Jerusalem, to be spent in erecting churches, entertaining pilgrims, and supporting the poor. His example was followed by the next Carlovingian kings, Louis the Pious and Louis the German. The Church of the Holy Sepulchre became a model for many in Germany, which were built under the patronage of the emperors, particularly those of Cologne, Fulda, and St Gall, and among these, the last most of all. Its influence is also traceable in many others, among them the Cathedrals of Spires, Worms, Mayence, the Sebaldus Church in Nuremberg, and in general, in those which were built between the eighth and the thirteenth centuries. It was felt also in the baptisteries, and in other internal arrangements. William of Tyre describes the

---

[1] Williams, *Entrance to the Church of the Holy Sepulchre*, vol. ii. p. 120. See Tobler, *Golgotha*, Pl. i. and ii. Comp. Williams, *Holy City*, ii. pp. 129-294; Willis' *Essay on the Architectural History of the Holy Sepulchre*.

[2] Robinson, *Bib. Research.* i. p. 390; Krafft, *Topog.* pp. 250-252.

[3] Eginhardi *Vita*, ed. I.; L. Ideler, p. 72.

Church of the Holy Sepulchre precisely as Arculfus does at the end of the seventh century: a rotunda with an open roof over the Holy Sepulchre, a chapel over Golgotha and the place where the cross was found. These were all brought together at the time of the Crusades by means of a great nave, thus giving to the church the form which it has retained up to the present time.[1]

DISCURSION VI.

THE WATER RESERVOIRS AND BURIAL-PLACES IN AND AROUND JERUSALEM.

Among the peculiarities which distinguish the neighbourhood of Jerusalem, and which depend in a measure upon the broad mass of limestone which underlies it, are to be reckoned, whether regarded in a topographical or historical light, the reservoirs and burial-places of the city. They serve as guides in antiquarian research, and date back, it seems probable, to the age of Solomon.[2] They would, however, be far more valuable to us as data, did we know the history of their origin and primary use, which, in consequence of the names and the legends which have been applied to them, often comes to us in untrustworthy form.

1. *The Cisterns, Pools, Water Reservoirs, and Fountains in and around Jerusalem. Rogel, Siloah, and the Spring of Mary.*

To what was said in the pages regarding the wells of the city, there remains something to be added under this head. In consequence of the dryness and the very great want of water in the entire neighbourhood of Jerusalem (there being but three springs in the valley of Jehoshaphat, besides the few wells which seem to be alluded to by Jer. ii. 13 as holding no water), the abundant supply which seems to have been always in the city (for we have no record of its ever giving out) is something quite peculiar to Jerusalem. Strabo even characterized Jerusalem in these brief terms: ἦν γὰρ

---

[1] Krafft, *Topogr.* p. 254.
[2] Ewald, *Ges. des Volks Israel*, iii. pp. 61–69.

Πετρῶδες, scil. Hierosolyma, καὶ εὐερκὲς ἔρυμα, ἐντὸς μὲν εὔυδρον, ἐκτὸς δὲ παντελῶς διψηρόν, κ.τ.λ. A subterranean Hierosolyma would disclose many thousands of cisterns,[1] generally very ancient, of which every house has one, several more than one, and some a large number. To these must be added the many covered fish pools, and the inexhaustible supplies beneath the Haram. All these must have received their supplies from the rain falling upon the roofs and the terraced sides of the city, or have been conducted from distant springs by means of aqueducts, in some cases running beneath the ground, or have been fed from springs which lie far beneath the surface, and whose existence is now unknown to us. These cisterns are of great antiquity, for we have no account of the construction of new water-works in modern times.

The largest cistern of all, to which fifty-two steps lead down, in the so-called Treasury Building or Hospital of Helena (the Akbet el Tekiyeh el Sahahira of Gadow), has been already referred to. Northward from it, in one of the streets near to the Damascus gate, Gadow speaks of a broad cistern in the garden of an English proselyte, which, according to the owner's statement, is never destitute of water, and whose narrow outlet cannot be climbed up to without danger. It contains also, besides the water, an old church, and the studio of a fresco painter. According to a Jewish tradition, the palace of Queen Helena of Adiabene once occupied the site. I have already spoken of the great cistern before the Damascus gate at the cave of Jeremiah. The same are to be found among all the great buildings, especially the numerous convents. The Latin Convent, says Scholtz,[2] can in a time of dryness supply all the Christian inhabitants of Jerusalem for a whole half-year out of its twenty-eight cisterns.

Of these concealed reservoirs, as well as of those connected with the baths on the west side of the Haram, I have already

---

[1] Robinson, *Bib. Research.* i. p. 301 et sq., 328 et sq.; Krafft, *Topogr.* pp. 183-190.

[2] Scholtz, *Reise in Pal.* p. 197.

spoken fully enough. Of the pools which are exposed to view on the north side of the city, I have mentioned the Birket el Hijeh, *i.e.* the Pilgrims' Pool, the Birket Hammam Sitti Marjam, generally dry,[1] and the Birket Israin, the last of which, lying within the city, has been universally called by pilgrims, but without sufficient reason, the Pool of Bethesda. In 1842, a part of it was filled up with earth dug up during the excavations around the Church of St Ann, in the course of which operations some halls came to light which had before been concealed. This little church lies north of Bethesda, beyond the grotto which is reputed to have been the birth-place of Mary. It dates back to the times of the Crusades, at which period a small nunnery was established at this point by King Baldwin I. Sultan Saladin afterwards converted it into a school for the sect of Shafites.

Robinson discovered, however, that there is no ground for believing that this evident relic of the fosse of the Antonia, often destitute of water as it is, is to be considered the five-porched pool of Bethesda of the Evangelists at the sheep gate (John v. 2-9), since it has received this name in the later legends, because the Gate of Stephen was held to be the sheep gate of Neh. iii. 1. The pool of Bethesda—a Chaldaic word, meaning house of pity—lay farther to the north of the Church of St Ann, but has been wholly filled up. The Birket Israin had no water in it as early as the beginning of the seventeenth century, and is at the present day covered with rubbish, and overgrown with weeds and bushes. Its depth is about eighty feet, its breadth a hundred and thirty, and its length from east to west three hundred and fifty-five.[2]

I have already spoken of the pool on the west side of the city, which has been called the pool of Bathsheba, but which has now entirely disappeared. So, too, of the Patriarchs' pools, or the so-called pools of Hezekiah. So, also, of the Dragon's well, to which Nehemiah alludes as being outside

---

[1] Robinson, *Bib. Research.* i. p. 232; Krafft, *Topogr.* p. 184; Schultz, *Jerusalem*, p. 32.

[2] Roberts, *The Holy Land*, Book, ii. : *Pool of Bethesda*.

the wall (ii. 13), but which is not named elsewhere. According to Robinson, Ewald, and others, it is only to be sought in the Gihon valley, on the west side of the city, where we have already become acquainted with the upper and lower Gihon pools,[1] or more correctly, the Mamilla and the Sultan's pools. The three springs of Jerusalem to which allusion has been made, lie all outside the city in the lower valley of Jehoshaphat. They are Rogel, Siloah, and the spring of Mary.

The well of Rogel,[2] *i.e.* of the Messengers, called also the well of Nehemiah and of Job, belongs to the oldest known localities of the land. Ewald says that the name is of Canaanitic origin.[3] It is a deep well, at the top of the junction of the valleys of Hinnom and Jehoshaphat. Robinson has given the most accurate description of it. He found it irregular, four-sided, walled in with great square stones, forming an arch on one side, and having the appearance of great antiquity. The depth he ascertained to be a hundred and twenty-five feet, there being fifty feet of water in it. This is sweet and wholesome, but not cold. Forty feet below the top there is an outlet; but during the rainy season the well overflows, and for sixty or seventy days during the winter these superfluous waters form a brook of considerable size. Schultz[4] thinks, however, with considerable probability, that the well itself does not overflow, but that its waters break out in its neighbourhood in the form of common springs, one of them farther south on the east side of the valley, the second one still farther south, and nearer the western wall. These neighbouring springs Schultz does not consider dependent upon, but independent of, Rogel; the more easterly one he terms the Almond Spring. Rogel seems to have suffered alterations at Mohammedan hands. Gadow observes, that close by this well is an old pool thirty feet square, above

---

[1] Robinson, *Bib. Research.* i. p. 347; *Christian in Palestine*, p. 147, Plate xlvii., and p. 170, Plate lxi.

[2] Robinson, *Bib. Research.* i. p. 331 et sq.; Roberts, *The Holy Land*, Book iv., *Fountain of Job*.

[3] Ewald, *Gesch. des Volks Israel*, Pt. iii. p. 64.

[4] Schultz, *Jerusalem*, p. 40; Scholtz, *Reise*, p. 136.

whose eastern wall there is built an arched room with niches. Round it there are drinking troughs protected with iron fences. On the 6th of March, Gadow found the top of the water sixty-four feet below the ground in the valley. Wilson asserts that the taste of the water is different from that in the Siloah spring, and suspects that Rogel receives its waters from the Kidron, the river filtering through the rock.[1]

It surprised Robinson that Rogel was mentioned by none of the earlier historians. Jac. de Vitriaco is unacquainted with it at the time of the Crusades, for he only mentions Siloah; yet Wilken cites Hugo Plagon, who wrote in 1184, and who speaks of discovering and clearing out an old well or fountain below Siloam, which yielded water profusely. This could have been no other than Rogel.

Robinson was unable to find a satisfactory explanation of the name Job's well,[2] Bir Eyûb, so common among the Arabs, although it was used by Mejr ed Din. It is employed, however, in the Arab translation of Joshua, dating from the middle of the tenth century, instead of the Hebrew name. But Gadow finds an explanation in the fact that the name Joab does not occur in the Koran, although the name Job or Eyub is one of great prominence in it. In the Jewish itineraries it is named the well of Joab, after one of the foremost of Adonijah's coadjutors. Krafft, too, concurs in the supposition that the name has crept in in consequence of the resemblance of Joab and Job; for at the time when David was in his last days, when his rebellious son Adonijah wished to make himself king, and to supplant his brother Solomon, and was celebrating the royal festivities incumbent on a newly made king, with Joab as one of his chief guests, we read in 1 Kings i. 9, 41, that the well of Rogel was the place selected for the feasting. Down to that deep valley the sound of the people's shouting came, as, in obedience to the injunctions of David, Solomon was anointed king. The rock Zoheleth, spoken of in direct connection with Rogel, may

[1] Wilson, *Lands of the Bible*, i. p. 497.
[2] Krafft, *Topogr.* pp. 94-96, 188; comp. Wilson, *Lands of the Bible*, as cited above.

still be seen in a projecting mass, a little above the wall, and on the north-east corner of the Mountain of Evil Counsel. It flings its shadow gratefully over the little plain which is found at the junction of the two valleys, the fairest and the most fertile spot around Jerusalem, the place of general resort for the people of the city. It bears the local name Wadi el Rubâb.[1]

The other name given to the well, that of the Messengers, may be traced to an incident connected with the history of Absalom's rebellion, recounted in 2 Sam. xvii. 17. We are told that the sons of Zadok the priest were concealed there, in order to prevent being followed. The name Nehemiah's Well has only been given since the time of Quaresmius, at the end of the sixteenth century, and seems to arise from a story, that he preserved the sacred fire in the well during the time of the exile, and on the return of the Israelites removed it from its hiding-place, and restored it to the temple.

The name Rogel occurs as early as the time of Joshua, that is, directly after the driving out of the Canaanite tribes, as one of the boundary points of Judah (Josh. xv. 8) and Benjamin (Josh. xviii. 16). From these citations it appears that Jebus then lay entirely within the territory of Benjamin. The boundary of the latter tribe began at the north-western corner of the Dead Sea, went westward through the mountain country as far as to en-Shemesh, *i.e.* the Spring of the Sun, probably the present Fountain of the Apostles, below Bethany, on the way to Jericho.[2] From there it ran to the well en-Rogel, and to the Valley of Hinnom, on the south side of Jerusalem: from that point to the highest part of the mountain tract, past the northern extremity of the Valley of the Rephaim or Giants. Thence it ran to the waters of Nephtoah, possibly the Yalo spring, in the Wadi el Werd. This statement, remarks Robinson, agrees in the most exact manner with the present position of Rogel; but should we

---

[1] Robinson, *Bib. Research.* i. p. 331; Schultz, *Jerusalem*, p. 79.
[2] Robinson, *Bib. Research.* i. p. 333; Keil, *Comment. zu Josua*, p. 282, Note 1, p. 283.

take, with him, not the Apostles' Spring as en-Shemesh, but one near the St Saba Convent, which Keil, however, considers too far south-eastward, the boundary would run along the lower Valley of Jehoshaphat to Rogel. Josephus, in his account of the rebellion of Adonijah, gives the precise locality of the place where the feasting took place, remarking that it was without the city, near the spring, in the royal garden.

The height of the water in this Nehemiah well gives a measure for wet and dry years in Palestine.[1] In 1814, 1815, 1817, 1818, and 1819, the water overran the top three times; in 1821, twice; in 1815 and 1821, the overflow was exceedingly profuse, and great fertility was the result. In 1816 and 1820 no perceptible rise in the water was noticed: that in the cisterns was speedily exhausted: a general scarcity of food was the result, and attendant sicknesses.

The fountain of Siloah—Siloam in the New Testament—i.e. the Sent,[2] is of much more account with Christians than it was held in the remote history of Jerusalem. It is mentioned but three times in the Scriptures. Isaiah speaks of the waters of Siloah, which " go softly," in contrast to the wild rushing tide of the Euphrates (viii. 6, 7). His meaning is, that the Lord in His time will avenge the scorn which the northern kings of Israel displayed towards His pious kings of Judah, by summoning the hosts of Assyria, and will overthrow the united hosts. Again, allusion is made to Siloah in Neh. iii. 15, who speaks of the rebuilding of the gate of the fountain by the hands of Shallum: " He built it, and covered it, and set up the doors thereof, the locks thereof, and the bars thereof, and the wall of the pool of Siloah by the king's garden, and unto the stairs that go down from the city of David."

The third allusion is in John's Gospel (ix. 7), where Jesus says to the man who was born blind, " Go to the pool of Siloam, and wash. And he went and washed, and came seeing." As there was in ancient times, as well as now, a spring and a pool of the same name, there arose in the later

---

[1] Scholtz, *Reise*, p. 138.

[2] Robinson, *Bib. Research.* p. 334 et sq.; Gesenius, *Comment. zu Isaias*, pp. 331, 332.

writers' allusions to the place a great deal of ambiguity. There was also a tower of Siloam mentioned in Luke xiii. 4, probably in the neighbourhood of the waters of the same name, and on one of the steep banks of rock near by. Its fall occasioned, as we learn from the words of the Saviour, the death of eighteen men. The passages cited, taken in connection with those in which Josephus alludes to Siloah, leave no doubt that the place corresponds with the Selwan of the Arabs of the present day, and which is located at the south-eastern extremity of the Tyropœon. Its sweet water flows no less abundantly than it did in ancient times. As early as 333, we have an allusion in the *Itinerar. Burdig.* to a pool of Siloah with four porches, and to another great pool, also beyond the city wall (juxta murum est piscina, quæ dicitur Siloah, habet quadriporticum, et alia piscina grandis foras).[1] But the same pilgrim also mentions a spring which runs for six days and six nights, but which rests on the Sabbath, running neither day nor night; whence Pliny, *H. N.* xxxi. 18: In Judæa rivus sabbatis omnibus siccatur. Jerome says with more exactness, in his commentary on Isa. viii. 6: The spring of Siloah lies at the base of Mount Zion: its waters do not flow regularly, but only at certain days and hours; and when they do, it is with a great rushing from holes and pits in the solid rock. In his commentary on Matt. x. 28, where he speaks of Gehenna, he says the idol Baal was set up at Siloah, near to Jerusalem, and at the foot of Mount Moriah. At the end of the sixth century, Antoninus Martyr,[2] in speaking of the Siloah spring, which had been shut out from the city since Justinian erected his church, alludes to the great fish pools in the Valley of Jehoshaphat, in which men, women, and lepers bathed daily, so far as the changes in the water allowed. From this it appears plain why allusion was made in the *Gesta Dei*, i. fol. 573, to a *natatoria Siloe*, although William of Tyre calls it an intermittent spring. This *natatoria Siloe* was said by the Lector of Ulm[3] to be at his time a

---

[1] *Itin. Antonin.* ed. Parthey, p. 279.
[2] *Itin. Anton. Martyr.* p. 19.
[3] Fabri, *Evagator.* ii. fol. 417.

dry spot, lying between walls formerly forming a reservoir, but in his day turned into a vegetable garden, by the side of which flowed the water of the Siloah spring. Robinson observes, that all the earlier historical notices up to Marin Sanutus (1321) relate merely to this one well, or its pools, in the Tyropœon, but that not one alludes to the spring of the Virgin Mary, lying farther north, in the Valley of Jehoshaphat, and with which the Siloah spring is connected. The pilgrims Tuchern and Fabri (1479) are the first to carefully discriminate between the two springs, at the same time without being aware of their connection. Fabri relates, that he with his companions was compelled to reach the waters of Siloam by working his way through a narrow cleft.[1] He says, moreover, that in some weeks the water ran only for three or four days, sometimes ceased entirely, and sometimes poured forth its supplies very abundantly. Sometimes he found the opening in the rock entirely dry, particularly during his visits in the cool of the morning to this wonderful fountain, which he thinks identical with the waters of Gihon, which were covered over in the time of Hezekiah. It was not till the first half of the eighteenth century that the false hypothesis arose,[2] that the Fountain of Mary is the true spring of Siloam, and the southern spring only the pool of the same name.

Robinson ascertained the distance of the Siloah spring from the eastern corner of Ophel to be two hundred and fifty-five feet, the depth of the water reservoir nineteen feet, the length fifty-three, the breadth eighteen. The western end is fallen into decay; several columns stand in the walls,— once supporting the roof of a chapel it may be, or, as Bartlett thinks, porches. At Robinson's visit the cistern in front of the entrance to the spring was dry: the stream from the well merely ran through it on its way to the garden. The smaller basin or mouth of the spring is a hollow excavated in the massive rock, in order to hold the water. Steps lead down on the inside to the water beneath the arch of natural

[1] F. Fabri, ii. p. 489.
[2] Robinson, *Bib. Research.* i. p. 334; Bartlett, *Walks, etc.*, p. 67, Tab. iv., p. 146, Tab. xliv.

stone. Close by, on the outside, is the reservoir, to which the water finds its way beneath the steps. This basin, only five or six feet wide, forms merely the end of the long and narrow subterranean passage, through which the water comes from the Spring of Mary, farther north. In the neighbourhood are several traces of more ancient cisterns.

Gadow[1] distinguishes here three separate water basins, which are entered on his map, as well as on that of Krafft, slightly different from Robinson's, while in that drawn by the English surveyors they are entirely passed over. There is the Ain Silwan, the reservoir described by Robinson, with a small pool lying before it twenty-three paces in length, ten in breadth, and only fifteen feet in depth. Steps descend into it at the north-west corner, and on the bottom may be seen fragments three or four feet high of four shattered columns. Farther southward, sixty paces away, there is a solitary pool, now in a garden where olive and fig trees grow: this is probably the *natatoria Siloe*. About double the distance from this, and to the south-east, there is a drinking fountain very much used, which receives its waters through a canal a hundred and twenty feet in length, running from the Ain Silwan. It lies directly above the royal garden, and at the confluence of the three valleys. On the north side of this drinking fountain there are traces of the cement used in the construction of cisterns. Thirty paces from the termination of the watercourse leading to the drinking fountain, there is an old wall twenty feet high, and running from north to south; between it and the drinking fountain there is an opening leading to a deep arch, whose cover is thin, and more recent than the structure which it covers. There seems to have been, according to the remains now existing, a reservoir about fifty feet square and twenty feet deep, once open, but afterwards covered, so as to save room. It appears to have been much deeper than the two farther north, which were fed from Ain Silwan. The Mohammedans lay great value upon the well of Siloah, and Mejr ed Din calls it and the well Zemzem the two fountains of paradise.

[1] Comp. Krafft, *Topog.* pp. 127, 174, 187.

The Spring of the Virgin Mary, Ain Sitti Mariam,[1] also called the well of Siloam, in contradistinction to the one farther south, known as the pool of Siloam. It lies only 1100 feet from the point of rock at the southern extremity of the Tyropœon, and owes to the investigations of Robinson the great interest felt in it, as possibly belonging to the very oldest sources of water supply in the city. It appears to be alluded to for the first time[2] in the fourteenth century; for Fabri[3] described it, in 1479, in his chapter headed *de Fonte Mariæ Virginis*. He discovered it before reaching the more southern spring of Siloam, and carefully discriminates between the two. Later writers fell into the habit of confounding the well of Mary with Siloah, and the name of Upper Siloah became current. Quaresmius (1639) cites the legend that Mary washed the swaddling-clothes of the child Jesus in its waters, thus giving pilgrims an excuse for paying their homage to it; but Fabri quotes still another legend, namely, that this retired spot served as a place for Mary to hide herself as she was fleeing with her babe from Herod, and sought to pass up the Kedron valley to the eastern gate, the Golden Gate, in order to consecrate him to the Lord, in accordance with the laws and customs of the Jews (Luke ii. 22).

Robinson,[4] who considers it improbable that the old east wall of the city should have excluded the only two living springs with which the city is supplied, and that they should thus have been allowed to fall into the hands of enemies, gives them a place within the wall, and holds, not in contradiction to Josephus, whose language on the subject is obscure, but in harmony with Nehemiah (ii. 14 and iii. 15), that the basin was the former King's pool, known also as the pool of Solomon; or, at any rate, that the water from it was used to supply these wells, in case they were subterranean, and within the walls. He has also completely convinced himself that the well of Mary is of great antiquity; that, at all events, it must be older

---

[1] Robinson, *Bib. Research.* i. pp. 232, 333 et sq.
[2] Krafft, *Topogr.* p. 187.
[3] Fel. Fabri, *Evagator.* i. pp. 415-417.
[4] Robinson, *Bib. Research.* i. pp. 311, 340.

than Siloah, since it sends its waters to it. I cannot pass by a remark of Schultz,[1] which seems, however, to have called no subsequent attention to the fact which he mentions. He says that the vault which encloses the well of Mary is very ancient, composed of very large stones (their size appears in Roberts' sketches), and bearing faint and illegible traces of an inscription. It would repay the efforts of some future observer, to see if something cannot be made of this inscription, at least to discover in what language it is written. All this, taken together, favours the view held very early, and accepted by Krafft, that this spring, although its origin is unknown to us, stood in connection with the great waterworks constructed by Hezekiah, and the lower outlet of the Gihon,[2] which, as Robinson remarks, may have received its characteristic name Siloah, the Sent (*missio aquæ*), from this fact. The subterranean communication would, according to that view, date back to the time of Solomon, and would owe its completion only to the labours of Hezekiah. Robinson's bold discovery, confirmed by the later researches of Tobler, has proved that the water of Siloah is derived from that of Mary's spring; but to ascertain whence the waters of the latter are derived, must be left to the efforts of future inquirers.

The water chamber[3] of the well of Mary lies very low: the outflowing stream follows the western wall of the Valley of Jehoshaphat. The chamber is cut in the solid rock. It is reached by going down a flight of sixteen steps: then comes a level spot of twelve feet extent, and then a second flight of ten steps. These steps are each about ten inches high, and they give a depth of not far from twenty feet below the surface of the ground. The water chamber itself, called by the Arabs Ain um ed Deraj, *i.e.* the Mother of the Steps, is some fifteen feet long, five or six feet broad, and six or eight feet deep. The bottom is strewn with small stones. The water

---

[1] Schultz, *Reise*, p. 171.
[2] Krafft, *Topogr.* pp. 127, 174, 178.
[3] Bartlett, *Walks, etc.*, p. 112; *Christian in Palestine*, p. 46; Roberts, *The Holy Land*, Book iii., *Upper Basin of Siloah*.

runs from it through a low artificial channel cut in the solid rock, and does not make its appearance till it reaches the Siloah spring. At present there is no other outlet but this, and according to all appearances, says Robinson, there never was any other. This fact, that there is a southern outlet, seems to indicate that there is an undiscovered subterranean channel leading to the Mary spring from the north.

This aqueduct cut in the solid rock, which conducts the water beneath the walls of Ophel to the spring of Siloah,[1] is alluded to in 1620 by Quaresmius, in terms which indicate that he was familiarly acquainted with its existence. There were some attempts made to trace its course, but they remained fruitless down to Robinson's time. People spoke as though there were no doubt that the two were connected, but there was no proof.

On the 27th of April, the water being very low, Robinson[2] and Eli Smith found not a soul at Mary's well, which was generally so thronged with people coming from the city to fill their bottles. They took advantage of the opportunity, and went down barefooted, with lights and measuring tape, in order to explore the well and its outlet. The water was nowhere more than the depth of a man's foot, and flowed away very gently: the bottom was covered with sand. They entered the outlet. It was generally two feet in breadth, and was cut out of the solid rock. The main direction was from N.N.E. to S.S.W., but there were repeated curves: for no long distance did it follow a straight line. For the first hundred feet it seemed to be fifteen or twenty feet high, the next hundred from six to ten, then only four, and became constantly less in this respect, till, after traversing a distance of eight hundred feet, they could advance no farther without lying down in the water. This they were not dressed for attempting; and the two bold explorers, after blackening their names with candles upon the roof, in token of the extent of their observations, came back to the mouth, hoping to make a second attempt at the other extremity, at the Siloah spring.

This trial took place three days' later, and brought the

[1] Robinson, *Bib. Research.* i. p. 232.   [2] *Ibid.* i. p. 239.

enterprise to a termination. They entered the subterranean canal at the last-mentioned spring, but found it impossible to proceed without tearing down some loose stones which were on the point of falling. The passage was so low and narrow, that it was with the greatest difficulty that they could proceed. It was only by overcoming formidable obstacles that they could get on at all, and with much loss of time. After meeting many windings of the channel from side to side, and passing many branches, plainly attesting the want of skill of the workmen, who were unable to follow the shortest course without making many of these deviations, they reached, to their great joy, after traversing a distance of nine hundred and fifty feet, the marks which they had branded three days before, and solved the debated problem. The whole length of the passage is 1750 English feet, a hundred longer than the straight line measured above ground; but that is easily accounted for by the repeated deviations from a direct course, occasioned by the rude skill of the artisans who constructed the work. It is probable that they began at both ends of the mine, and it must have been a task of great difficulty for workmen so primitive to meet in the middle. Robinson and Smith found the bottom of the passage very slightly inclining, giving the water a very gentle and uniform flow. There was nothing of that rushing mentioned by earlier writers. It is not at all probable that there is that difference in the taste of the water from the two wells which has been alluded to by some of the older pilgrims; and the intermittent jetting up of the waters, about which so many fables have been told, may be common to both. That there is a regular rise and fall of the water, although not at regular intervals of three or seven days, or twenty-four hours, etc., Robinson and Tobler have satisfactorily proved. The former found the rise of the water in Mary's spring to be about a foot: it lasted about ten minutes, and covered the low steps. The washerwomen there told him that there is a similar gushing up two or three times daily during the summer: sometimes, they say, the spring is entirely dry, and then suddenly the water breaks out among the rocks. The popular method of explaining this phenomenon is, that there is a great dragon in the well: it runs only when he is asleep, and when he is awake he keeps the water all to himself. An Arab told Robinson that the water comes from the spring under the great mosque; but whether conducted by an artificial passage or by a natural channel, is yet unknown. The expression, John v. 2–7, regarding the pool of Bethesda, into which an angel went at certain times and "troubled the waters," seems to imply that the mere water was not invested with any healing properties; for the first one of the sick lying around who went down into the pool after the *troubling of the waters*, was made perfectly whole. This expression "troubling" seemed to Robinson to indicate the intermittent flow of the spring, and therefore to correspond well with the New Testament description of the pool of Bethesda, whose position had been placed farther north by the prevailing legend.

Tobler visited Mary's spring[1] very often during the winter of 1845, choosing the hours of early morning and late evening to make his observations, in consequence of the great number of persons resorting thither in the day time to procure water. In March 1846 he succeeded in passing through the entire length of the rock channel from Mary's spring to Siloam, and the results which he attained fully substantiate the conclusions gained by Robinson.[2] Even the measurements made in the tunnel are substantially the same in length, breadth, and height. Tobler observed a number of brown stripes along the sides: these seemed to him to indicate the altitude reached by successive floods of water. He suspected at one place that he might have touched the opening of a second channel leading to the Siloah spring, but the extinguishing of his light while in the very middle of the whole course leads to the belief that he was misled in this by one of those numerous windings to which Robinson referred. The con-

[1] Tobler, *über Siloah*, in *Ausland*, 1848, Nos. 52, 53, pp. 205–211.

[2] Barclay, the American missionary, has since made an attempt to follow Robinson and Tobler in this perilous adventure. His success was not equal to theirs, as he did not pass through the entire length of the canal.—ED.

struction of this subterranean mine seemed to him to have been effected with less expense and pains than would have been undergone if it had been made in the form of an open canal between the two springs, not to speak of its being kept better from the reach of an enemy. The water was cooler too; it was brought nearer to another portion of the population of the city, and could be easily collected in a pool below Siloam, and used for the watering of the gardens and the cultivated valley. But happy as was the conception of carrying out the plan of connecting the two fountains, the work was executed nevertheless in so rude and unskilful a manner, without an exact grading, without a straight direction, and with a repeated loss of the true line, that its construction may with great probability be placed before the time of Solomon, whose architects were men of great skill. Tobler remarked, as did Robinson, the changes in the height of the water, and states them as commonly ranging over a scale of two inches. On the 21st of January he observed the very unusual phenomenon of a marked rise in the height, which attained to an altitude of four and a half inches, and was accompanied by a slight wave. On the 14th of March a similar rise lasted an hour and a quarter at the greatest altitude; then subsided gradually to its natural proportions. This time it rose six and a half inches above its habitual level. The flood was usually observed about three o'clock in the afternoon. The temperature of the water at the beginning of the increase was 13° Reaumur, later 14° +: shortly before, Tobler had ascertained the warmth of the water in the Ain esh Shefa, or Baths of Healing, to be 15°. The temperature of the spring of Mary at all other times during the winter was steadily maintained at 14° R. Tobler says that the water is sweeter, and with less of a salty taste, in winter than in summer; and at the time of the accession of the new supplies at the time of flood, the taste is less salt than in the Baths of Healing. Tobler does not propound any theory in connection with this; but it seems to be very natural to believe that the spring of Mary receives its waters regularly from a cooler basin, which, on account of its greater depth or greater quantity of water, has a proportionately lower temperature; but that this is raised when the waters of the healing baths, which are nearer the surface and the air, flow in. This seems to happen only under certain physical conditions, on which depend the flood of the water in Mary's spring, which continues as long as this side stream enters, and then gradually subsides, and receives its regular supply from some cooler source, perhaps beneath the Haram itself. At any rate, there can be no doubt regarding the connection of the spring of Mary with other great basins; for it is proved by the temperature, the taste, and all those peculiarities which show a common origin for the waters which are found in the *Hierosolyma subterranea.*

2. *The Necropolis around Jerusalem. The Rock Crypts in the Valley of Hinnom; the Rock Chambers and Mausolea in the Valley of Jehoshaphat, as far as to the Grave of Mary; the Rock Graves north of the City; the Tombs of the Prophets, the Judges, of Helena, the Kings, and Herod.*

A second kind of memorial, and a very important one in helping us to settle the ancient topography of Jerusalem, is found in the ancient graves of the city, because, being mostly subterranean, and cut in the rock, they are capable of little displacement, and in this respect are unlike the structures on the surface, which have not only been subjected to complete destruction, but have passed into entirely new forms, their parts being in many cases removed to a distance, and transformed into the materials which make up other buildings. To attempt to trace all such changes must lead to error. The fragments which remain of the three walls mentioned by Josephus, despite the appearance here and there of blocks of quarried stone, pillars, arches, and other ornaments, are very difficult to trace, and must be given over mainly to hypothetical conclusions. With the exception of a few main points, all that belongs to the early Jewish condition of the city, and to its character when held by Gentiles, is very obscure. So, too, the Christo-Byzantine and Arabian periods down to the time of the Crusades, are subject to the same difficulties, with regard to its architecture, of which the Church of the Holy Sepulchre is an example. There the subterranean grave has undergone no changes, although its surroundings must have been wholly made over; and yet it must be confessed that careful study is able to detect in the remains many features which date back to the construction of the original church bearing that name. In less important buildings there is less opportunity to trace the architectural changes, and no particular attention has as yet been paid to them. This is a task which it is for the future to attempt. There are objects in the city which present some encouragement to explorers: the Kasr ed Jalud, for example, the ancient Goliath fortress, in the north-west part of the city. Scarcely less promising are the old wall, with its Frank tower; the Burj Jebel Chani, east of the Herodian gate, in the north-eastern part of Bezetha. But many of the edifices once here, and whose names are met in the narratives of the ancient pilgrims, have wholly disappeared: among them the Byzantine Churches of Chariton, Ægidius, John the Baptist, John the Evangelist, those of Maria Major et Minor, and of Maria de Latina.[1] Scholtz and Tobler[2] are the two men who have most clearly indicated in their researches what remains to be done in examining those architectural remains which appear at all promising. The last mentioned alludes particularly to the proudly aspiring Mosque Maulawiyyeh, the ancient Church of St John, situated near the Damascus gate. On its inner walls may be perceived even now fresco figures, lightly whitewashed over. Above, at the Street of the Herodian gate Bab es Saheri, there are still to be seen marked relics of the Church of Mary Magdalen, called by the old pilgrims the house of Simon the Pharisee, and by the Arabs Mamunijeh. In this edifice fresco paintings dating back to the Frank occupation of Jerusalem are to be seen.

[1] Compare Robinson's admirable *History of Jerusalem*, in *Bib. Research*. i. p. 365 et sq.
[2] Comp. Scholtz, *Reise*, pp. 171–177; Tobler, in *Ausland*, 1848, No. xviii. p. 74.

South-east of this lies the deserted Church of St Ann el Salchiyyeh, which was given by Saladin to the Shaafites for a school—the only building which has been restored by the Turks. Still south-east of this Church of St Ann lie, on the north side of the so-called pool of Bethesda, the ruins of a former nunnery, called also by the name of St Ann, the reputed mother of Mary—Haret attiseh Hannah. It stands in the street Sucket Bab el Hotta.

In the neighbourhood of the Church of the Sepulchre there are still to be seen distinct ruins of the former Hospice of the Knights of St John, which at the time of Benjamin of Tudela's visit gave shelter[1] to four hundred of this order, whose duty was to care for the sick, while four hundred others, in a second hospice, were always in a state of readiness for war. The place is marked by the white marble clock-tower, which was not thrown down as it now is by the Mohammedans, but partly felt the natural influence of time, and also in the year 1719 was despoiled of its top, for fear that it should fall over and injure the great dome of the chapel close by. The minaret known as Muristan, standing just south of it, was built in 1465 in the spirit of Ishmaelitic defiance. East of the Church of the Sepulchre, says Tobler, and close by the court of the Abyssinian Convent, there are interesting architectural relics of the Frankish possession of the city: there stood 'the edifice where the musicians lived, whose duty it was to sing in the adjacent church; and westward, *vis-a-vis*, resided the Patriarch of Jerusalem, in a house also adjoining the church. The architecture of it is striking even in its present form. When one thinks of the rich gifts which have been made to all the religious foundations of Palestine and Syria, and recalls the minute particularity with which such transactions have been recorded in countless documents, in the court records, in the Codex Diplomaticus of the Church of the Holy Sepulchre and of the Hospital of St John, all of which can be used to great advantage in studying the topography of Jerusalem at the time of the crusaders, it will not seem too sanguine an expectation to

[1] See B. von Tudela, ed. Asher, p. 69.

look confidently forward to a time—not distant, let us hope—when the careful use of these materials will give us new data towards solving difficult problems relating to the early topography of the Holy City.

With regard to the Hospice of the Knights of St John, of which only the lower storey remains, filled with earth and rubbish, and forming a garden, Scholtz[1] cites some particulars which seem to warrant the conclusion that careful inquiry would yield valuable results. At the time of the Crusades this building seems to have lain between the Bazaar and the Holy Sepulchre, and to have been three times as large as the Armenian Convent, or five hundred paces long. It was nearly as broad as it was long, and appears to have had much the appearance of a fortress. When Saladin had availed himself of treachery, and scaled the walls of Jerusalem, the Christians fled to this hospice, and defended themselves obstinately there for a long time: they were obliged to yield at last, however, and were all massacred. Saladin took up his residence there. The Mosque ed Demah was erected by his nephew in 1216, the minaret in 1417: the latter was destroyed by an earthquake in 1459, but was rebuilt in 1465. Scholtz says that walls are lacking in the interior, and that the whole is deserted at the present time. Little huts and booths are set up on the southern and eastern sides: these are the property of the Patriarch of Jerusalem. Scholtz tells us of one of those patriarchs who fell in love with a Turkish girl, gave up his faith, accepted the Koran, and left behind him a numerous family, forty branches of which were in the city when he was there: these live on the income of the houses in the immediate neighbourhood of the old hospice, and in the region where Schultz, Krafft, and others claim to have discovered architectural relics dating back to the most ancient times in the history of Jerusalem.[2]

It only remains for me to make such mention of the remarkable ancient graves, which exist in uncounted numbers on all sides of Jerusalem, as may throw light upon the topo-

---

[1] Scholtz, *Reise*, 168, 169; comp. Williams, *Holy City*, i. Suppl. 47.
[2] Schultz, *Jerusalem*, pp. 31, 61; Krafft, *Topog.* p. 26 et sq.

graphy and history of the city. In dealing with them, we must follow the guidance of the most earnest inquirers and the most accurate students into the character of the belt of sepulchres which girds this unique centre of the Christian world. We find that in the early Hebrew literature, for example (2 Kings xxiii. 6), there are allusions to the graves of the common people in the Valley of Kidron; for when king Josiah wished to purify the temple of all idolatry, and to burn the grove which the priests of Baal had transplanted there, he caused the ashes to be thrown upon the burial-places just mentioned. Family vaults and tombs were held in high estimation by the great men of Israel from a very early day, as can be seen from the pious care for his dead taken by Abraham in Hebron, from the sepulchres of the kings on Mount Zion, and from the words spoken by Isaiah at the time of king Hezekiah. In the twenty-second chapter, from the 15th to the 17th verses, we have the words of rebuke spoken by the prophet to Shebna[1] the treasurer, who was just on the point of hewing out a rock sepulchre for himself and his family: "What doest thou here, and why dost thou hew out a sepulchre in the high places, and make a resting-place for thyself in the rock? Behold, Jehovah shall cast thee out," etc. Such burial-places were a universal necessity of Israel in its earlier days.

I have on a previous page alluded to the graves of Christians who have died within modern times: they lie, the reader will remember, upon the southern slope of Zion, and outside the city walls. The Mohammedans bury their dead in three different places,[2] but mainly on the east side of the city, under the shadow of the Haram walls: the second spot is north of Jerusalem, near the cave of Jeremiah; and the third on the western side, in the upper valley of Gihon, and near the Mamilla pool, whose name they interpret etymologically Ma min Allah: The one come from God. The Jews' burying-place lies on the western slope of the Mount of Olives, above the old graves in the Valley of Jehoshaphat, where they await,

---
[1] Gesenius, *Comment. zu Isaias*, p. 694.
[2] Krafft, *Topog.* p. 221; Schultz, *Jerusalem*, p. 28.

in common with the ancient people of their nation interred there, the last day and the great judgment. All other graves are memorials of more ancient times, but of the most varied dates: they have slight, and only slight, differences in their internal structure and their external form, unless they be distinguished by sculptures and inscriptions. The uniform type of these[1] is a door cut in the solid rock, leading to a chamber, and that to others, of equal height with the door; the whole hewn out of the rock, but separated from each other by walls equally old with the tomb itself, and utterly without decoration. In these walls there are niches for the reception of the bodies; but they are so set in, that only one can lie in each wall, making it necessary to excavate many chambers for a family, and so rendering this kind very expensive. Or we have them so that the niches admit of the bodies being laid, not parallel with the main entrance, but at right angles,—an arrangement which allows far more interments within a small space, and which therefore put this kind of tomb at the disposal of the less wealthy. It was a great advantage, and therefore usual, to make use of existing quarries, and to convert them into tombs; and it made it still easier, when, at the corner of a great mass of rock, there were already two entrances, to connect them, and provide walls and niches, converting them into spacious tombs. All this gave rise to many diversities in the outward form.

Really artistic *mausolea*, such as those found in many of the great heathen capitals, are found to be very rare at Jerusalem: those which remain are mostly the ones in the Valley of Jehoshaphat and on the north side of the city, whose builders are now quite unknown to us. There are no graves in those parts of the city which, before the capture by the Romans, were included within the walls,—for example, on the north side of the Valley of Hinnom at Ophla, and on the south slope of Zion,—since all Jewish burial-places could only lie outside of the city, the graves of the kings alone excepted.[2]

[1] Robinson, *Bib. Research.* i. p. 352; Schultz, *Jerusalem*, p. 97.
[2] Robinson, *Bib. Research.* i. p. 360 et sq.

During recent years Krafft and Tobler have called more particular attention to the sepulchral monuments than they had ever received before: they have collected their inscriptions, and given minute sketches of their interiors.

1. *The Necropolis in the Valley of Gihon, and Ben Hinnom.*

In the Valley of Gihon, on the west side of the city, and around the Mamilla pool, there are no graves of importance mentioned, which would throw light upon the history of Jerusalem: the number, too, seems to be very small. They are all in a state of ruin, and are little visited. Below[1] the lower pool of Gihon, or the Birket es Sultan, however, the remarkable succession which follows the southern wall of the Valley of Hinnom, begins the unquestionably modern burying-ground of the Karaites.[2] A wall rises to a height of thirty or forty feet, composed of successive strata of blue limestone, overshadowed by a cluster of dark olive trees. It runs on, all the way becoming richer in tombs, to the junction of the Hinnom and the Jehoshaphat valleys, where the deeper-lying grounds of the King's Garden are surrounded on all sides by a wall full of rock-hewn sepulchres. They extend their course northward through the whole length of the Valley of Jehoshaphat past the Church of Mary, and to the northernmost trace of the Kedron, near the tombs of Helena, Simon the Just, and the Judges, the last of which are wrought with some art, and are ascribed on archæological grounds to the time of the Roman possession. At the western commencement of the lower valley of Hinnom, within an ancient quarry, is the burial-place of the Karaites, a sect of Jewish separatists, who receive only what is written, and pay no heed to tradition. They have long, flat, semicircular gravestones, with Hebrew inscriptions of a recent date. Hard by is a rock sepulchre, over whose entrance are some very rudely executed Hebrew[3] words, which used to be considered Phœnician, but which have been shown to relate to the Karaites just alluded

---

[1] Scholtz, *Reise*, p. 177.
[2] Krafft, *Topog.* p. 190.   [3] *Ibid.* pp. 179, 191.

to. There are given the date of birth and of death, and a brief prayer to God.

Not far from that place are found some graves hollowed out of the rock wall, and near them the words inscribed—

ΤΗΣ ΑΓΙΑΣ ΣΙΩΝ,

which Dr Clarke,[1] who discovered them, interpreted as implying that here was Mount Zion, and the place where Jesus was buried,—a view which has long since been disposed of. The words, says Krafft, signify nothing more than that these old Jewish sepulchres were used again in the more recent Christian times, and that the new occupants belonged to the Church of the Apostles worshipping on Zion, which was known as the hallowed Zion as long ago as when Willibald wrote (A.D. 786). Joh. Phocas, writing in 1185, confirms Willibald in this, and terms the place ἁγία Σίων.

Farther eastward, and close by the path leading from Zion through the Valley of Hinnom to the Mount of Evil Counsel, there lies a tomb over whose entrance stands a Greek inscription, stating that ten Germans, probably pilgrims, are interred there.[2] These and many other of the rock chambers lying farther east, some of which are furnished with pilasters, others with hewn crosses, and with brief inscriptions such as ΜΝΗΜΑ, served, although dating back to the remote Jewish times, in subsequent centuries as the burial-places of Christians, and show that at all times there has been there a crowded necropolis for the always large population of Jerusalem.

Still farther east, the number of graves increases in Aceldama, the field of blood of tradition, according to Matt. xxvii. 7, 8, and Acts i. 19, but which was formerly called the potters' field. It is a fact that even at the present time there is found there a not unimportant layer of white clay, still used for potters' purposes. From Jer. xix. 1, 2, 11, we learn that the house of the potter to whose earthen vessel the

---

[1] E. D. Clarke, *Travels*, iv. p. 326 et sq.; Robinson, *Bib. Research.* i. p. 353; Krafft, *Topog.* p. 192.

[2] Krafft, *Topog.* p. 193.

prophet was directed, lay in the valley : he was to break it before the potters' gate in the Valley of Ben Hinnom, before the eyes of the people and their elders, for a sign and a threatening that Jehovah would in like manner break His rebellious people who served Baal in this valley and on the heights around, and brought their own children in their arms as an offering to the image of Moloch. The valley was thereafter no more to be called Tophet, the high place of sacrifice, nor Ben Hinnom, the children of Hinnom, but the Valley of Slaughter.

According to Matthew, the thirty pieces of silver which were paid for the betrayal of Jesus, and which the traitor threw into the treasury of the temple before he hanged himself, were not suffered by the high priest to be retained, but were used as purchase money to buy the potters' field as a place to bury pilgrims in. This it has remained during the long succession of centuries from that time to this. In the year 1143, the church in Aceldama, in which strangers were buried, became the Hospital of the Knights of St John.[1] Eusebius and Jerome even allude to it. In the fourteenth century it was in the possession of the Latins; then of the Armenians, who paid a high sum for the privilege of being interred there, under the delusion that if they should be put in that place after their death, it would atone for the sins which they had committed during life. It is only since the beginning of the eighteenth century that the place has ceased being used for sepulchral purposes. The same delusion, or rather the opinion, that in the dry clay of the potters' field bodies would pass into a state of perfect decay sooner than anywhere else, was the occasion for the Pisanese in 1218 to remove seven shiploads of it to their city, and to place it in their Campo Santo : some of it is said to have found its way to Rome also. The present graves in Aceldama[2] are not excavated in the clay, but in the rock, and are arched over: they probably occupy the site of former quarries, and their walls are profusely covered with crosses hewn in the rock.

---

[1] Sebast. Pauli, *Codice diplomatico*, i. p. 23, No. xxii.
[2] Robinson, *Bib. Research*. i. p. 395; Schultz, p. 39.

Still farther east, there is met an excavated chamber with the inscription cited above as seen elsewhere, τῆς ἀγίας Σιών, showing that the graves of members of the Church of Sion extended as far eastward as this point, and indeed yet farther, for the same words are found farther on.

The ornamented tomb, with its four chambers and its many niches (according to Tobler's plan of the city, No. 10, the Graves of the Apostles in the Valley of Hinnom), dating probably from the time of Herod, is held, considered in connection with some adjoining ones, to be the Latibula of the Apostles, in which, according to the legend of the middle ages, eight of the twelve were concealed during the time of Jesus' imprisonment. In the sixteenth century it was restored by the Franciscan monks at the expense of the Catholic king Philip, and decorated with images of the saints. Here, too, according to the tradition, was the grave of Ananias the high priest.

Farther eastward, Krafft[1] describes an entrance to a subterranean vault, which was bordered by carved work, access being effected by turning a large stone[2] upon its edge, while a second large one stood just in front in a place prepared for it, reminding him of the arrangement connected with the sealing of the stone at the grave of Jesus (Matt. xxvii. 66, xxviii. 2). The interior of this half-closed tomb he found richly ornamented. At the entrance appeared a chamber twelve feet square, arched over in the form of a domed vault, and having several small and simple pilasters; at the side are other chambers with niches at the side, and stone coffins. Opposite to the entrance there is a second door leading to a chamber lying still deeper, furnished also with niches,—an arrangement which corresponds to the descriptions of Jewish graves given in the Talmud. Here Krafft discovered an inscription which shows that the tomb was used by Christians during the middle ages. It runs thus: The place of interment of ten men, superintendents of the Convent of Benas of George. This, however, does not

[1] Krafft, *Topogr.* p. 197.
[2] Comp. Ritter, *Erdkunde*, xv. p. 380, etc.

throw any full light upon the question, since there were several foundations named in honour of St George. It may be that the one referred to here was the little Church of St George el Chuddr, whose ruins lie on the west side of Hinnom, north of the fallen Arabian village Abu Wair, west of Birket Sultan. This is probably the one indicated by No. 1 on Tobler's plan ;[1] a rock grave given in outline, extending in a south-easterly direction. Among the numerous crypts which belong to this group of tombs is one with this inscription: "The grave of several men from Rome, connected with the holy Zion," making it probable that people of various nations, going to Jerusalem to perform their pilgrimages, were buried there. The rock graves which are found still below the junction of Hinnom and Jehoshaphat cover a considerable extent of surface, and are in some cases by no means small, but they are destitute of any architectural grace.

2. *The Necropolis in the Valley of Jehoshaphat, from Ben Hinnom northward along the Kedron to the reputed Grave of Mary, and those of the Prophets at the Mount of Olives.*

From the Fountain of Rogel and the King's Garden, at the confluence of the three valleys (including the Tyropœon), the one bearing the name of Jehoshaphat runs northward as far as the Gate of St Stephen, and then widens considerably, and extends farther in the same direction, making the circuit of the north side of the city. Both of its sides are full of tombs.

At the Gate of Stephen, the western wall under the terrace of the Haram is fully a hundred feet high. Here it is traversed by a road leading westward out of the city, and conducting over the generally dry brook Kedron, and to Gethsemane, by a bridge.[2]

The reputed grave of Mary, which is there, is surmounted by a church, half of it, however, subterranean. The entrance

---

[1] Schultz, *Jerusalem*, p. 38.
[2] Robinson, *Bib. Research.* i. p. 476; Bartlett, *Walks, etc.*, p. 99; Williams, *Holy City*, ii. p. 431.

to the sepulchre is very similar to that on the south side of the Holy Sepulchre, and dates apparently from Byzantine times. Before this is a small sunken court (probably once a quarry), from which a staircase leads down to the church. Tradition ascribes the erection of it to the Empress Helena; others, like Brocardus, Mar. Sanutus, consider the grave much older, and involved in the destruction and ruin effected by the Romans when they took the city. The earliest mention of it is made by Arculfus in the year 705, and by Willibald in 786. In the *Itinerar. Burdig.* of 333 there is mention neither of this grave of Mary nor of Gethsemane, although the valley is repeatedly mentioned by pilgrims to the Mount of Olives and Bethany. We have no historical information regarding the end of Mary, and there is an equal lack regarding the place of her sepulture; yet the situation near Gethsemane, the scene of her Son's greatest sorrows and inward trials, seems not at all unnatural, and very satisfactory and comforting to the mass of believers. The localizing of her grave in some distant part of western Asia, to which she is said to have been sent by the Apostle John, has no more historical ground to rest upon than that which places her grave here in the Kedron valley, and at the foot of the Mount of Olives.

Arabian authors and the present native Arabs call the church el-Ismaniyeh, *i.e.* Gethsemane, taking the name from the small olive garden scarcely a hundred steps away, surrounded by a simple stone wall, and square in its shape. This wall encloses the celebrated group of eight very old olive trees, which are surrounded by heaps of stones. These trees may date back to the time of Helena (326), and very probably to the time when Jerusalem was taken possession of by Omar and his Arab hordes. Even the Turks respect the trees, and allow no injury to be done to them. Bové, the experienced botanist,[1] measured the circumference of these trees, whose height is usually from thirty to forty feet, and found it to be generally from eighteen to nineteen feet: he thinks them at least two thousand years old, allowing the

[1] Bové, *Recit.* in *Bullet. de la Soc. Geogr. Paris*, 1835, iii. p. 382.

growth of a half millemetre for each year. There is no tradition extant concerning any visit to these trees before the time of Helena. Eusebius, who wrote several years later, speaks of the Gethsemane of the Scriptures at the Mount of Olives (Matt. xxvi. 36; Mark xiv. 32), a place of prayer for believers. The Bordeaux pilgrim speaks of a stone standing at the Mount of Olives, at which Jesus was betrayed. A hundred years later, Jerome sets Gethsemane at the foot of the mount, and says that a church was then erected at the spot. This edifice Theophanes alludes to in *Chronic.* A.D. 863 as still standing; but of it no trace is now to be seen, unless it be the sepulchre and chapel hard by, known now by the name of Mary.

The peaceful solitude of the Kedron valley has had its suitable consecration in the last struggles of the Saviour on earth, who withdrew thither at nightfall from the tumult of the city, in anticipation of His great sufferings, and in preparation for His betrayal. During the day He taught in the temple, says Luke (xxi. 37); but at evening He went out and spent the night at the Mount of Olives. In Luke xxii. 39, we are told that He went out, according to His custom, to the Mount of Olives; that His disciples followed Him thither, and that there He kneeled down and prayed. From John xviii. 1 we learn that Jesus went out with His disciples over the brook Kedron, where was a garden, in which Jesus tarried with His followers, and in which also His betrayer found Him. No pilgrim can ever pass this solitary valley without quickened thoughts and heightened emotions, looking upon this gloomy grove where Jesus contended with the terror of death in the presence of His disciples, while they left Him to tread the wine-press alone (Isa. lxiii. 3, 5), to bear the burden of His pains (Isa. liii. 5, 11), while they slept.[1]

I have already spoken of the path leading from the bridge to the Mount of Olives, but must here speak somewhat more minutely of the bridge itself, and the valley which it traverses. It is erected upon a kind of dam terrace which crosses the valley; its southern wall is perpendicular, but its northern

[1] Von Schubert, *Reise*, ii. p. 518.

edge is no higher than the masses of rubbish which have been heaped up in the valley to such an extent as sensibly to raise its level above the bridge. The open arch is seventeen feet above the water-bed; but a pair of subterranean canals, one of which issues from the Church of Mary near by, serve to convey the rain-water under this bridge, whose erection is ascribed to Helena by the older pilgrims. The whole breadth of the valley is narrowed at this point to about four hundred feet: it becomes narrower as it extends southward, and then first discloses traces of a regular water-bed, which, however, is often dry for years. There follows a second bridge with one arch, at whose eastern side are seen those tasteful *mausolea* which have their own special names. These are followed by Jewish graves on both sides, and by the village of Silwan, opposite the space between the fountains of the Virgin and of Siloah. It lies not directly in the valley, but about half-way up the side of the eminence. Here may be traced the beginnings of that luxuriant growth of figs, pomegranates, olives, and other fruits and vegetables, which particularly characterizes the junction of the valleys, making the place always a marked one in the topography of Jerusalem — the King's Garden of the ancient times, and also the seat of that worship of Moloch and Baal which gave it the name of Tophet, the Unclean Place, and caused it to be shunned by the later Jews on account of the fires kindled in the sacrifices to those gods, and to be called Hell, the place of perpetual damnation, and Gehenna.[1]

The present village of Silwan is for the most part built out of an ancient city of tombs : the fellahs have either used the vaults themselves for houses, or have built slight structures in front of them out of the loose stones lying around.

On coming to the end of these abodes of the living in the houses of the dead—a sight which is artistically not without some picturesqueness—we come to a stone monument hewn out of the solid rock,[2] which in its form reminds the observer

---

[1] Robinson, *Bib. Research.* i. p. 274. Comp. Roberts, *The Holy Land*, Book iv.; Bartlett, *Walks*, etc., p. 110, Tab. xii.

[2] Krafft, *Topogr.* p. 198.

of the pyramids on the plain of Gizeh. It is in the form of a pylon, *i.e.* a small pyramid with an oblong base, whose flattened top sustains a fluted tore; a small door leads to the interior, and opens towards the valley.

As we pass down to the deepest place in the gorge, stepping over the countless gravestones which cover the Jews buried there, reaching higher up the valley than the upper part of the village of Silwan, and bearing inscriptions which are but a few centuries old at the longest, we arrive at some architectural monuments of great interest on the east side of the valley.

These *mausolea* stand in a row running from south to north, and have received at the hands of a comparatively recent tradition the name of Zacharias, James, Absalom, and Jehoshaphat.[1] Those designated by the first and third of these appellations are monoliths, real rock monuments; the two others are only grave caverns with doors. They lie in the narrowest part of the valley, close by a perpendicular wall of rock, out of which they seem to have been hewn. The bed of the Kedron is hard by. The style of their architecture appears to be a peculiar mixture of Persian and Greek, like that found in Syria, or more eminently that to be seen in the *mausolea* of Wadi Musa, the former residence of the Nabathæans and Idumæa-Arabian possessors, who are not to be assigned to a period prior to that of Christ. It is the style of the Herodian age, which the showy monarchs of that line loved to indulge in, and is explained by the marriages which they contracted with the princes of other nations, leading as they did to the incorporation of the manners, and even architectural styles, of those nations. These *mausolea* cannot be ascribed to the age of Hadrian, for they lack the purity and the elegance of the Roman sculpture and architecture left by that monarch. They show on the one hand an excess of ornament, and on the other a certain rudeness and lack of unity.

[1] Robinson, *Bib. Research.* i. p. 349; Krafft, *Topog.* pp. 199-202; Williams, *Holy City*, ii. pp. 157-160; Bartlett, *Walks*, p. 114, Tab. xiv., p. 145, Tab. xli.

Who the Zacharias was, who is commemorated in the southernmost corner of these monuments, is unknown. The legend ascribes it to the one who was stoned between the temple and the altar, the son of Jehoiada, a contemporary of king Joash (2 Chron. xxiv. 21, Matt. xxiii. 35). It is a massive cube with a pyramidal point above, and is hewn out of the solid rock. It is thirty feet in height, and eighteen feet square. At the side stand some pillars of the Ionic order.

Just as uncertain and untrustworthy is the authenticity of the tomb of James, to whose front chamber a low doorway, enclosed between two Doric columns, leads. Tobler gives a sketch of the interior, and states that there are six different rooms and several niches for the reception of the dead. The legend asserts of it that it was the place of refuge to which the Apostle James fled, and where he remained hid in the interim between the crucifixion and the resurrection of Jesus.

The so-called grave of Absalom is sixty steps farther north, and close by the lower bridge over the Kedron. Tobler has given a good plan of it. Williams has availed himself of the drawing of the architect Scole, and published in his work an elegant view not only of its external, but also of its internal arrangements. It is a square solid mass of rock, twenty feet on each side, and is decorated on the outside with small Ionic pilasters. In the interior there is a vault recently discovered, with niches for the reception of the dead. For a long time it has been the universal custom for Jews to cast stones at this tomb, and to spit on it in passing, in accordance with a custom which they share with the Arabs, and which is intended to express horror towards those who array themselves against God. In this case the object is to keep in perpetual loathing the memory of Absalom and his rebellion against the authority of his father David. In 2 Sam. xviii. 17, 18, we have the following account of the burial-place of Absalom : " And they took Absalom, and cast him into a great pit in the wood, and laid a very great heap of stones upon him : and all Israel fled every one to his tent. Now Absalom in his lifetime had taken and reared up for himself a pillar, which is in the king's dale : for he said, I have no son to keep my name in remembrance; and he called the pillar after his own name: and it is called unto this day, Absalom's Place." In harmony with this passage, Josephus states that Absalom erected for himself in the king's valley a marble monument, and called it his "hand:" it lay two stadia from Jerusalem, a position corresponding with that part of the Kedron valley which is below the king's valley, since, according to Josephus, the Mount of Olives was six stadia distant from the city. Williams holds[1] the monolith which bears the name of Absalom to be the one originally raised by the rebellious aspirant for the crown, and traces its external ornaments to the later hand of some Idumæan prince; and even Wilson[2] thinks that the primitive form is from the old Jewish times, afterwards modified and made more elegant by the addition of the refinements of modern art. Yet although we must confess that it is hardly to be admitted that this monument dates back so far as the time of David, and grant that it is the product of the period when Idumæan power was largely influential in Palestine, it must be allowed that there was some ground conceded by Josephus for the Jewish legend, and some valid reason for the Jews to continue to give it the name of Absalom. The first mention of the place by the pilgrim writers is that made by Benjamin of Tudela,[3] who does not call it the grave, but the monument of Absalom. He also alludes to the grave of king Uzziah as being close by. The Burdigala pilgrim speaks in 333 of two monolithic monuments standing in the same neighbourhood, and calls them the graves of the prophet Isaiah and of king Hezekiah. The statements of subsequent pilgrims are exceedingly confused. The Arabs of the present day call Absalom's grave Tantur Faraon,[4] *i.e.* the Horn of Pharaoh.

The so-called grave of Jehoshaphat,[5] which, like the last

[1] Williams, *Holy City*, ii. p. 456.
[2] Wilson, *Lands of the Bible*, i. p. 488.
[3] Benj. v. Tudela, ed. Asher, i. p. 71.
[4] Wolcott, in *Bib. Sacra*, 1843, p. 34.
[5] Robinson, *Bib. Research*. i. p. 350; Krafft, *Topog.* p. 201; Williams, *Holy City*, ii. pp. 449, 451.

mentioned, is a monolith, is ascribed by the Jewish tradition to the pious king of that name; but we learn from 1 Kings xxii. 51 that he was buried with his fathers on Mount Zion. A three-cornered gable crowns the entrance to several rock chambers with niches, whose walls bear traces of ancient fresco pictures of saints.

In the second of the chambers, of which Tobler gives four or five, Krafft found twelve obituary stones of Jews ranged regularly upon the floor, and probably still used. It was here that in the winter of 1842-43 the parchment roll conjoined with the Hebrew Pentateuch was found, which at first attracted great attention, and which was sent to the library of the Vatican, where it was judged by Schultz[1] to be a modern manuscript in Babeli characters. The Jews have the custom, he says, of hiding away every roll of the law that seems to be in the least injured; and this seems to have been the case with this roll recently discovered.

Between these sepulchral monuments and the Kedron is the present burial-place of the Jews, bearing the name the House of Life.[2] Between this and the middle summit of the Mount of Olives, about half-way up the declivity, and about a hundred paces south of the Chapel of the Ascension, belonging to the Latins, lies a separate burial-place known as the graves of the prophets, the Kubur el Umbia[3] of the Arabs, regarding whose origin, and the reason of its unique internal arrangement, we know very little. The entrance is on the north-west side to a round subterranean vestibule, from which you pass to two concentric semicircular passages, provided with countless niches, hewn out of the soft limestone. Other less regular passages run deeper into the mountain, and form a miniature labyrinth, of which the lowest parts are full of rubbish and earth, and hard to trace. They have probably been examined with care by Tobler. The comparison of this monument with the Peristereon of

[1] Wilson, *Lands of the Bible*, i. p. 489.
[2] Williams, *Holy City*, ii. p. 452.
[3] Schultz, *Jerusalem*, pp. 41, 72; Williams, *Holy City*, ii. p. 447. Comp. sketches in Tobler, and Krafft, *Topog.* p. 202.

Josephus, and with a columbarium (according to Krafft and Schultz), is shown by Robinson to rest on no satisfactory grounds. The description which Wolcott gives of the interior is the following.[1] Through an opening in the rock above, you descend into a semicircular chamber in the rock, about twenty feet in diameter, with an arched roof: to this there is a side entrance. Two passage-ways go out from this chamber; a third one appears to be closed up: the two which are open extend for thirty feet. Between them two concentric passages form galleries, an inner one and an outer one. These would be, were they free from dirt and earth, ten feet high and six feet wide: they are arched, and overlaid with stucco. The outer gallery has a length of a hundred and fifteen feet, with thirty-two niches, and two small chambers with six niches. A narrow excavation leads down from the northernmost passage, and ends, after extending on for a distance of more than a hundred feet, in a clayey soft soil, which was the reason, it may be, why the galleries were not carried farther. The hypothesis that that singular labyrinth was a temple of Baal has no other basis than the conical arch of the circular vestibule; and not less unsatisfactory is the theory which makes it the sepulchres which the Pharisees erected in honour of the prophets whom they had stoned (Luke xi. 47).[2]

3. *The Burial-places in the northern part of the Valley of Jehoshaphat, and on the north side of Jerusalem.*

In the northern part of the Valley of Jehoshaphat, the graves and tombs continue on in an unbroken series around the north-eastern, northern, and north-west portions of the city, as far as to Scopus, whose waters flow into the Kedron, running over the Nabulus road. In part they have been changed by later quarrying operations to great holes and

[1] Wolcott, in *Bib. Sacra*, 1843, pp. 36, 37.

[2] In addition to what is given here, the student should not fail to consult the later pages of the American Barclay, the French de Saulcy, and the Swiss Tobler, all of whom have diligently studied the graves north of the city.—ED.

stone-pits, and have become unrecognisable as graves; in part they have been grown over with low stunted bushes, and have gained thereby in artistic beauty; and in part they have been allowed to become reservoirs for water. The most distant and the most remarkable group of these burial-places is one lying a half-hour's distance north of the Damascus gate, at the height of the watershed which supplies the Kedron, and at the place where the path begins to descend towards the Beit Hanina. It bears the name Tombs of the Judges. The whole neighbourhood, the beginning of the slope towards the Mediterranean, is thickly sown with graves. Those to which I particularly allude are remarkable for their architectural decorations, and for their inner arrangements, and remind one of the fine catacombs of Egypt.[1] Tobler has taken a sketch of them, according to which there seems to be first a vestibule, then a large central chamber with three or four side chambers, in which he counted sixty-eight niches. The portal at the entrance is remarkable, like that leading to the tomb of Jehoshaphat, for its rich decorations. The place has been considered to be the burial-place of members of the Sanhedrim, because their number (seventy) so well agrees with the number of the niches; others have held it to be the tombs of the righteous mentioned in Matt. xxiii. 29, where the hypocrisy of the scribes and Pharisees was made especially manifest. The place is not mentioned in the legends till the sixteenth century. The Graves of the Judges, as they are generally called, are reputed among the Jews to be especially holy; the tract around them is very attractive.

The tomb of Simon the Just (Kaber Sadik Simun), farther towards the south-east, lies in the midst of the broad valley of the upper Kedron, known as Wadi ed Jos. It is a place of much resort for the Jews on the thirty-third day after Easter, as they repair thither to celebrate the memory of the son of Onias, who was high priest at the time of the Egyptian supremacy of Ptolemy Soter.

The catacomb of Queen Helena of Adiabene, which lies directly south of the one last named, on an adjacent plateau

[1] Robinson, *Bib. Research.* i. pp. 240, 269; Krafft, *Topog.* p. 204.

of rock, and which is mentioned by all travellers, is often known, but incorrectly, as Robinson [1] has shown, by the name Tombs of the Kings,[2]—an appellation which only comes into use among the pilgrims of the sixteenth century. The oldest visitors to the Holy Land do not mention the place at all, for it was not in the first centuries after Christ considered as a hallowed resort: Marin Sanutus refers to it a few times, but only casually, as a *Sepulchrum Helenæ, reginæ Jabenorum*. Before him, Eusebius had said with regard to Queen Helena that she had erected some celebrated *stelæ* or *cippi* over her tomb, which before had been in another part of the environs of the city. The locality of this tomb is, however, stated with great exactness to have been on the north side of the city, in the passage in which he speaks of Paula's coming to Jerusalem from the north, and passing the tomb of Helena on the left, *i.e.* on the east. The main road over the Scopus is now the same that it was then, and the situation is defined clearly by these few words. Pausanias gives another proof of the identity of this monument: he speaks particularly of the remarkably fine stone doors of the tomb, and compares it to that of Mausolus in Caria. This catacomb of Helena lies about a quarter of an hour's distance north of the Damascus gate, on the right of the Nablus road, where it begins to run down into the Kedron valley. Two great square courts, open at the top, are sunk eighteen feet into the rock, of which the longer or more eastern one forms the vestibule to the second. These two courts serve as reminders of the graves of private citizens as well as of priests at Thebes. The passage from the outer to the inner one is under a portal, which formerly rested upon two pillars, which are now broken down, as well as the pilasters at their side. Yet the whole of the horizontal piece above remains entire, and is supported at the ends: its length is twenty-seven feet. The pillars which formerly supported it divided the passage-way into

---

[1] Robinson, *Bib. Researches*, i. p. 357 et sq. See Catherwood's *Sketches*. Comp. also Williams, *Holy City*, vol. ii. p. 519.

[2] Krafft, *Topog.* pp. 211-217; Bartlett, *Walks*, etc., pp. 127-132; Roberts, Book i.; *Christian in Palestine*, p. 153, Tab. li.

three tolerably equal compartments. The frieze above is remarkable for its fine sculpture, which reminds the observer of similar forms in Wadi Musa: great bunches of grapes between flowers and garlands, and fruits, and horns of plenty, are found along the whole length of the portal. At the side, too, there are reduced copies of Doric triglyphs. The representation is not full, indeed: it lacks completeness; but the impression left is of a work admirably executed, and corresponding, like the tombs already referred to in the Valley of Jehoshaphat, to the Herodian epoch. At the southern extremity of these vestibules, which are now full of dirt and fragments of stone, making it almost impossible for the visitor to force his way, two passages lead to a rock chamber in the south, and one to a similar apartment in the west, from both of which several other passages diverge and lead to other chambers, giving the whole a truly catacomb-like aspect. These are all supplied with niches, to receive the bodies of the dead. Robinson and others have described the ruins minutely. They belong unquestionably to a tomb intended for royal personages, though their interior presents nothing now but a bare, tasteless wall. The floor is covered, as I have just said, with fragments of rock and with dirt. The passages from the vestibules have attracted much attention from antiquarians, because they were once closed by stone doors with hewn panels: these were closed from within. They are now broken, and their fragments strew the floors. At the time of Maundrell's visit (1697), one of these doors was still hanging on its stone hinges, which were fastened[1] into the natural rock—the same kind of stone, indeed, out of which the great side door was made, which he has described as a great curiosity to him. Towards a solution of the problem how these doors were arranged, he noticed that the upper hinges are of twice the length of the lower ones, and that the lower door did not touch the under sill, but was two inches above it, making it plain how the door would play freely. In one of the innermost and lowest rooms there are three great niches at the side, in which sarcophagi of white

---

[1] Maundrell, *Journey from Aleppo to Jerusalem*, p. 77.

marble once stood, adorned with beautiful carved work. The fragments now strew the floor, and show that there was the place of the greatest honour—the room appropriated to the heads of the family owning the tomb. The magnificence of this sepulchral monument, although not vying with those of Greece, caused the belief to be long prevalent, that this was the resting-place of the kings of Israel. Although it is probable that there was no such extended use of the word Zion as would warrant us in accepting this view, yet it has so far found credence, that these catacombs north of the city have been very generally visited by pilgrims as the tombs of the kings.

Robinson has laid down all the grounds[1] which he thinks afford a reasonable degree of certainty (despite the objections of Wilson, which seem rather trivial) that this was the tomb of Queen Helena, wife of King Monobazus of Adiabene, who, having gone over with her son Izates to Judaism, lived in Jerusalem, and built a mausoleum there. Josephus speaks three times of it, and in his account he states that it lay three stadia distant from the city, and opposite the north gate, which agrees with the distance, and the position of the tomb now to be seen. He also states that the mausoleum was crowned with three pyramids. These, which seem to have been similar in appearance to those which are to be seen at Petra, have long since been thrown down; yet traces of their existence are to be seen in the three divisions of the main portal already spoken of, to each one of which a pyramid probably corresponded. Each one of these was erected, it would seem, in honour of one of the royal family; and the three marble sarcophagi already spoken of appear to have been reserved for the father, mother, and son. We know, moreover, that the seven heads of the Maccabee family were designated by seven pyramids. We have, in addition to this body of evidence, the fact that Pausanias makes a very laudatory allusion to the tomb of Helena, comparing it to that of Mausolus at Caria, and stating that these two are the

[1] Robinson, *Bib. Research.* i. p. 610, Note xxix.; comp. Wilson, *Lands of the Bible,* i. p. 428.

finest that he had ever seen. He makes particular mention of the doors of solid stone, and of the mechanism by means of which they were moved, although in this his praise seems to have been a little exaggerated. The Jews of the present day, according to Krafft, consider this to be the grave of a very wealthy and benevolent Jew, Kolba Sebuah, who is often mentioned in the Talmud; unquestionably a later tradition, but perhaps conveying[1] a remote hint and faint remembrance of the liberality of Queen Helena, of whose large gifts to the Jews of Jerusalem Josephus speaks in warm eulogy.

Of the cave of Jeremiah, lying south-east of the tomb of Helena, very little is known with any certainty.[2] It seems as though it must have been originally intended as a resting-place for the dead, although its appearance is rather that of a stone quarry. Being now used as a Mohammedan burying-place, it has not been examined with great care. There is much probability that in this so-called cave of Jeremiah, in which name there is no allusion whatever to the prophet, we are to look for the monument of Herod, which Josephus speaks of in three different places, and which he locates between the tomb of Helena and the fuller's grave. This Herod is probably the builder of the third wall, Herod Agrippa, after whom also the Bab es Zahary, the Porta Villæ Fullonis of the middle ages, received its name of Herod's Gate. Many other monumental remains are to be still seen north of the city, but I will close my account with those already described.

DISCURSION VII.

THE CLIMATE AND THE SOIL, THE PLANTS AND THE ANIMALS, OF JERUSALEM, JUDÆA, AND PALESTINE.

Regarding the natural history of Jerusalem and its neighbourhood, our information is at present scanty.[3] Ob-

---

[1] Wilson, *Lands of the Bible*, i. p. 427.
[2] Krafft, *Topog.* pp. 217–219.
[3] Robinson, *Bib. Research.* i. p. 428 et sq.

servers have thus far confined their researches to what pertains to and illustrates the study of man, and have paid little heed to other matters which have great interest in themselves, and which tend not a little to throw light upon the past.

The elevation of the city some thousands of feet above the Mediterranean, as well as above the valley of the Jordan and the level of the Dead Sea, must contribute much to lessen the temperature beyond that enjoyed by places not far distant. Von Schubert[1] gives the mean temperature of Jerusalem as about $13\frac{1}{2}°$ Reaumur, which is not so high as that of Naples. Although Jerusalem lies in the same parallel with Morocco, the palm, though growing very tall, brings no fruit to perfection. The shrub-like cotton plant and other tropical growths which flourish in the neighbourhood of Jericho do not thrive here; but, on the other hand, at Jerusalem, as at Bethlehem and Hebron, there is produced a fiery wine, like that made on the Greek islands and on the west coast of Asia Minor, which have a climate like that of the mountain-land of Judæa, notwithstanding their more northern situation. The olive tree, the fig, the walnut, and the pistachio, have their true home here, and yield their fruit in abundance. The coolness of the winter-time projects itself into the spring, and the heat of summer into the autumn, more markedly than in districts farther west. The mean summer temperature von Schubert estimated as from 23° to 24° Reaumur, but it sometimes rose as high as 32°—a dry, parching heat. Even the nights, with their prevailing east and south-east winds, bring little relief; and the crusaders used to complain bitterly of the discomfort, and to seek protection in the caverns and pits, there being a great and universal want of shade. The cold north winds of winter, on the contrary, make the use of furs very comfortable.

At Easter the ground is generally covered with the young green corn and grass. In May the heavens are cloudless; at the end of the Egyptian Chamsin touches of this hot wind are felt as far north as the Mount of Olives. Early in June the wheat and barley harvest begins, and then follows the

[1] Von Schubert, *Reise*, iii. p. 104; Schultz, *Jerusalem*, pp. 27, 28.

fierce summer heat. With the rise of the Nile in August there are seen in Judæa thin fleecy clouds coming up from the south-west, and floating high above Jerusalem: a heavy dew then falls, refreshing indeed, but too late to do much good. During September and October the land longs for rain; the heat of the autumn months is very great. In October the first drops fall, to the great relief of the land; then follow the periodical and very heavy rains, continuing on until December. Everything then is beautiful and green, though the last month of the year is often wet and cold. Still, as a general rule, Christmas-time is one of the most delightful of the whole year, and January a genuine spring month. It has to yield, however, to the cold and the storms of February and March, after which we have a second spring. It must not be denied, however, that a considerable degree of cold is sometimes experienced in January, just as is the case in Rome; but in both places, Judæa and Rome, snow is not at all permanent, and ice is rarely formed. The distant mountains, however, are often seen for days together capped with white. This is the period when all the cisterns and reservoirs of the city must be supplied with water for the next six months. The last months of the year are therefore the most unfavourable for a visit to Judæa.

The universally distributed limestone, with the beds of chalk which underlie it, are the components of the soil immediately around Jerusalem. Among the masses of these two are often found ferruginous red strata, enclosing encrinites: they lie very deep, and reappear on the east side of the Jordan and the Dead Sea, and may be traced as high as the tops of the highest of the mountains there. Marl, and a loam admirably adapted to tillage, according to Bové, are found in the most sunken hollows, and richly reward cultivation. Sand districts seem to be entirely lacking, and von Schubert states that he first met them beyond the Lebanon. Everywhere there is an abundant supply of holes and caverns,[1] which may perhaps partly account for the fact that Palestine has suffered less from earthquakes than the neighbouring

[1] Von Schubert, *Reise*, iii. p. 110; Schultz, *Reise*, p. 140.

countries, where basalt and other formations are more frequently met.

Bové found,[1] in the Valley of Jehoshaphat, only a meagre gain for his herbarium; for the late months of spring, themselves dry, were followed by intense heat, which burnt up every living thing. Notwithstanding this, however, he found in the dense shade of the olive and fig trees of Aceldama, many varieties of plants; and the King's Garden yielded abundant returns of cabbage, parsley, artichokes, melons with a green meat (the Maltese variety), pumpkins, cucumbers, aromatic shrubs; and fruits, mainly pomegranates, plums, apples, pears, cherries, figs, mulberries, jujubes, and pistachio nuts. Where there is a supply of water, there is an ample harvest.

Scholtz has given[2] the indigenous Arabic names of many plants which he found in Palestine; but he has not coupled them with the terms recognised in systematic botany. Von Raumer, following Rosenmüller, Klöden, von Schubert, and others, has given an instructive catalogue of the plants in their relation to the Bible.[3] I follow von Schubert,[4] however, not only as a contemporary observer, but as one who has done the great service of retaining with strict care the Arabic names.

The olive tree, according to von Schubert, was, and is still, the prince among the trees of the country: it seems to have its natural home here; for seldom are very old olives met which yield so good an oil as those of this district. When the Koran swears by the fig and the olive, it is as if it swore by Damascus and Jerusalem.

The fig tree is universal, but is found in its greatest profusion around Jabrut, a day's distance from the Jewish capital, and on the hills of Bir and Sinjil. As one goes towards Samaria, tracts are seen so extensive, that the eye

---

[1] Bové, *Naturaliste, Recit.*, in *Bulletin de la Soc. Geogr. Paris*, 1835, T. iii. p. 383.
[2] Scholtz, *Reise*, p. 140.
[3] Von Raumer, *Pal.* pp. 85–91.
[4] Von Schubert, *Reise*, iii. pp. 112–117.

cannot take them all in at once: the fruit (Tin Bershuwy, to distinguish it from Tin Jimmayz, the sycamore fig; Tin Serafendi, the Zyziphus fig; and Tin Shuke, the Opuntia fig) is remarkable for its delicate aromatic flavour, but it is smaller than the fig of Smyrna.

The vine is met only in a few places. As in Hebron, so all through Palestine, its cultivation is accompanied with the preparation of dibs and wine: that made in the Lebanon keeps longer than that of any other part of the country.

The almond tree (*Loz*) blossoms before the commencement of the cold days of February; around Bethlehem and Hebron, however, not till March, when also the apricots, apples, and pears bloom. The purple blossom of the pomegranate comes later, and is contemporaneous with the white vesture of the myrtle: later still we have the rose, the fragrant jessamine, etc.

The lofty cypress is cultivated around Jerusalem only in gardens; the azerole grows wild in the hills; as do also the walnut, the arbutus, the laurel, the pistachio, the terebinth, and the evergreen oak. Besides these, we have varieties of the Rhamnus, juniper, thymelæa, firs on the hill-tops, the sycamore and carob tree, the mulberry and opuntia, generally growing on slopes and in hollows. The two last mentioned have been but recently introduced. Oranges and citrons are very seldom to be seen in gardens.

Of the cereals,[1] there are many varieties in several parts of the land, but especially upon the plain of Jezreel, and on the uplands of Galilee, growing in great profusion: probably only the wild successors of the grains once regularly raised in those places: proofs, says von Schubert, of the remarkable fertility of Palestine at a former day, and its admirable adaptation to corn. On the roads to Nazareth and Nablus, Hänel[2] found wild oats growing in abundance. Besides wheat and barley, von Schubert found among these wild crops, rye, in appearance very like that of Germany: he fell in with the latter mainly on the slopes of Tabor and on

[1] Von Schubert, *Reise*, iii. pp. 115, 201.
[2] G. Hänel, in *Zeitsch. d. deutsch. Morgenl. Ges.* ii. p. 432.

Esdraelon, and found that it outstripped in height even the bearded wheat. The occasional spots which are now tilled with any care, and rescued from the waste which everywhere prevails, are sown with the common Egyptian varieties of grain: summer millet (dura gaydi), the common millet (dura sayfa), the autumn millet (dura dimiri), together with varieties of Holcus sorghum. Wheat (kumh), spelt, and barley (shay-in) flourish everywhere. Among leguminous plants, we have the chick pea, or hommos (cicer arietinum), the Egyptian bean, the fuhl (vicia faba), the gishungaya (Phaseolus mungo), the gilban (Lathyrus sativus), lentiles, abs, and peas (bisfilleh).

Among the vegetables, the varieties of the tubiscus stand pre-eminent, especially the great favourite bamia towileh (Hibiscus esculentus), bamia beledi, and wayka (Hib. præcox). Potatoes (kolkas franshi) are only here and there cultivated, and then by the Franks. Lettuce (khus) and artichokes (karshuk) are very often met in the convents: those raised[1] in the King's Garden are of especial excellence, and are not to be confounded with the Jerusalem artichoke, which is Helianthus tuberosus, and is known also as gira sole. The water-melon (batikh) and the cucumber (khiar) grow in moist places. Flax (etam) is little attended to; hemp (bust) is more; cotton (qotn) as well as madder are cultivated in favourable localities: neither of these, however, do well in the neighbourhood of Jerusalem, but require a less elevated position. Where we leave the few well-watered lowland tracts, and come to the higher and dryer regions, we meet an abundance of aromatic shrubs, such as the Syrian marjoram (Origanum syriacum), the rosemary, germander (Teucrium ros marinifolium, not the T. creticum, as was formerly supposed), and many other fragrant plants of similar character. On the dry Mount of Olives the pilgrims are in the habit of collecting the little blood-immortelle (Gnaphalium sanguineum): from Carmel and Lebanon they carry away with them the great eastern immortelle (orientale). The Gnaphalium sanguineum, also known as the blood-drops of

[1] Wilson, *Lands of the Bible*, ii. p. 33.

Jesus, is not found in Egypt, but is entirely peculiar to Palestine; it is a flower of remarkable beauty,[1] very similar in appearance to the yellow Gnaph. arenarium of Germany, only the parchment-like leaves are not golden-yellow, but blood-red: the time of their blooming is in May. The splendour of the varieties of lilies, tulips, hyacinths, narcissus, anemones, is a particular charm of the spring-time: even the wild leek flowerets are not devoid of a certain beauty. The fruit of the mandrake of Palestine (Mandragora autumnalis) is in request among the oriental Christians, as well as among the Mohammedans, because in the neighbourhood of Jerusalem it is invested with a certain great potency. It is rarely met here, however, but is far more common south of Hebron, and around Tabor and Carmel. The upland flora of Palestine cannot be compared with that in the lower valley of the Jordan, from the Dead Sea up to the source of the river in the Anti-Lebanon,—a short distance, but one in which, according to von Schubert,[2] more varieties of plants can be seen within the course of a few days than in most other parts of the world would require months to reach.

The statements of von Schubert[3] relate not to Jerusalem alone, which in itself offers but a limited field, although its neighbourhood may afford something of value relating to the fauna of Palestine.

Despite the abundant patches of grass, beef cattle are seen but seldom: the bullock in the parts around Jerusalem is small and ugly-looking: veal and beef are not often eaten. In the upper Jordan valley, on the Tabor, and around Nazareth, the bullock thrives better, and still better on the east side of the Jordan, and towards Damascus. The buffalo (Gamus) is found very unfrequently along the sea-shore: in size and strength he is very similar to the Egyptian variety; but the number of these is very small, when we take into account the amount of pasturage, though in the upper Jordan valley they are numerous. The heavy taxes upon large cattle prevent the raising of them: the herds of sheep

[1] Sieber, *Reise*, p. 32.
[2] Von Schubert, *Reise*, iii. p. 116.   [3] *Ibid.* pp. 117–121.

and goats not being subject to heavy rates, are much larger. The native sheep of Palestine show a tendency to fat tails. The hair of the long-eared goat of Syria is tolerably fine, but appears to be inferior to that of some of the varieties of Asia Minor. Of stags, von Schubert met with but one variety: he saw it near Tabor, and not far from the spot where Hasselquist, nearly a hundred years before, had observed it. On the other hand, there are several kinds of antelopes. The training of camels is only made a thing of importance in the Jordan valley, on its eastern bank, in the neighbourhood of Baalbec, and in the south of Judæa. The rearing of horses, too, is much neglected; and it is only here and there that a fine Arabian steed is seen. The ass is considered in Palestine a nobler animal than the horse; and so too are the mule and the donkey, both of which are the best creatures in traversing the wild mountain roads of the land.

Beasts of prey have become very rare. The wild boar (khanzir) is only met in the richly-watered country around Hermon and Tabor, down the Jordan valley, and in the plain of Jezreel: the lion (assed) is now only known there in untrustworthy stories and traditions: of the common panther (nimr) there are only faint indications here and there: the wower or wubber (Hystrix syriacus), the klippdachs, which makes its nest among the rocks on both sides of the Jordan, and particularly in the neighbourhood of the Convent of St Saba, are not now much seen, excepting in the places indicated: the dog (abul hhosseyn) is common, however; so, too, are the fox (taleb), the hare (erneb), the jackal (dib), the great enemy of the herds, and the porcupine. The hedgehog, met near Bethlehem, says von Schubert, is not the long-eared Egyptian variety, but that of Europe: the mole and the common rat are common. Bears are found in the Lebanon and Anti-Lebanon, particularly in the neighbourhood of Damascus; but on Carmel they appear to be extinct.

Among the birds of prey, the carrion kite and the common kite (hedy) are the most numerous. Flocks of pigeons, very different from the European variety, are everywhere the inhabitants of the caves and clefts. Other European birds, too, are met, such as the lanner and the crow. Snakes are very rare. The chameleon is met south of Hebron; and turtles are not rare in the neighbourhood of Bethlehem—the same variety which is met at Rome and near the Mediterranean. The country appears to be rich in insects, but mosquitoes give little trouble: the best known of all the insects of Palestine, says von Schubert, is the bee.

DISCURSION VIII.

THE INHABITANTS OF JERUSALEM—ITS POPULATION—THE MOHAMMEDANS—THE ORIENTAL AND OCCIDENTAL CHRISTIANS, AND THEIR SUBORDINATE SECTS—THE JEWS.

Jerusalem is divided into several quarters (Hareth), which I have heretofore alluded to: as el-Jahud, or that of the Jews; el-Arman, that of the Armenians; el-Nussarah, or that of the Christians; el-Mugharibeh, or that of the Africans, the smallest of all; and el-Muslimin, or that of the Moslems, the largest of all. These again are subdivided in some cases into various sects; the Christian quarter, for example, which comprises Latins, Greeks, Syrians, Copts, Abyssinians, Georgians, Maronites, Nestorians,[1] etc. Only the Armenians are unbroken, for even the Jews' quarter is subdivided among the Sephardim, Ashkenazim, and Karaites. When we add to these the variety of people with their manifold tongues, from all parts of the world, who are represented by these different religious persuasions—the Hindoos,[2] for example, who form a part of the Moslem quarter, and strictly adhere to their national peculiarities; and when we take into account the extraordinary concourse of people who come together at Easter from India, from Persia, from Russia, and all Europe, even from North America, it will be seen that it would be hard to find a place which offers so great and so numerous diversities of national life. It must be said, indeed, that this

[1] Robinson, *Bib. Research.* i. p. 418 et sq.
[2] Wilson, *Lands of the Bible*, i. p. 446.

is marked at Easter, as hinted just above; for during the greater part of the year there is little life in the city excepting what comes in connection with the bazaar: the streets are deserted, and the neighbourhood of the city solitary. At such times, were it not for the passing of peasants with their laden asses over the main roads to the city, and for the women carrying their water-skins to the wells and reservoirs to get water, and to give to the flocks of sheep collected there; and were it not for the white-veiled Moslems wandering around the graves of their fellow-believers, or reposing in groups upon them, the city and the adjacent country would be sunk in desolation and in the stillness of death; and the only life that would be apparent in it would be that which centres in mosques, churches, and houses for the poor.

The number of the population is, as usual in the East, very difficult to determine with exactness, and the increase and decrease are exceedingly irregular. Robinson reckoned the population of the city to be 11,000: 4500 Moslems, including 1150 men; 3000 Jews, including 500 men; and 3500 Christians, including 850 men. The number of adult male Christians he estimated to be composed of the following subordinate divisions: 460 Greeks, 260 Latins, and 130 Armenians. Williams admits the justness of Robinson's census.[1] Schultz, who examined into this matter with great care some years later,[2] reckons 5000 Moslems, 3400 Christians, and 7120 Jews; in round numbers, some 15,500 souls. To these must be added the Turkish garrison, 1000 strong, and some hundreds of persons standing in connection with the various consulates and missions, giving an aggregate of not far from 17,000. Among the Christians are estimated 2000 Greeks, who make up here, as in all Palestine and Syria, the greater number: there are, besides, 900 Roman Catholics, 350 Armenians, 100 Copts, 20 Syrian, and as many Abyssinian Christians. There are about 600 Jews, who are Turkish subjects; the Sephardim are mostly of Spanish origin; the other foreign Jews, generally Poles and Germans, are known as the Ash-

[1] Williams, *Holy City*, ii. p. 548, and App. Note 4, pp. 613, 614.
[2] Schultz, *Jerusalem*, p. 33.

kenazim, and are generally under the protection of the various consulates. These two divisions, taken together, number about 1100. The number of pilgrims who come together at Easter is estimated at the highest at 10,000: in one year there were but 5000, in another but 3000. The number is small, compared with that in former years. The population of Jerusalem seems to have been subject at all periods to great changes, owing to a more or less tyrannical government, and the spread of the plague. Only eight patrician or effendi families now claim to trace their genealogy back to the companions of Saladin, while among the Jews there is not a single old family. The Christians who do not belong to the churches, convents, and other religious institutions, are the traders of the bazaar,[1] and ply each some petty industry: they make soap; they weave; they manufacture, as in Bethlehem, rosaries, crosses, wax tapers, images, etc. The possession of landed property, as well as the various rights of the population, are so little established under the Turkish law,[2] that there is no living in security. In the courts the testimony of Mohammedans alone is taken against a Mohammedan; that of Mohammedans and Christians against the latter; and against a Jew, Mohammedans, Christians, and Jews.

The greatest part of the landed property in the city is the so-called Wakf, *i.e.* the property of mosques, convents, and public institutions: so much is taken up in this way, that only a little is left for private possession. Wakf el Haram is the largest of these Wakfs: it embraces most of the houses of the Jews' quarter; and since there are no private owners of the houses there, they have been suffered to fall into the wretched condition in which they are now seen. Besides this one, there is the Wakf el Tekijjeh, the former Hospital of Helena, now a Moslem house for the poor and sick. A Wakf Franji is the property of the Latin Convent, a Wakf Rumi of the Greek one, and a Wakf Arman of the Armenian Convent. The name given to private property is Mülk Maukuf (*manus mortua*): this always

[1] Wilson, *The Lands of the Bible*, i. p. 453.
[2] See Wolff's *Reise* for the fullest specification of such matters.

falls to the religious institutions where there are no male heirs. The Mülk or private property forms, therefore, but a very small part of the city. The value of each lot is always divided into twenty-four parts, which are rarely held by a single individual. Almost every piece of ground has several possessors; and it becomes very difficult to purchase land, since the owner of a single one of the twenty-four parts may prevent the formation of a contract. This difficulty was a very serious one at the time when the land was purchased for the Anglo-German church, and other similar occasions. Even at the epoch of the Crusades, says Schultz, it was not easy to obtain a title to land in Jerusalem; but now, with the additional drawback of Moslem jealousy, it is much more difficult.

The life of the Mohammedans of Jerusalem displays no marked dissimilarity from its character throughout the East; but it is a matter greatly to be regretted, that since 1840, the Egyptian sway, which at least assured security for life and property throughout the land, has been exchanged for the Turkish, with its feebleness and its confusion; so that now it is necessary to come back to the employment of a band of armed Beduin, if one would not be plundered, and perhaps killed; and this despite the fact that now there is a pasha resident at Jerusalem, whereas the city was formerly included in the pashalic of Damascus. The taxes have, indeed, been made somewhat less than they were during the Egyptian supremacy, but the miserable system[1] of hiring them out has been made universal in Palestine,—a system which bears particularly hard upon the Christians of Jerusalem, who[2] are obliged to pay 105,680 piastres, or not far, in round numbers, from a thousand pounds sterling. The little colony of Hindoo Mohammedans, discovered in Jerusalem by Wilson, deserves a moment's notice. Many of this sect, who have made the pilgrimage from India to Mecca and Medina, have come as far as Jerusalem, and in many

---

[1] Schultz, *On the Taxation of Palestine*, MS. account.
[2] Gadow, *über gegenwärtige Besteuerung einiger Districte des Paschaliks Jerusalem*, in *Monatsb. für Erdk. in Berlin*, vol. vi. 1850, pp. 2–7.

cases have married and settled in the Moslem quarter. Such are gladly welcomed by the Mohammedan authorities, since they often bring presents with them, rice among the rest. When Wilson[1] visited them, they numbered about twenty-five, and were strictly English subjects, and protected by the English consulate. A much larger number have settled in Damascus; and these two little colonies form a remarkable complement to those which have been found farther west, and which have continued to hold fast to their faith—such, for instance, as the community which Pallas discovered in Astrachan. In answer to the question why they came so far west to live, the Jerusalem Hindoos answered, that "wind and water are good in Ind, but in Sham they are bitter: Ind is the land of the unbelievers, Sham the land of the believers." Sham is Syria, and the West to be extolled as highly by the Moslems as the Hedjas by the Arabs.

The missions and the Protestant bishopric established by the united Churches of England and Prussia, have already done enough to justify the hope that the future will witness a marked improvement in the life of the nominal Christians in Jerusalem. The incessant internal strifes of the various sects; the unchristian manner in which the sacred festivals of Easter, Whitsuntide, and others are celebrated; the ignorance, superstition, jealousy, selfishness, which are displayed, added to a tendency in one sect to conciliate unbelievers in order to awaken their hatred against other sects, have been displayed in Jerusalem no less than in Bethlehem, commented on by almost all travellers, and condemned in round and worthy phrases. The thing is so well known and so much lamented,[2] that I need not speak of it minutely here. The evil is heightened by the fact that it tends to harden the Moslems in the conviction that their faith is the true one, and to fill them with scorn towards the men who thus quarrel over the places most sacred in their eyes; and so long as the Turks retain their authority in the land, it is to be feared

---

[1] Wilson, *Lands of the Bible*, i. p. 446.
[2] Schultz, *Reise*, pp. 192-225; Wolff, *Reise*, p. 96; Williams, *Holy City*, ii. p. 531 et sq.; Wilson, *Lands of the Bible*, i. p. 449 et sq.

that the claims of Christianity will find little sanction in their eyes by the conduct of most of the Christians who resort to Jerusalem. The oriental adherents to Christianity are divided into the following sects: Greeks, Georgians, Armenians, Syrians or Jacobites, and Copts. The occidental Christians are the Latins, and the people attached to the American congregation and to the Anglo-German mission.

1. *The Orthodox Greek Church of the Orient.*[1]

Four patriarchs stood at the head of the Christian community of the East during the early centuries, but without any special claims. Of these four, the bishop of Cæsarea, where Eusebius once resided, was selected by the Council of Nice to be Metropolitan of Jerusalem. Later a patriarch was established there, and Cæsarea and Scythopolis were taken from the northern diocese, and Rabbath Moab, Petra in Arabia, from the Egyptian patriarchate, and a new one formed which comprised sixty-eight bishoprics, and had twenty-five suffragans under its jurisdiction. Of this diocese, which extended from the Lebanon to the Dead Sea, and from the Mediterranean to the Desert east of the Jordan, only a wreck remains. Although its wide domains comprise Phœnicia, Judæa, Galilee, Samaria, Idumæa, and Arabia Petræa, yet only fourteen episcopal residences are distributed over it; and of these many are merely nominal, and cannot be occupied by bishops. In case they are (as they generally are) *in partibus infidelium*, they reside generally in the great Greek Convent at Jerusalem which was founded by Constantine; and when the country is peaceful enough to allow it, they go out to visit their respective flocks. Only the bishops of Ptolemais or Acre (to which Nazareth belongs) and Bethlehem are accustomed to reside statedly at the episcopal capital.

The patriarch of Jerusalem, for whom a stately palace has been recently built between the Church of the Holy Sepulchre and the Latin Convent (the old palace of the patriarch of Jerusalem at the time of the Crusades is at pre-

[1] Williams, *Holy City*, ii. pp. 539-549; Wilson, *Lands of the Bible*, ii. pp. 451-479.

sent the residence of the superintendent of the Sherif, the Nakib el-Ashrak), had for a long time resided in Constantinople, because it was thought that there he could be more influential in favourably affecting the Turkish Government than if he were to live in Palestine. His vicar bears the title Wakil. The late patriarch Athanasius, an old man of ninety, for many years lived in great retirement on an island in the Sea of Marmora, while a synod composed of a hundred and fifty men, most of them bishops and priests from the Greek islands, managed the business of the patriarchate. In 1843 he died. According to the right of his position, he had before his death appointed his successor, the bishop of Tabor, who was then acting as his minister at St Petersburg; but the divan of the sultan, who had to act upon the selection, rejected it, and chose a stranger. Meanwhile the two members of the synod who had remained in Jerusalem, the prominent bishops of Lydda and Petra, would not admit the right of the Porte to interfere: they nominated the former of the two. The sultan accepted the nomination, and he was inaugurated as patriarch in 1845. He at once set new movements on foot: he selected his treasurers, sacristans, custodians, and various subordinates, and gave them active occupation in the Greek Convent in caring for the numerous pilgrims, and in seeing that every one enjoyed the hospitality of the place for a day and a night. On the day after their arrival the gifts of the pilgrims were received, their names registered by the authorities, and they were apportioned to the various convents of their church, to be taken care of, body and soul, during their stay in Jerusalem. There are twelve monasteries for the reception of the men, and five nunneries for the women of the Greek Church, who come as pilgrims to Jerusalem. These institutions are indeed very small; yet the pilgrims are all of them able to attend mass in the chapels, and all of them receive their directions where to find the stations; and thus a good degree of system is preserved, notwithstanding the numbers to be provided for. Six married and resident priests are entrusted with the spiritual care of the pilgrims: they preach in the Greek language, and alone receive the sacrament; all others, even the ignorant monks themselves, being excluded from partaking of it, in order to avoid the scandals of former times. But even these six priests are destitute of all culture: they have no seminary whatever; and the establishment of one in Jerusalem would be of great advantage to them. According to the statement of the present patriarch, who seems to be more deserving of praise than his predecessors, there reside permanently six hundred orthodox Greek Christians in Jerusalem: in his patriarchate he numbers 17,280, the most of whom live in villages mixed with the Mohammedans.

In the Convent of St Demetrius the main body of the Greek monks live; at the Church of the Holy Sepulchre there are at least thirty, who discharge the duties of public service in rotation. Syrians are not received into the order of the monks; Greek Christians of Arabian extraction have been for two hundred years excluded from these convents, in consequence of their wild and vagabond character; only natives of the islands of the Greek archipelago are admitted to the priestly order. The Greeks hold the largest part of the Church of the Holy Sepulchre, but all other communions have their own, although often a very limited and jealously guarded one.

2. *The Georgians*.[1]

Their church in Jerusalem is at present very much overshadowed, but its great antiquity and its early importance ought to give it a very prominent place. Iberia, Albania, and the lands on the south side of the Caucasus, adopted Christianity after its presentation there by the apostles Andrew and Simon, and renewed their faith when St Clemens preached there, who was banished to those regions by Hadrian, and when the voice of St Ninna had been heard there, whose uncle was a bishop in Jerusalem during the third century. The first Christian king of Iberia, Miriam, while on a pilgrimage to Jerusalem in his eightieth year, received a present from Constantine which was intended to be used for founding a church there. This gift was

[1] Williams, *Holy City*, ii. pp. 549-554.

subsequently largely increased, and made more adequate to meet the end proposed. Thus far we have to rely on the legend. But this story is confirmed by history; for Procopius tells us that the Emperor Justinian placed the church of the Iberians and that of the Lazi, a branch of that body, in the Hermitage at Jerusalem. The princely family of Bagrathion, which traced its lineage back to the house of David, and which at a later day as the kings of Georgia[1] waged its thousand years long war with Persians and Turks in behalf of Christianity, left their home at Jerusalem, according to the statement of Constantine, and established their throne in Georgia under Justinian II., whose reign extended from 685 to 711. Living there, they have always been prominent donors to the institutions of their old home, and have done much for the Holy Sepulchre. Their labours only ceased at the time of the gradual decline of their kingdom during the seventeenth and eighteenth centuries, and with its ultimate absorption, in 1801, in the Russian Empire.

There was once a time when the Georgians held no less than eleven churches and convents in Jerusalem: their king, Vachtang VI., was able to send a gift of two thousand tomans to the Holy Sepulchre. At the commencement of the sixteenth century they had more rights and privileges than any other of the Christian confessions; for, says the pilgrim Baumgarten, in the year 1507, they went with great pomp, with arms and banners, through every part of the Holy City, entirely free from tribute and other civic burdens; and their chivalrous wives even were equipped like Amazons ready for battle. They were then in possession of the Chapel of the Holy Cross. During the many wars with Persians, Turks, and Caucasians, Georgia fell into a decline: Armenians and Greeks gradually gained possession of the impoverished church, convents, and other religious foundations of the Georgians, and only a single convent remained to them— probably the one which had been built by their king Tatian in the fifth century, upon the land given by Constantine to Miriam. It remains known to the present day by the name

[1] B. Dorn, *Erster Beitrag zur Geschichte der Georgier*, pp. 7-119.

of the Convent of the Wood of the Holy Cross—Deir el
Masallabeh. It lies outside of Jerusalem, twenty minutes'
walk westward from the city, in a side valley leading off to
the left from the road to Jaffa. It is there, as the legend
says, that the wood grew, out of which the cross was made.
The massive walls of the building, with its iron gate and its
low entrance, show how dangerous the neighbourhood used to
be considered, and how necessary that it should be guarded
strongly, although the building had in the seventeenth century
two hundred cells, and might be supposed able, from har-
bouring so many monks, to have offered a stout resistance.
Yet one of the last superiors was murdered in his bed by the
Arabs. During late years there have been only three or
four monks there, under the superintendence of an archi-
mandrate. The church, despite its ruined state, has something
striking in its position, and many fine mosaics, but seems to
be but little visited: a rebuilding of it would be well worthy of
the heir of the domain once held by the Emperor Heraclitus.
Tischendorf,[1] who visited the place in 1846 on his way to
the Convent of St John farther west, says that it lies about
three-quarters of an hour from the city; and that he found
there a library with many Georgian, Armenian, and Arabic
manuscripts, and some Greek parchments strewing the floor;
and convinced himself that much that is valuable matter was
there, although the greater number of Georgian manuscripts,
and many others which the learned Scholtz had discovered
twenty years before, were far from being of value. In the
church he found still remaining some rich fresco paintings,
executed in a devout spirit: these Krafft supposes to have
been the gift of the older Georgian kings, who followed
Miriam's example in making the pilgrimage to Jerusalem.

3. *The Armenians*,[2]

the oldest cognate race with the Georgians, and in common
with them very early recipients of Christianity, abandoned

---

[1] Tischendorf, *Reise in der Orient*, ii. p. 69; Krafft, *Topogr.* p. 263.
[2] Schultz, *Reise*, pp. 215-223; Williams, *Holy City*, ii. pp. 554-560;
Wilson, *Lands of the Bible*, i. p. 452, ii. pp. 479-506.

the orthodox church at the Council of Chalcedon in 491, and were afterwards viewed by their old friends with great bitterness. They are the chief representatives of the monophysite heresy. Their union with the orthodox Greek Church is earnestly sought by their general head, the Catholicos or Patriarch at Etjmiadzin, and by the Russian Government; for if it were accomplished, the united body would present a formidable front to the Romish Church. Of Catholics in alliance with Armenians, there are none in Jerusalem, although their number is very large in Aleppo and other Syrian cities. The number of the schismatic Armenians in Jerusalem is small, at the highest three hundred and fifty souls; but they are in the possession of the greatest wealth, an assured position, the most attractive convent in Jerusalem, and the richest church—one which in size, too, is larger than that of the Holy Sepulchre, and which, moreover, is surrounded by the finest gardens. It, as well as the convent, is consecrated to St James the son of Zebedee, whose name is given as that of the first bishop of Jerusalem, and who is revered as a martyr. They both were the property of the Georgians; but the latter, too poor at length to pay the heavy taxes which the Moslems laid upon them, made them over in the sixteenth century, in connection with the reputed house of Caiaphas on Zion, outside of the gate, to the Armenians, reserving the title, however, to themselves. The church was built by George I. Curopalata in the eleventh century. Up to the present day the very ancient archives of this foundation remain unexamined. The extensive trade which the Armenians carried on during the long centuries of their banishment from their old home,—and which they, as the medium of communication between the East and the West, Vienna on the Danube, and Calcutta on the Ganges, the Neva, the Euphrates, and the Nile, knew how to appropriate to themselves,—has, in connection with their great fairness in dealing, brought them to the highly honourable place which they hold not only in Jerusalem, but throughout the East. The great apostle Gregorius Illuminator is the bond that holds all the scattered communities of this body in unity: the

Catholicos is the officer who now holds his place and wears his dignities. His two vicars, who are rather to be called his political agents, are the Armenian patriarchs, who have taken up their residence at Jerusalem and at Constantinople. The first of these laid aside the title of Bishop of Jerusalem in favour of that of patriarch in 1310; the Bishop of Constantinople took the same step in 1461. The diocese of the latter, which was subject to the Catholicos, extended over the whole Turkish Empire, Palestine excluded. The Patriarch of Jerusalem is independent, and enjoys unlimited control in his diocese, which extends over Palestine and Cyprus. He is a man of much note there, in the city of his residence. There are several hundred Armenians dwelling in Jerusalem, the most of whom are merchants and agents.

4. *The Syrians or Jacobites.*[1]

The Syrian people, once one of the most cultivated on the globe, lost its political independence with the extension of the Persian power over western Asia, and has never since attained to national importance. Early converted to Christianity, the Syrians carried the gospel with great zeal from Antioch on the Orontes, and as Nestorians, from the Euphrates at the same time, not ceasing in their efforts till they reached China. The Syrians who remained on the west bank of the Tigris attached themselves—since the Nestorians, as a heretical sect, were excluded from the orthodox church—to the monophysite doctrine, like the Copts in Alexandria and the Armenians, with whom they were in alliance. The views of the Syrians, however, were but slightly different from those which were held to be orthodox. They called themselves Jacobites, after a certain Jacobus Baradæus, a heretical monk, a disciple of Severus of Antioch, who in the first half of the sixth century called into new activity the already weakened party of the Monophysites. His followers, under the name of Jacobites, formed the

---

[1] Schultz, *über syrisch-jacobitische Christen in Asien*, in *Monatsb. der Ges. für Erdkunde in Berlin*, 1850, pp. 267-281; Wilson, *Lands of the Bible*, ii. pp. 506-519.

largest part of the Copts of Egypt, who, so long as the Emperors of Constantinople exercised influence there, were constantly persecuted and oppressed.

These Syrian Christians, who were also known as Surjani, retained their Asiatic locations, particularly in the country of the Kurds, on the upper Tigris and Euphrates, where the chief dignitary, the patriarch, resided. He formerly lived in Amida, but more recently his home has been in the Convent Deir el Zafaran, near Mardin. From this point he sends metropolitans to each of the twelve bishoprics under his supervision, to Mosul, Diabekir, Bitlis, Damascus, Aleppo, Jerusalem, and even to Malabar. Still, these have so little coherence and interdependence, that the Roman Catholic propaganda has attempted by means of French priests to take from them six of their churches and convents. In Jerusalem, the only possession which remains to them of their former possessions, is the little Church and Convent of St Mark,[1] on Mount Zion: the others have either all been seized by the Turks in default of the non-payment of taxes, or have been appropriated by their reputed protectors, the Armenians.

The head of the convent in Jerusalem a century ago was a bishop of Orfa, Abd el Nur, a very celebrated man: he was transferred to Damascus, and left in his place a vicar to attend to the business of the convent, as well as to the wants of the little community in Jerusalem, consisting of only three families. These seem to have at length disappeared, and the great Abyssinian Convent, which claims supremacy over all the monophysite institutions, has taken charge of the relics of the old and desolate Jacobite Convent, including its furniture and documents, among which it is possible that there are some of considerable historical importance.

5. *The Copts and Abyssinians.*[2]

The largest portion of the population of Egypt, which, according to Macrizi, bore the name for centuries of Nile

---

[1] Williams, *Holy City*, ii. pp. 560-562.

[2] Wilson, *Lands of the Bible*, i. p. 452, ii. pp. 519-543; Williams, *Holy City*, i. pp. 562-567.

Copts, and was made up of Egyptians, Nubians, Abyssinians, and Jewish renegades, adopted the monophysite doctrine of their patriarchs in Alexandria. These, as long as the Byzantine power was felt in Egypt, were so oppressed by the stronger orthodox Greek Church, that at the time of Omar's expedition they even made common cause with the Mohammedans, and fought against the Greeks in the field. The patriarch Benjamin, the head of the Jacobites in Alexandria, paid tribute, and was considered a dependant during the time of his patriarchate,[1] A.D. 640. The conversion of the Abyssinians to the gospel, which began at Alexandria, continues to be recognised up to the present time, by the appointment of the highest dignitary of the Abyssinian Church by the patriarch of Alexandria.

The Copts, who number about a hundred in Jerusalem, attach themselves—as do their fellow-believers the Jacobites, and the Abyssinians of the Holy City, all of whom have long been opponents of the Greek Church—to the Armenians, since these are able, in consequence of their high position, to afford them protection, and from their wealth to give them substantial aid in their need. Of their special peculiarities little is known. It is only within recent years, since there has been an interest awakened in Abyssinian affairs, that much attention has been devoted to them. In the year 1842, the Copts in Jerusalem possessed only six poor houses, from which they received a very slight income in way of rent; yet even these were claimed by intriguing factions; and since there could be no documents received from the region of the Upper Nile without considerable delay, the Copts were compelled to advance a sum equal to a hundred and five pounds sterling, in order to bribe the Turkish cadi, for this was the worth of the houses on the point of being confiscated. The use of a cistern, which had always before been at their command, was denied them by a Turkish neighbour; and in order not to perish with thirst, they were obliged to pay forty-five pounds more. They used to own a little convent of St George, near the so-called pool of Hezekiah; and at the time when Ibrahim

---

[1] Macrizi, *Gesch. der Copten, aus dem Arabischen*, p. 51.

Pasha controlled the destinies of the country, they were putting up a new building to serve as a convent, or a caravanserai for the accommodation of their pilgrims. After the Turkish Government was restored, however, they were driven out from their own building, then well advanced, and it was converted into Turkish barracks.

The main property at present consists of the great Coptic convent which stands close by the east side of the Church of the Holy Sepulchre, above the great cistern which bears the name of Helena's Treasury. It dates from a time when the body claiming it were in much more favourable circumstances than they are at the present time, probably from the period when the Egyptian Mamelukes had the control of Palestine, and took them under their protection. The advantage of this ceased, of course, in the year 1517, in which the Sultan Selim I. took possession of Jerusalem. In the convent the story is still told, that the Coptic secretary of one of the Mamelukes was allowed, in consideration of his faithful service, to ask for anything which he might choose, and preferred a petition in favour of the dilapidated convent near the Holy Sepulchre. The Copts, in honour of their patron, gave the restored convent the name Deir es Sultan. This seems to have taken place but a short time before the capture of the city by the Turks, and there was no opportunity for the institution long to enjoy [1] its newly gained advantages. A token of the valuable protection afforded by the Egyptian sultans, and of their liberality, is seen in a heavy iron chain which is fastened into the wall close by the gate, and which has proved strong enough to secure the possession of the convent to the Copts up to the present day.

A prior stands at the head of it, a married man, as in other Egyptian convents. The convent is served by some poor Coptic and Abyssinian priests, and is khan or hospice [2] as well as convent. The number of pilgrims who visit the place is, however, very small. In the year 1816, Mr Banks, in company with his companion Buckingham,[3] called upon an

[1] Krafft, *Topog.* p. 263. [2] Macrizi, as cited above, p. 85.
[3] Buckingham, *Palestine*, i. p. 329.

Abyssinian prince who was then lodging there. They found the building to be made up of several little rooms or cells surrounding a court, from which the dome of the Church of the Holy Sepulchre could be seen. The prince was in one of these little rooms with five or six attendants, besides women, and three pretty maids. Banks saw there some finely written books in the Amhara language, executed on parchment, and with illuminations. This establishment has been, so far as the King of Shoa could go in the matter, recently transferred to the direction of the evangelical bishop, and the change promises to be of great advantage to the hitherto neglected Abyssinian priests. It is to be hoped that the excellent and untiring Bishop Gobat will be able to kindle the almost extinct spark of African piety to a brighter flame than it has ever before displayed.

Of western Christians, the Latins, or those belonging to the Roman Catholic Church, are the most numerous, not only in Palestine, but also in Jerusalem. They are native-born Arabs, speak the Arabic language, and are about a thousand in number. The most elaborate account of them has been given by Scholtz,[1] formerly professor in Bonn. After the time of Jerome, hermit and monastic life began to be very general throughout all Palestine; and the tendency of Europeans to travel thither, in order to adopt one of those two forms of self-sequestration, was much heightened by the inroads which were sweeping away the population of Europe, as well as by the princely gifts which the Byzantine emperor was making to the Holy Land in founding convents, churches, and hospices. These, however, were for the most part swept away when the caliphs had gained possession of the country. The merchants of Amalfi were able, however, to retain their Convent of St Maria de Latina, in the neighbourhood of the Church of the Holy Sepulchre, down to the time of the Crusades: the building became subsequently an hospital of the Knights of St John. After the occidental Christians had been dispossessed of their estates and ecclesiastical build-

[1] Scholtz, *Reise*, pp. 193–230; Wilson, *Lands of the Bible*, ii. pp. 569–579.

ings by Saladin, the Latins, under the guidance of the order of St Francis, took up their abode in the Cœnaculum on Mount Zion, where King Robert of Sicily erected for them a Pilgrims' House, which was very much used by Europeans in the following years, till, in 1560, the Christians were forcibly ejected from it, under the plea that it was wrong for them to make so common use of the burial-place of David. They then looked for a home elsewhere, and found one in the Convent of St Salvador, which had formerly belonged to the Georgians, but which, like most of their buildings, had fallen much out of repair: it was bought by the Franciscans in 1569. It is supposed to be the same Iberian convent which king Vachtang founded, and which Justinian subsequently restored.[1] The privileges which were enjoyed by the Franks through the protection of the kings of France, and Louis XV. in especial, were shared by the Latin monks of this convent, which, to use Scholtz's expression, formed, in respect to matters both ecclesiastical and secular, a *status in statu*. Their privileges were, however, much reduced during the French Revolution and the invasion of the East. Charles IV. of Spain received, in 1793, from Sultan Selim the title of Protector of the Sanctuaries and of the Priests of the Holy Land, which had formerly belonged to the kings of France. Napoleon also received the same, but it led to no practical results. The Latin or Franciscan Convent is at present in the possession of twelve or fifteen monks, mostly Spaniards or Italians, under the direction of an officer chosen by the Pope every three years, and called the Guardian of Mount Zion. His rank is higher than that of the heads of the other twenty Franciscan convents in the Holy Land, and he styles himself the Custos Terræ Sanctæ. Formerly he exercised an episcopal jurisdiction: this continued till the year 1847, when a titular patriarch took up his residence in Jerusalem. In the convent there is a new Arabic printing-house. The Latins own the Garden of Gethsemane, as well as a small convent in the neighbourhood of the Church of the Holy Sepulchre, and the Church of Scourging in the Via Dolorosa.

[1] Williams, *Holy City*, ii. pp. 567-572.

Their stately hospice near the main convent is the usual place where pilgrims from Europe are lodged. With regard[1] to the efforts of the Roman Catholic Church to induce the Georgians, Armenians, Syrians, Copts, and Abyssinians to come over to their profession, Wilson has given the fullest account in his *Lands of the Bible.*

With regard to the founding of the recent evangelical bishopric in Jerusalem, there has been published a full and authentic account,[2] to which I refer the reader, in case he wishes to inform himself regarding a project by no means destitute of significance, and full of hope for the East. In relation to the present condition of the Protestant Episcopal Church in Palestine, I must ask the reader to consult Strauss' *Sinai und Golgotha* (last edit. 1865), Tischendorf's *Travels in the East,* Krafft's *Topographie Jerusalems,* and Wilson's *Lands of the Bible,*—all the works of eye-witnesses. It must appear surprising to the reader, to find this subject presented in the Rev. Mr Williams' circumstantial work under the one-sided heading, "The English Mission."

The Church of England was in possession of a parsonage merely on Mount Zion, when King Frederick William IV. of Prussia felt himself constrained to give to the National Church of Prussia a place side by side with that of England; since, according to his convictions, the cause of evangelical Christianity in the East, and especially in the Holy Land, has no hope of accomplishing permanent good, and of influencing men to any considerable extent, unless there be all possible unity in the church itself. To reach this high goal, the "first evangelical bishopric" was established on the part of both England and Prussia, endowed with the same sums on both sides, and with the understanding that the bishops should be selected alternately from each nation. After the death of Alexander, the first bishop, Samuel Gobat, a German,[3] was chosen. Since his election the church on

---

[1] Wilson, *Lands of the Bible,* ii. pp. 585-600.

[2] *Das evangelische Bisthum in Jerusalem,* Berlin 1842.

[3] *Die Jahresfeier der evangelischen Stiftung in Jerusalem,* No. 6, 1851, pp. 22-25; Krafft, *Topogr.* p. 259.

Mount Zion has been erected and consecrated; evangelical schools have been established in Jerusalem, Nazareth, Nablus, and Szalt in Gilead, the last of which has been taken in charge by the Greek bishop, after he had become a co-worker with the English and Germans; an industrial school is in flourishing operation; an hospital has been opened, out of the promptings of Christian love, for Jews, Mohammedans, and Christians, and has been fruitful in good influences; a deaconesses' institution has been founded; a Brothers' Mission has been begun by the Bâle Society; the Wupper Valley Colony, near Bethlehem,[1] is in progress; and the Abyssinians, who usually make a two years' stay in Jerusalem, have been put by the King of Shoa under the spiritual direction of Bishop Gobat, and have already begun to carry back to their countrymen some of the blessings which instruction in the Bible and fellowship with enlightened Christians have imparted to them. The English mission has already, under Gobat's wise direction (he formerly laboured in connection with it), made common cause with the bishopric of Jerusalem, to the evident advantage already of the Greek Church in the East. The most salutary results are to be hoped from it, if God continues to give His blessing to the enterprise. The grain of mustard seed has been sown, which, under God, has already begun to bear fruit in the north of the Holy Land, on the formerly inaccessible side of the Jordan, and even in the neglected church of distant Abyssinia.

With regard to the American mission, whose labours are largely enjoyed in Jerusalem,[2] although since 1821 Beirut has been the centre of operations, the fullest consecutive account is given in its organ the *Missionary Herald*, where the reader must look to find the most valuable and instructive documents which have been sent home by the agents of any society, and where a rich store of scientific, historical, and antiquarian details may be seen by any inquirer into the condition of the Holy Land and its inhabitants. For many interesting facts, too, I am indebted to the contributions of

---

[1] *Church Missionary Intelligencer*, vol. ii. 1851, p. 191.
[2] Wilson, *Lands of the Bible*, ii. pp. 281-284.

Drs Robinson and Eli Smith to the *Bibliotheca Sacra* and to the *Oriental Herald*. Although these sources indicate so unmistakeably the extremely valuable nature of the labours accomplished by the agents of the American Missionary Society, that no partisan statement can mislead us, and awaken a doubt regarding the great, permanent, and practical value of labours so strongly seconded by American wealth, yet as there have been attacks made upon the efforts of the American missionaries, I will cite a word or two from Robinson, which will set in the clearest light the purposes of these men, of whom, in respect of honourable aim, enlightened minds, purity and genuine nobility of character, even their opponents speak in praise.[1] The object of the American[2] mission is by no means to entice the members of the oriental churches from their own communions, and to lead them to Protestantism: its purpose solely is to reawaken in them a living faith in the old and simple truths of the gospel, as they existed when they were first spread abroad, and before they were obscured in the hearts of believers. To this end alone the missionaries direct their united energies, in the hope that they may so powerfully affect individuals with the contagion of their own spirit, that while remaining in the bosom of their own church, they may become the means of diffusing the same life and love, reawakening an interest in the old and faded words and symbols of truth, and causing superstition and error in all its varied forms to disappear. The results of their strivings are slowly reached, it must be confessed; yet the last twenty years show some fruit, even though the number of baptized converts be small. These men work in perfect stillness; their channels of influence are those of domestic, social, and religious life; they found schools, and preach sermons in the native languages, translate the Bible into the same, and issue these and school-books from their own presses. Thus it will be seen that they put no secondary value upon scientific culture; and as physicians, religious advisers, and benefactors

---

[1] Williams, *Holy City*, on American Congregationalists, ii. 512–519.
[2] Robinson, *Bib. Research.* i. p. 225.

of the people in all possible ways, they have won the universal confidence of Christians, Mohammedans, and Jews, even through all Syria. They have even overcome in a great measure the aversion of the Druses. All this they have done in spite of the strong antagonisms in religious parties and in political factions. God grant that in the future they remain equally successful in avoiding the perilous rock of oriental politics!

There yet remains something to be said regarding the Jews in their own city, in addition to what has been said regarding their depressed condition elsewhere—in Safet, Tiberias, Hebron, and other places. Wilson,[1] who, like the missionary Ewald before him, has paid the most careful attention to this subject, and has studied the condition of the Jews not in Palestine alone, but throughout the East, from Egypt to Bombay, and who may almost be said to have made this the special interest of his life, is our best guide.

The Hareth el Jehud with its 7000 inhabitants, occupying the very narrow tract which forms the depression between the Haram and Mount Zion, has the most wretched houses, and narrow lanes, and dark corners, full of filth and rubbish. In order to show more markedly the hatred which the Turks cherish towards the Jews, they have transferred their shambles thither. The Jews have divided themselves into three parties —the Ashkenazim, Sephardim, and Karaites or Separatists.[2]

The Ashkenazim—the so-called German, Russian, and Polish Jews—are divided again into sects: the Peroshim, *i.e.* the Pharisees; and the Khasidim, *i.e.* the Puritans. They are, for the most part, natives of Jerusalem; but Jewish devotees have come to them from every part of Europe, in order to die there. The most of these are Polish and German, very few English. They are, as a general rule, poor, and live on alms which are collected for them all over Europe. Still there are some well-to-do families among them, who live without any ostentation. The foreign Jews are, for the

---

[1] Wilson, *Lands of the Bible,* ii. pp. 661-686.
[2] *Ibid.* i. pp. 454-461.

most part, under the protection of various consulates, and have very little to do with the Turkish Government. They are without any head, since they represent so many nationalities, and look up to so many different consuls, the five chief European powers being represented in Jerusalem.

The Peroshim have two synagogues, one of which lay long in ruins, and has only been restored within recent times by contributions from all the countries of the East and West. They comprise all the Jews who belonged to the Ashkenazim or so-called German Jews up to the time when the modern sect of the Khasidim arose in Galicia. These Peroshim estimate their number at about six hundred souls.

The Khasidim are far less numerous, embracing but about a hundred individuals, among whom persons of Safet and Tiberias, who are connected with their fanatical sect, are accustomed to enrol themselves. In their religious worship they usually shout and gesticulate with a vehemence and an enthusiasm which run to a wild excess: they hold this to be one of the special conditions of piety, and do not regard any festival complete unless it witness displays of this sort, which are unknown among other bodies of Jews. Their spiritual head, whom they call the Zadik, *i.e.* the Righteous, they hold to be a saint, and suppose him to be in direct connection with supernatural beings. They believe in a transmigration of souls, like the Hindus, study the Cabala, and regard the Sohor as their supreme authority. The high rabbi, Moses ben Aaron, writes magic formulæ for them, summons angels, and the like. Their mode of life is, however, free from blemish: they have two synagogues and a good printing department. The latter has, within recent years, issued the first volume of a geography of Palestine, by a learned Bavarian Jew, Rabbi Joseph Schwartz.

The number of the Sephardim, the second class, is much the largest in Jerusalem, amounting, according to their own statement, to seven hundred families and three thousand souls. The most of them trace their origin to ancestors who lived in the various provinces of the Turkish Empire; but many of them have lived as families for several generations in Jerusalem. Many trace their lineage back to the Jews who were driven from Spain by Ferdinand and Isabella at the end of the fifteenth century: they are hence often termed Spanish Jews. To a great extent they are subject to the Porte. The chief rabbi bears the title of Hakim Pasha, and enjoys a certain degree of civil authority: he has a guard before his house, who assist him in enforcing the taxes: his influence extends over all the Sephardim in Palestine. He himself lives in a very neat, comfortable, and well-appointed house, wholly, however, in the eastern style. Wilson found him very friendly and communicative: his library is valuable, and particularly rich in Hebrew manuscripts. He praised as particularly valuable the topography of Palestine written in 1322, and called the *Khafthor va-ferach*. In two synagogues, which stand near together, the Jews, who have come from such widely scattered regions, meet for public worship.

The sect of Karaim or Karaites has but few representatives in Jerusalem: Wilson found only five men who claimed to belong to it. These called themselves interpreters, though the name Jerusalemites is given to them also, because they make public lamentation, with great apparent sincerity, over the fall of Zion. The Sephardim despise them, and call them Sadducees,—a title which, however, they solemnly disclaim: nor does there seem to be any good reason why it should be applied to them.

The Jews who come to Palestine are led almost invariably by purposes of devotion: they despise, therefore, every kind of mechanical occupation, and do very little trading even: a few of them keep little retail shops, and one is a stone-cutter. The most, however, prosecute studies in the Jewish law,[1] and offer prayers for their brethren in the great world, since, according to the Talmud, the world will revert to its primeval chaos, unless there be offered prayers, at least twice every week, in the four sacred cities of Palestine—Jerusalem, Hebron, Safet, and Tiberias. On this account the pious Jewish scholars are supported by the gifts of their brethren in the faith, which are collected by agents from the Holy

[1] Krafft, *Topog.* p. 264.

Land, but mainly by the head rabbi of the Spanish Jews.
Krafft says that the youngest of these remained four years
out of Palestine, and returned with the sum of 46,000 francs,
of which two-thirds were contributed by Spanish and one-third
by German Jews. The wives of these pious Hebrews show
themselves all the more industrious in household occupations,
as if in compensation for the idle life led by their husbands:
they care well for their children too, and stand upon a much
higher plane of culture than the Moslem women. This the
Jews are very proud of, and oftentimes their pride takes the
form of the sumptuous apparel of their wives. Their chief
business is to visit the Wailing Place on the west side of the
Haram—the Mount of the Holy House, as they call it; and
the stones which are there the Jews universally consider the
work of their own distant ancestors. The Mosque of Omar
is to them identical with their own temple: el-Aksa they call
the Midrash (*i.e.* the college) of Solomon; and the Aurea
Porta on the east side of the Haram they term the Gate of
Grace.

NORTHERN JUDÆA.

CHAPTER I.

THE REGION IMMEDIATELY ADJOINING JERUSALEM.

DISCUSSION I.

BETHANY AND ABU DIS, ON THE EAST OF THE CITY—THE WILDERNESS OF JOHN THE BAPTIST—AIN KARIM, THE CONVENT OF ST JOHN, AND DEIR EL MASALLABEH ON THE WEST.

OVER the Mount of Olives there runs a shallow wadi for a distance of three-quarters of an hour to Bethany. The road would not measure two Roman miles, and corresponds well with the fifteen stadia or furlongs[1] which are mentioned in John xi. 18. On the west-north-west of the village lies a hill; a little south is a very deep narrow wadi or gorge running towards the east. Beyond this, and higher than Bethany, lies, at a distance of twenty minutes, the deserted village of Abu Dis, the Betabudison of Tobler,[2] and according to Schubert, the Bahurim of 2 Sam. xvi. 5-13. This place Brocardus[3] mentions as a mere castle lying on a high hill about two bow-shots west of Bethany. The latter is now an impoverished village of only about twenty huts, with here and there large bevelled stones strewing the ground, indicating that once it was a place of importance. The monks still point out the house of Mary and of Simon the leper, as well

---

[1] F. W. Krummacher, *Evangel. Kalender*, 1851, pp. 75-84.
[2] Tobler, in *Ausland*, 1848, p. 79; v. Schubert, *Reise*, iii. p. 70.
[3] Brocardi *Terr. Sctæ*. Descr. in *Novus Orbis Sim. Grynæi*, fol. 512.

as the grave of Lazarus. The latter has given the name el-Aziriyeh (the Arabic form of Lazarus) to the village. The reputed tomb is a cellar-like excavation in the centre of the village, hewn out of the limestone, and entered by a flight of twenty-six steps. Robinson remarks that it has by no means the aspect or the shape of ancient graves. Besides, the testimony of John xi. 31, 38, condemns the monkish legend conclusively, for the tomb is distinctly said to be beyond the limits of the village. Regarding the identity of Aziriyeh and the ancient Bethany there is not a doubt: the modern name has been commonly supposed to be of Moslem origin; but Tobler[1] has shown, from the *Life of Euthymius*, that the place as early as in the sixth century bore the name Lazariotæ. This word passed in the course of time into the present Arabic form which designates the site of the ancient Bethany. The author of the *Life of Euthymius* (Cyrillus) mentions the village Abu Dis under the name Bet-Abudison, as having been the home of the saint about the year 500. It was first visited by Scholtz in 1821, and subsequently by Tobler.

The vault in which Lazarus was buried is mentioned in the *Bordeaux Itinerary* as early as in 333; and only seventy years subsequently, Jerome speaks of a church which was built over the vault, and which Antoninus Martyr[2] calls a *Monumentum Lazari*. After the crusaders had possessed themselves of this region, the Church of St Lazarus was conveyed to the canon of the Holy Sepulchre. Queen Melesinda[3] founded a large monastery at Bethany, and set her sister Júveta over it as abbess. Bethany being very much exposed to Beduin attacks, in the year 1142 a large tower was erected there at great expense, and a strong garrison established. In 1254 the Pope conveyed this castle to the Knights of St John. The large hewn stones which are still seen may be the remains of the tower erected by Melesinda, it having been, according to William of Tyre, *turrim muni-*

---

[1] Scholtz, *Reise*, p. 210; Tobler, MS. communication, 1847.
[2] *Itinerar. B. Antonini M.* p. 13.
[3] Willermi Tyr. *Archiepisc. Historiæ*, lib. xv. fol. 887; Sebastiano Pauli, *Codice diplomatico, etc.*, Lucca, p. 443.

tissimam quadris et politis lapidibus. A church, which Felix Fabri found existing at Bethany in his day, was subsequently transformed into a mosque, but at present is a heap of ruins.

Tobler has visited[1] several places, both on the east and on the west sides of Jerusalem; but the progress of discovery in the immediate neighbourhood of the city has not advanced much beyond what it was when Robinson confessed that hardly a commencement had been made in this quarter, even the biblical Arimathæa and the Emmaus of Jesus' last walk with the disciples not being identified with absolute certainty.

Among the places touched by Tobler in his explorations is the Convent of St John, lying one and a half hours west of Jerusalem, and half as far as Deir el Masallabeh, or the Convent of the Holy Cross. The latter bears the usual name of Ain Karim,[2] and lies in a fertile valley surrounded by pleasant hills, the building being the conspicuous centre of the landscape. Russegger says[3] that here grow the finest olives and grapes, and that they have a reputation through the whole land. Tischendorf[4] visited the place, and found it inhabited by Franciscans, all Spaniards: he says that the convent owes its present prosperity to the kindness of Louis XIV., and that it is the finest of all the Latin religious houses in Palestine. The church connected with it stands, according to the legend, directly over the spot where John the Baptist was born; and in this wilderness, and directly around the fine spring which here breaks from the ground, the preacher is believed to have prepared himself for his career in the district beyond the Jordan. The walls of the church are richly inlaid with marble, and adorned with gold and silk: a marble staircase descends to the caverns where John is said to have first seen the light. Over the altar is the inscription, "Behold the Lamb of God!" Skilfully wrought bas-reliefs adorn the walls of the room, and illustrate the history of John from his infancy to his behead-

---

[1] Tobler, in *Ausland*, 1848, No. 20, p. 79.
[2] Wilson, *Lands of the Bible*, ii. p. 267.
[3] Russegger, *Reise*, iii. p. 113.
[4] Tischendorf, *Reise*, ii. p. 70.

ing. The most costly decoration is over one of the altars, and is a painting of John, executed by Murillo. Buckingham alludes particularly to the roses which grow in the valley in which the convent stands;[1] and Sieber declares the spot to be one of the loveliest in the neighbourhood of Jerusalem.

DISCURSION II.

PLACES DIRECTLY NORTH OF JERUSALEM.

Although much remains to be done in this direction, as well as on the east and west of the Holy City, yet the masterly method introduced by Robinson and Smith, so fully appreciated by Olshausen[2] in his *Critique*, makes the confession only just, that an excellent beginning has been made in this direction. I have already alluded to the great obstacles which stood in the way, to the mass of idle legends which had to be annihilated, to the confused accounts of preceding hasty and ill-informed travellers[3] which had to be rectified, before the two Americans could have a clear field, and begin what might be called a satisfactory, and certainly a thoroughly scientific, investigation of the country north of Jerusalem.

Directly north of the city, and at a distance of twenty-five minutes, is the high land known as Scopus,[4] whence is gained even now an imposing view of Jerusalem, with its towers and minarets. Only a quarter of an hour towards the north-east from this point may be seen in the valley the small and little known village of el-Isawiyeh: northward, in the distance, and east of the road to Nablus, er-Ram (the ancient Ramah) may be descried. Going northward it is necessary to cross Wadi Suleim before reaching the high land on which lies the

---

[1] Buckingham, *Travels*, p. 355; Sieber, *Reise*, p. 82. Comp. Richter, *Wallfahrt in Morgenland*, pp. 36, 37; and Robinson, *Bib. Research.* ii. pp. 325, 333, 469, ii. 2, 10.

[2] Olshausen, *Rev.* above quoted, p. 145; and Robinson, *Bib. Research.* ii. 1 et sq.

[3] Robinson, *Bib. Research.* i. p. 435.

[4] *Ibid.* i. pp. 275, 579; Wilson, *Lands of the Bible*, ii. p. 35.

village of Anata, at a distance of an hour and a quarter from Jerusalem. This corresponds with the statement of Josephus, that Anathoth was twenty stadia or furlongs from the city. This was the place where the prophet Jeremiah was born, and lay within the domain of Benjamin (Jer. i. 1, xxxii. 8).

The time for the destruction of Jerusalem had not fully come when Isaiah was able to announce that the Assyrian army, which was approaching from the north, should be driven from their camp by the hand of pestilence (Isa. xxxvii. 36). The approach of the army is announced, though not in strategical language, by the prophet; and his words, though uttered in the manner of a seer,[1] have a great deal of topographical value, and aid much in giving a clear view of the district north of Jerusalem, where the Assyrians had pitched their camp (Isa. x. 28-33): "He (*i.e.* the Assyrian) is come to Aiath, he is passed to Migron; at Michmash he hath laid up his carriages: they have taken up lodging at Geba; Ramah is afraid; Gibeah of Saul is fled. Lift up thy voice, O daughter of Gallim; cause it to be heard unto Laish, O poor Anathoth. Madmenah is removed; the inhabitants of Gebim gather themselves to flee. As yet shall he remain at Nob that day; he shall shake his hand against the mount of the daughter of Zion, the hill of Jerusalem."

Of the places here mentioned, Anata, the ancient Anathoth, displays in the wall and some great hewn stones the traces of its former importance: even a few pillars were seen by Robinson among the ruins. The present village shows but a few huts, and shelters scarcely a hundred inhabitants. On the hill above the town there are corn-fields, surrounded by fig and olive trees: in the neighbourhood a kind of building stone is found, which is very much prized in Jerusalem. A fine view is gained from this place over the whole eastern part of the territory of Benjamin, the course of the Jordan, and the north end of the Dead Sea. The Ramah alluded to by Isaiah as crying out in terror lies on a cone-shaped hill towards the north-north-west, where the village er-Ram is now found; Gibeah of Saul (now Tell el Ful), with its

[1] Robinson, *Bib. Research.* i. p. 463 et sq.

high heaps of stones, is more to the south, while Geba (the modern Jeba) lies directly northward, where the Assyrian army encamped. The places mentioned by the names of Madmenah, Gebim, and Nob, lay south of Anathoth, and therefore in the immediate vicinity of Jerusalem: the site of Nob is indicated with great exactness by the expression, "He shall shake his hand against the mount of the daughter of Zion, the hill of Jerusalem." These places have not been identified either by Robinson[1] or by Wolcott,[2] who have carefully gone over the same ground. Nob is the place mentioned in 1 Sam. xxi. 1 and xxii. 19, and the place which in the time of Saul was consecrated by the presence of the tabernacle; the spot, too, where the sword of Goliath was preserved; the town to which David fled for refuge, and in which the priests were put to death by Saul, in consequence of the kindness of Abimelech to David.

1. *From Anata by way of Häsmeh to Jeba, the Geba of Isaiah, and not the Gibeah of Saul: the Graves of the Amalekites.*

From Anata the road runs across the next two wadis to the village of Häsmeh, which lies south of Wadi Farah, on an eminence very similar in general character to that of Anata, but not so high. The village has massive houses, but at the time of Robinson's visit they were empty, the people having fled for fear of the conscription to the wilderness of the Jordan. The whole district consists of a succession of deep, rough valleys, running down to the valley of the Jordan, with broad ridges between them, which sometimes terminate on the east in high, precipitous cliffs. The village of Häsmeh or Hizmeh has been conjectured to be the same as the Asmaveth of Neh. vii. 28, which is mentioned in immediate connection with Anathoth.[3] The road taken by Robinson runs from the latter place northward, but does not follow

---

[1] Robinson, *Bib. Research.* i. pp. 464, 579.
[2] Wolcott, *Jerusalem and its Environs,* in *Bib. Sacra,* 1843, pp. 37, 38.
[3] Comp. Krafft, MS. communication; and Bartlett, *Walks,* etc., pp. 240-243.

the height of the ridge, which extends onward as far as to Samaria, but takes a course lying east of the true watershed, and crosses the upper course of a succession of wadis, whose sides are generally so steep that it was impossible to pass up and down without dismounting. The stone found there is invariably limestone; but the level uplands are fertile, although it is only here and there that cultivated fields are seen. It is farther eastward that the fearful desert begins, which extends down towards the Jordan valley. It was in this that the Parah of Benjamin,[1] mentioned in Josh. xviii. 23, may have lain, a trace of which name Buckingham discovered in this neighbourhood.

After crossing Wadi Farah, one comes to the height of land on which stands Jeba.[2] The place is small, and lies in ruins, the hewn stones of which indicate the importance of the site in ancient times. North-east of Jeba may be seen the village of Rummon, lying in a very striking position upon the summit of a cone-shaped hill, where the last remnant of the tribe of Benjamin took refuge when they had been pursued from Gibeah by the combined tribes (Judg. xx. 45, 47, and xxi.). This Jeba is unquestionably the Geba mentioned in the vision of Isaiah; but it is not the Gibeah of Saul, whose name might indeed be easily interchanged with it, but which must be sought nearer Jerusalem.

Although south of er-Ram there is no place discovered which bears a name kindred in form to that of Gibeah, yet there is no doubt, on grounds considered valid, that the Gibeah which was the birth-place of Saul is to be recognised in the modern Tuleil el Ful or Fulil, a good hour's distance north of Jerusalem. These grounds have been well stated by Robinson[3] in the *Bibliotheca Sacra*. The objections which have been brought by Gross against the identity of the Gibeah of Saul and Jeba have been fully confirmed by the discoveries of the American traveller.

---

[1] Rödiger, Review above quoted, No. 71, p. 564.
[2] Robinson, *Bib. Research.* i. p. 440.
[3] *Bib. Sacra*, 1844, vol. i.; E. Robinson, on Gibeah of Saul and Rachel's Sepulchre.

Josephus alludes twice to Gibeah of Saul, once as being twenty, once as being thirty, stadia from Jerusalem. But Ramah, according to the same authority, was forty stadia away from the capital: Gibeah must have been much nearer than Ramah, and Jeba lies only a half-hour's distance eastward from er-Ram.[1] The route which Jerome states was taken by Paula indicates that she passed by Gibeah; but the place was so near to Scopus, that it could have been no other than that now occupied by Tell el Fulil.

Josephus, in his account of the approach of Titus, speaks of a village called Gabath Saul, thirty stadia from Jerusalem, and says that here the work of reconnoitering went on over the high plain of Scopus. This plateau extends northward to the hill bearing the name at present Tuleil or Tell el Ful. On the west side of this cone-shaped eminence there is a gentle decline towards a low plain, which extends away towards the north and east as far as the hill on which lies er-Ram, and towards the west to the broad and high plateau where el-Jib or Gibeon is situated. Where the roads running north and north-west come together is a λόφος, as at Josephus' time, *i.e.* a cone-shaped hill, called Tuleil el Fulil, quite isolated and alone. It marks unquestionably the birth-place of Saul. A little to the west are to be seen old foundation walls, and some massive hewn stones, the remains, it would seem, of an ancient city. Captain Newbold[2] was led to suppose in 1846 that the place was one known by the Arabs as Kabur ul Amalikeh, or Graves of the Amalekites. The resemblance to the most ancient graves in the Egyptian pyramids led him to this conjecture, for he connected them with the Hyksos, or shepherd kings, the so-called sons of Amalek.[3] A later authority has identified their builders with the Sheta, a branch of the Rephaim, who, after their invasion of Egypt, removed to Palestine.

[1] Robinson, *Bib. Research.* i. p. 440.
[2] Asiatic Soc. Ap. 21, in *Athenæum*, 1849, No. 1124, p. 491.
[3] F. Corbeaux, in *Proceedings of the Royal Soc. of Literature*, 1850, No. 21, pp. 319, 221.

2. *Route to Mukhmas, Michmash; the rock pass of Wadi es Suweinit.*

From Jeba, the Geba of Isaiah, the road runs northward to the village of Mukhmas,[1] in the name of which the Michmash of the ancient Israelite history has perpetuated itself. Before reaching it, however, it is necessary to pass down the steep slope leading to Wadi Suweinit, which forces its way in an imposing manner between high walls of rock. This wadi comes from the north-west, running south-eastward from Beitin and el-Bireh, and becoming constantly more grand and wild, till it unites with Wadis Fuwar and Farah, and forms the Wadi Kelt spoken of in a previous part of this work, which issues into the Jordan valley near Jericho.

This steep valley is almost unquestionably the rocky passage which Isaiah mentions as existing between Michmash and Geba: the place corresponds well with the prophet's few but significant words. In the valley, on the left of the route which Robinson took, there lie two isolated hills, cone-shaped, and with rocky sides, which once may have had much sharper tops than at present is the case. They seem to indicate the Bozez and Seneh where the Philistines took up their post, and near which Jonathan accomplished that deed of heroism which is recorded in 1 Sam. xiii. 23, xiv. 1, 4, 5. The Philistines had advanced as far as to Michmash, and pitched their camp there, and from that point three of their bands had issued to desolate the land; one on the road to Ophra, another on the way to Beth-horon, and still another on the route which runs to the desert valley of Zeboim (1 Sam. xiii. 16-18). They were all driven back, in consequence of Jonathan's daring achievement, to Aijalon (1 Sam. xiv. 31). This valley appears to have been in earlier times the barrier between the tribes of Ephraim and Benjamin: for Geba, on the south side, was on the frontier between Judah and Benjamin (2 Kings xxiii. 8); while Bethel, on the north side, but farther west, lay on the southern boundary of Ephraim (Josh.

[1] Robinson, *Bib. Research.* i. p. 441 et sq.

xvi. 1, 2, xviii. 13). From this Wadi Suweinit, a quarter of an hour conducts one to the slope on which Michmash lies: the village lies about three and a half hours from Jerusalem. Eusebius and Jerome state it to be almost nine Roman miles from the city. Once it seems to have been well fortified, but now it is more ruined than Anathoth.

3. *Road to Deir Diwan. Efforts to discover the site of Ai: Ruins of Medinet Chai or Gai (Ai), and of Geba, on the steep wall of Wadi es Suweinit.*

From Mukhmas the road runs northward over rolling land, and through a wadi which passes from Deir Diwan southward, and enters Wadi Suweinit. This valley is full of graves; and as it narrows in approaching Deir Diwan,[1] it begins to display traces of human habitations. Robinson and Smith hoped to be able to discover in this neighbourhood some indications of the once celebrated city of Ai, and were led to believe that the district around the village of Kudeirah was the place best adapted to institute investigations. They were unable to discover anything which could confirm their hopes, and at length abandoned the attempt to discover the site of Ai. Wilson[2] was equally unsuccessful in subsequent attempts.

Deir Diwan is itself a large and prosperous village, lying in an uneven rocky basin, and surrounded by hills, on which grow corn, olives, and fig trees. There is not a trace of antiquity to be seen there, however.

After his return from Jeba, Robinson learned, but too late to make personal investigations, that east of this place there are important ruins.[3] Dr Krafft[4] made them the object of subsequent investigation, and believes that he has discovered there the site of the stronghold of Ai. The place now bears the name Chai or Gai: it lies forty minutes east

[1] Robinson, *Bib. Research.* i. pp. 442 et sq., 573 et sq.
[2] Wilson, *Lands of the Bible,* ii. p. 287.
[3] Robinson, *Bib. Research.* i. p. 440.
[4] Krafft, MS. communication, 1848; Strauss, *Sinai und Golgotha,* p. 393.

from Jeba (the Geba of Isaiah).[1] This place he thinks answers to the requisitions[2] of Josh. vii. 2, where Ai is said to have lain upon a height east of Bethel; and to those of Josh. viii. 1–35, where the account is given of the capture of the city by stratagem. Robinson, however, has replied to Krafft's conjecture, and pronounces strongly against it. He calls attention to the fact that the two places Bethel and Ai were so near together, that Abraham erected a tent in the valley between them (Gen. xii. 8). At the time that the first force was sent up from the Jordan to attack Ai, and was driven back, the people of Bethel all joined in the pursuit. Still they were not so near as to prevent Joshua from sending a large force to the rear of Ai without being perceived by those of Bethel. Robinson[3] shows that the site of Ai conjectured by Krafft is at least eight miles from that of Bethel, and is, moreover, connected with it by a circuitous way. And it must be confessed that the hypothesis of Robinson, propounded with great modesty and caution, that Kudeirah near Beitin is the site of Ai, is much more probable than that advanced by Krafft.

At a subsequent period Ai was rebuilt, for in Ezra ii. 28 we learn that two hundred and twenty-three men returned to Bethel and Ai from the captivity; and Nehemiah (xi. 31) alludes in like manner to the children of Benjamin, who lived in Michmash Aija, *i.e.* Ai and Bethel. At the time of Eusebius and Jerome, a few scattered ruins of this place could be seen not far to the south-east of Bethel.[4]

---

[1] Krafft, *Topogr.* Pref. p. ix.
[2] Von Raumer, *Pal.* p. 177.
[3] Robinson, in *Bib. Sacra*, vol. v. No. xvii. Feb. 1848, p. 93.
[4] Robinson has discussed Krafft's conjecture ably, though somewhat curtly, in *Bib. Research.* iii. 288, rejecting it totally; and there is now no uniformity among travellers in their judgments respecting the location of Ai. Van der Velde (in conjunction with Mr Finn) decides confidently that its site is Tell el Ajar, s.e. from Beitin, and on the Wadi el Mutyyah (*Mem.* p. 282). Stanley conjectures that it was at the head of Wadi Harith (*S. and Pal.* 202).—ED.

4. *From Deir Diwan to Taiyibeh :*[1] *conjectural site of Ophrah.*

Beitin lies only an hour north-west from Deir Diwan, Taiyibeh a little farther, an hour and a half towards the north-east. The steep sides of Wadi el Mutyah must be descended, and the valley, itself three hundred feet deep, be crossed, in order to reach Wadi el Ain, which comes down from the north, and which takes its name from a fine spring on the west side. It is an hour's distance from this spring to Taiyibeh. The village forms the crown of a cone-shaped hill resting on a high ridge of land. On the summit, the highest thus far reached north of Anathoth,[2] lie the ruins of a tower, as in so many similar villages, under whose protection the houses seem to stand, while olive and fig trees cast their shade around. A broad panorama is seen from the tower, comprehending the Ghor of the Jordan, the gorges of the Zerka, and even the heights of the Bilkah. Southward the Frank Mountain can be descried. The landscape is in the main destitute of greenness, but it makes a deep impression upon the observer. Only three hundred paces from the village, on a point of the hill, lie the ruins of the small Church of St George. All the inhabitants of the place are Christians, and though exempt from military conscription, are compelled to pay a very heavy tax to the Turkish Government. This amounts to no less than about 1900 Spanish dollars yearly. The entire population comprises but three or four hundred souls, of whom only twenty-five are ablebodied men, who are liable to tax.

Robinson thinks that the situation of this place is so admirable, that it could not fail to have been the site of some important place in past times, and conjectures that it was the Ophrah of the tribe of Benjamin, of which Eusebius and Jerome say that it lay five Roman miles east of Bethel. This seems to correspond quite as closely with the situation of Taiyibeh as with that of the place thought by Krafft to be Ophrah, a half-hour south of Gai.

[1] Robinson, *Bib. Research.* i. p. 443 et sq.
[2] Strauss, *Sinai und Golgotha,* p. 395.

The situation of this place is no less interesting to the student of the New Testament than of the Old, if it is to be considered, as many suppose, identical with that city of Ephraim which Jesus chose as a place of refuge when the high priests were pursuing Him and wishing to put Him to death (John xi. 54). This city of Ephrem or Ephraim is justly considered identical with the Ephron or Ephraim of 2 Chron. xiii. 19,—a place which Abijah king of Judah wrested from Jeroboam, together with the city of Bethel, which was not far away. Josephus tells us that Vespasian came from Cæsarea to the mountains, and conquered the ruler of Gophnah and Acrabah, together with the small cities of Bethel and Ephraim. This is doubtless the Ephraim or Ephron which Jerome locates twenty miles north of Jerusalem. In the book of Joshua (xviii. 23) a similar name (Ophrah) is mentioned in Benjamin, which corresponds, however, to the Ophrah mentioned in 1 Sam. xiii. 17, five miles east of Bethel. Both seem to refer to the same place. These places —the Ephraim or Ephron and the Ophrah of the Old Testament, and the Ephraim of the New—all appear, therefore, to have been in the wilderness, about five miles east of Bethel and twenty north of Jerusalem, and therefore to confirm Robinson's[1] conjecture that the site is now occupied by the village of Taiyibeh. From this place runs the road which Jesus took as He crossed the Jordan for the last time, and entered the land of Peræa (Matt. xix. 1; Mark x. 1).

5. *From Taiyibeh to Beitin or Bethel.*

From Taiyibeh Robinson turned back south-westwardly, reaching Beitin,[2] two hours distant, and on the main road from Jerusalem to Nablus. The only place on the way is el-Alya, lying on a high plateau, a village of only a handful of houses, near a fountain known by the same name.

Through the basin here seen surrounded by hills there run two valleys from the north, which unite farther south-

---

[1] Robinson, *Bib. Sacra*, vol. ii. 1845, pp. 398, 400.

[2] Robinson, *Bib. Research.* i. pp. 447 et sq. and 575; Bartlett, *Walks, etc.*, pp. 242, 243; *The Christian in Palestine*, Tab. xxix. p. 122.

ward, and pursue then a common course towards the south-east. Between these two valleys, on the high ridge of land, lie the extensive ruins of Beitin. The houses are in ruins, and deserted, and only serve as a protection at night for herds and flocks. The prospect from the place is a limited one: at the south-west only Bireh (Beeroth) is to be seen, while southward the most remote village is Shafat. Only eight minutes' walk to the south-east of Beitin are the ruins of a small castle, Burj Beitin or Makhrun, a square fortress of hewn stone, with a Greek church in the middle, and with scattered pillars here and there, out of one of which a cross is hewn. Ten minutes farther south-east, on the highest part of the ridge, there is a second and larger church, in the greatest state of desolation, however. The ruins of Beitin occupy the whole upper part of the eminence, which slopes off towards the south-east, where several roads meet. They cover a space of three or four acres, and consist mainly of foundation walls; partly, however, of real remains of ancient structures. On the highest point stand the ruins of a castellated tower; at the southernmost extremity of the walls there may be seen the ruins of a Greek church, which appears to have been constructed of the stones of a still larger building, within which it stands, and whose line of walls may still be traced. In the western part of the city Robinson discovered the remains of a cistern, the largest (those of Solomon excepted) which he had seen in Palestine, being three hundred and fourteen feet long, and two hundred and seventeen wide. Wilson[1] states that the construction is very similar to that of the tanks of India.

Notwithstanding that the site of the once so renowned Bethel (formerly called Luz), on the primitive boundary of Ephraim and Benjamin (Josh. xviii. 13, xvi. 1, 2), was completely unknown at the beginning of the present century, yet there is no doubt concerning the identity of Bethel with the modern Beitin. Instances of *el* passing over into *in* are very common, as in Jibrin from Jibril, Israin from Israil, Zerin from Jezreel. The situation and the distance from marked

[1] Wilson, *Lands of the Bible*, ii. pp. 40, 287.

points confirm in the fullest manner the identity which has been clearly shown by Robinson. Yet previous to his visit the missionaries Nicolayson and Elliot made the discovery in 1836 that Beitin occupies the site of Bethel. The place was one of the most sacred throughout the whole Jewish history: its fate is pictured strongly by Amos v. 4, " Seek not Bethel, nor enter into Gilgal, and pass not to Beersheba; for Gilgal shall surely come into captivity, and Bethel shall come to nought."

In Beth-el (the house of God) Abraham pitched his tent upon the high land, where even now excellent pasturage is found, and Jacob erected an altar to the Lord. Yet the opportunities for grazing were not sufficiently extensive to allow both Abraham and his brother Lot to remain in the same place, and hence they parted in the manner described in Gen. xiii. 9. The prophet Samuel went every year to Bethel to judge the people. In order to make his subjects turn away from their worship of Jehovah, Jeroboam selected this place as the site of the homage to be paid to the golden calf. Bethel belonged subsequently to Judæa: it was destroyed, but was inhabited again by the Jews on their return from exile. At the time of Eusebius and Jerome it was only an unimportant place, whose situation fell into entire oblivion at the time of the Crusades. Yet the remains of churches show that the place was never destitute of population. In the centuries after the holy wars, the site of the ancient Bethel was sought elsewhere, in the neighbourhood of Shechem; and even in Robinson's time the monks of Jerusalem did not suspect that Beitin occupied its site. All preceding tourists, therefore, passed this little village by without remark.

6. *From Beitin to el-Bireh (Beeroth).*[1]

From Bethel the road runs an hour's distance south-westwardly to el-Bireh, passing by the spring el-Akaba, and a cavern supported by two pillars, which serves as a reservoir, fed, it would seem, by an internal subterranean source. El-Bireh lies on the crest of a ridge running east and west, and

[1] Robinson, *Bib. Research.* i. p. 451 et sq., p. 565.

forming at the same time the watershed between the Jordan and the Mediterranean. It lies so high that it can be seen from a great distance. Yet its houses are built in a manner by no means conspicuous, and lie half-buried in the ground; so that the road itself runs[1] in some places over their roofs, where these are contiguous to the wall of rock.

Many great stones, massive remains of walls, and also a large square building, are the traces of the former importance of this place, which was once in the possession of the Knights Templar, the builders unquestionably of the fine church on the crest of the mountain, whose altar, sacristy, and walls are still standing. At a spring near by can be also seen the remains of an earlier time. Bireh had in 1838 a hundred and thirty-five taxable inhabitants, sixty of whom were impressed as soldiers, and the whole population can be safely estimated at about seven hundred.

From el-Bireh the distance is some two and a half hours to Jerusalem. The place is, according to Robinson, in all probability the Beeroth of the Old Testament, belonging to the territory of Benjamin, and yet not a place of great importance in biblical history. It is mentioned in 2 Sam. iv. 2, and xxiii. 37. At the time of Jerome it could hardly be identified, and its site remained unknown till Maundrell[2] recognised that the el-Bireh of the present time is the Beeroth of the Bible. The name he derives from a very profuse spring, Bir,[3] which was also noticed by Otto v. Richter at the southern entrance of the city.

7. *From Bireh to Jerusalem by way of Atara (Ataroth), er-Râm (Ramah of the Prophetess Deborah), Tuleil el Fûl (Gibeah of Saul), and Shafat.*

From el-Bireh the road runs south-westward past the spring Ram Allah, crossing the great watershed, from which the great coast plain and even the blue line of the sea can be descried just beyond the white sand-dunes around Jaffa.

---

[1] Otto v. Richter, *Wallfahrten*, p. 53.
[2] H. Maundrell, *Journal*, p. 64.
[3] Wilson, *Lands of the Bible*, ii. pp. 39, 287.

Passing Beit Unia, north-west of el-Jib (Gibeon),[1] and Neby Samwil (Mizpeh), the Jaffa route leading to Beit-ûr (Beth-horon) is soon reached. Of this place I shall have to speak subsequently. I have already discussed Gibeon, Mizpeh, and the western slopes of the Judæan mountain-land. On the direct road running from Jerusalem to Nablus, it remains that I shall speak of Atara and er-Ram.

The village of Ram Allah[2] lies upon the crest of the high watershed: the wadi in which Beit-ûr lies runs directly west from it, down towards the sea-coast. The village is inhabited by eight or nine hundred Christians, who appear to be tolerably well conditioned: their houses are all well built and new; no traces of antiquity are discoverable in the place; the soil around is productive, and yields abundant crops of corn, olives, figs, and grapes. This place, like Taiyibeh, belongs to the great mosque of Jerusalem, and is compelled to pay a heavy tribute to it yearly. The population is distinguished for energy and industry.

Following the line of watershed southward from Bireh,[3] some old walls called Suweikeh are first passed, and then Atara is reached, where the upper Wadi Hanina has its commencement. This wadi has been elsewhere spoken of as the natural barrier between the mountain districts of Judah and Ephraim.

In Atara there are to be seen some extensive ruins with arches, and above the same two ancient water-cisterns, a hundred feet long and forty wide, probably marking the remains of a former place bearing the name of Ataroth. In the book of Joshua two places of this name are mentioned, which both lay upon the confines of Ephraim and Benjamin. The one mentioned in conjunction with Beth-horon in Josh. xvi. 5 is probably the one found here, as the other mentioned in ver. 7 lay nearer Jericho. It is the same place which is subsequently alluded to in Josh. xviii. 13, although this might be supposed to refer to the Ataroth in the tribe of Ephraim,

---

[1] *The Christian in Palestine*, p. 124, Tab. xxxiv., *Gibeon from Neby Samwil.*

[2] Robinson, *Bib. Research.* i. p. 453.   [3] *Ibid.* i. p. 575 et sq.

which lies farther north, near Shechem, near Jilgilia, north of Jifna, where Robinson[1] discovered another Atara. The two Ataroths which are mentioned in the *Onomasticon* give no solution of the Old Testament question regarding their respective locations.

The site of er-Ram[2] is more important, which lay south of Atara on a high hill, about ten minutes' distance east of the great Jerusalem highway, and at a spot where the Wadi Farah begins its course eastward to the Jordan, and the Wadi Ram, one of the first north arms of the Wadi Hanina, begins its westward course towards the Mediterranean. Ram is a pitiful village of only a handful of houses, between which, however, several massive stones are still seen, among which are some pillars which indicate the nature of the architecture once standing there. The little mosque, once a Christian church, is built out of these remains. Er-Ram lies in a very conspicuous position; but it is to be discriminated from the Ramah of Samuel, the one near Hebron, and other places of the same name. It is, however, unquestionably the Ramah which lay upon the route of the Levite who is mentioned in Judg. xix. 13 as journeying from Jerusalem by way of Gibeah and Ramah in Benjamin to the mountains of Ephraim. According to Jerome, it lay six or seven Roman miles from the Jewish capital.

It was between Ramah and Bethel, upon the mountains of Ephraim (which unquestionably extended within the territory of Benjamin[3]), that, according to Judg. iv. 4 and v. 12, the prophetess Deborah sat under the palm trees and judged Israel. King Solomon, according to Jerome, re-erected Ramah and Beth-horon, and converted them into cities of some splendour. Ezra (ii. 26) and Nehemiah (vii. 30) name the place subsequently to the captivity, but during the dark ages it was completely forgotten. It is hardly alluded to by the writers of many hundred consecutive years. Even quite recent travellers did not know where this important place lay, although the monks repeated their

---

[1] Robinson, *Bib. Research.* ii. 265. Comp. Keil, *Com. zu Josua*, p. 309.
[2] Robinson, *Bib. Research.* i. p. 516 et sq. [3] *Ibid.* ii. p. 9.

traditions about its site; for even Schubert[1] passed close by it without suspecting the fact, as did also Richardson, Scholtz, Monro, and others. Its discovery is one of the many things which we owe to the unwearied exertions and sharp-sighted inquiry of Robinson and Smith. Jeba, the Geba of Isaiah, lies a half-hour's distance east of er-Ram, but cannot be seen thence. A little more than a half-hour's distance westward is el-Jib, or Gibeon; and about as far southward is Tuleil el Ful,[2] the Gibeah of Saul. Gibeon, Anathoth, Mizpeh, and Michmash, are all to be seen from the lofty site of er-Ram.

Leaving this pitiful village, we reach in ten minutes southward Khuraib er Ram, *i.e.* the ruins of er-Ram,[3] where eight to ten shattered arches lie parallel with the main road: they seem to have belonged to a khan once stationed here.[4]

Still farther south we bear eastward from the main road, in order to reach the high Tell or hill on which lies Tuleil el Ful, with its heaps of massive stones.[5] There seems to have once stood there a great square tower fifty-six feet in length, and forty-eight feet in width, composed of hewn stones, from whose remains there is still afforded an extensive prospect, only surpassed by that from Neby Samwil, the ancient Mizpeh. There are no other architectural remains found at this place, which, on grounds already stated, has been identified with the ancient Gibeah, the birth-place of Saul. South of this spot is the little village of Shafat, only fifty minutes' walk from the Damascus gate of Jerusalem. This place is probably identical with the hamlet of Sharifat, mentioned by von Richter,[6] who noticed, as did Robinson, the signs of a large population once inhabiting this region, now all deserted and barren.

---

[1] Von Schubert, *Reise*, iii. p. 124.

[2] Wilson, *Lands of the Bible*, ii. p. 38; *Christian in Palestine*, p. 124, Tab. xxxii.

[3] Robinson, *Bib. Research.* i. 579; Wilson, *Lands of the Bible*, ii. 38.

[4] Comp. v. Schubert, *Reise*, iii. p. 125.

[5] Wilson, *Lands of the Bible*, ii. p. 36.

[6] O. v. Richter, *Wallfahrten*, p. 53; Robinson, *Bib. Research.* i. p. 580.

CHAPTER II.

THE MOUNTAIN ROADS, WITH THEIR PASSES WESTWARD TO THE COAST OF THE MEDITERRANEAN:

THE PLAIN OF SHARON, AND THE TOWNS OF RAMLEH, LYDDA, AND KEFER SABA (ANTIPATRIS).

ROM Jerusalem westward to Jaffa, its port, the ancient Joppa, ten hours away, there are several paths, all of them having their beginning in the rocky land of Judæa, and ending in the level plain along the coast.

Heavy articles of merchandise[1] are now, as they always have been, transported by way of the Beth-horon pass, which lies north-west of Jerusalem, where an upper Beit-ûr and a lower Beit-ûr indicate the line of descent, after passing which the road runs on to Ludd (Lydda) and Jaffa.

A valley lying south of the Beth-horon pass, Wadi Suleiman by name, is merely a side arm of the same, separating from it in the high land (near el-Jib), and joining it again near Jimzu. This is commonly considered the easier route, and travellers usually choose between these two. It is probable that this was always the case, although we lack full evidence on this point.

There is a third route lying farther south, which does not make the great bend to the north-west which the two just specified do, and this is the most direct road from the capital to its port. It runs by way of Kulonieh and Kuryet el Enab through Wadi Aly to Lydda, where it joins the other two,

[1] Robinson, *Bib. Research.* ii. p. 252.

and the three form thenceforth a single road to Jaffa. This route, too, may have been much used in antiquity, but there is no decisive evidence whether it was or was not. The older writers used to be too neglectful of details which might throw light on such points as this. The mountainous tracts which lie between the three routes just detailed, although the distance is so short across, have never been thoroughly explored, and much yet remains to be done before it can be asserted that the topography of this region is accurately known. Many valleys must first be traversed, heights taken, and the sites of ancient convents visited. I will first speak of these three routes in detail, and then pass to the coast, and the places upon it.

DISCURSION I.

THE SOUTHERN ROUTE, BY WAY OF KULONIEH, KURYET EL-ENAB, AND WADI ALY.

The earliest direct reference to this route, according to Robinson, seems to be the statement of Eusebius and Jerome, that Kirjath-jearim lay nine miles from Jerusalem, on the road to Diospolis, *i.e.* Ludd or Lydda. If the present Kuryet el Enab is identical with Kirjath or Kirjath-jearim, *i.e.* with the ancient border city of Judah and Benjamin, to which the ark of the covenant was carried from Beth-shemesh, there can be no doubt that this route was one which was taken at the time of Jerome.

Robinson did not himself pass over it. He gives, however, the statements of two travellers who did: one,[1] the account of Dr E. Smith, who, in company with ladies, leisurely journeyed from Jaffa to Jerusalem; and the other that of the missionary Lanneau, who gives the measurements more fully, but who went by way of Ramleh instead of Ludd. Both accounts are valuable, for most tourists hasten over this part of their journey, taking little note of the objects on the way.

[1] Robinson, *Bib. Research.* ii. p. 243, and Note to p. 528.

1. *Eli Smith's Itinerary from Jaffa to Jerusalem.*

		Hours	Min.
1. From Jaffa to Yazur,	. . . . .	1	0
2. To a village,	. . . . .	1	0
3. ,, Ludd,	. . . . .	1	35
4. ,, er-Ramleh,	. . . . .	0	45
5. ,, Kubab, on the first rising ground,	. .	2	10
6. ,, Latron, beginning of Wadi Aly,	. .	1	0
7. ,, Saris, highest point,	. . . .	2	30
8. ,, Kuryet el Enab, in a valley,	. .	0	30
9. ,, Jerusalem,	. . . . .	3	30
	Total, .	14	0

2. *Lanneau's Itinerary from Jerusalem to Jaffa.*

		Hours	Min.
1. From Jerusalem to Kulonich,	. . .	1	30
2. To Kuryet el Enab,	. . . .	1	30
3. ,, Saris,	. . . . .	1	0
4. ,, Bab el Wadi,	. . . . .	1	0
5. ,, Latron,	. . . . .	1	0
6. ,, Kubab,	. . . . .	1	0
7. ,, Ramleh,	. . . . .	2	0
8. ,, Surafend,	. . . . .	0	30
9. ,, Beit Dejan,	. . . . .	1	0
10. ,, Yazur,	. . . . .	0	30
11. ,, Jaffa,	. . . . .	1	0
	Total, .	12	0

With horses or with mules the journey is usually accomplished in the same time, it taking nine hours between Jerusalem and Ramleh, and three between this point and Jerusalem.

Wilson[1] asserts that this route is the shortest: the same was taken by Fabri, and in the present century it was the one traversed by Turner and Buckingham in 1815, von Prokesch in 1829, and by von Wildenbruch in 1843.

Ramleh lies in the plain, running along the coast northward as far as to Mount Carmel, and bearing the celebrated name of Sharon. Just east of this place, however, the level sandy tract comes to an end, and the common hard lime-

[1] Wilson, *Lands of the Bible*, ii. pp. 263-268.

stone of the mountain-land begins to appear. At the village of Anabah the ascent commences: then the road passes by the pitiful village of Fineh, which appears to be built upon ancient ruins, near which is a spring known as Job's fountain.[1] The hills which are next encountered towards the south-east are generally cone-shaped, with valleys or wadis lying between, or simply basin-shaped depressions. The broadest one of these is the very fruitful Merj Ibn Omeir,[2] which reaches away towards the south-west. On the southern border of this lie el-Kubab, Beit Nuba, and Yalo. The strata of chalk are not at first marked in their character: they appear now horizontal, now sloping at various angles. The nearer to Jerusalem they are examined, the more marked is the terrace shape which they assume. Shortly before reaching Jerusalem the declivities increase, and the more desolate is the surface. Many of the small places passed on the way have not yet been named in the accounts of the tourists.

The village of Yalo has been already shown to be the ancient Ajalon, in the territory of Dan (Josh. xix. 42). The doubts respecting the identity of the two have been shown by Jerome to be unfounded, whose $Aλώμ$ must have lain about two miles from Nicopolis (Lydda), and could have been no other than this Yalo; and another confirmation is, that Ajalon is not merely mentioned in connection with places lying farther to the south (by which its location would be somewhat doubtful), as with Beth-shemesh, Hebron, Lachish, Gath, Bethlehem, and others; but in another passage cited by Robinson (2 Chron. xxviii. 18),[3] it is mentioned in immediate proximity to the cities of the plain which the Philistines possessed: the list closes with Jimzu, a place south-east of Lydda, and showing that Ajalon must have been in the immediate neighbourhood. The identity of Yalo with Ajalon and Ailom may therefore be considered as established.

---

[1] Von Wildenbruch, *Reiseroute in Syrien*, in *Monatsb. der Berliner geog. Ges.* 1843, Pt. i. p. 229.

[2] Robinson, *Bib. Research.* ii. p. 252.

[3] Wilson, *Lands of the Bible*, ii. p. 266.

The three villages, el-Kubab, Beit Nuba, and Yalo, have not been visited by travellers. Wilson, the only one who mentions them in their connection, says that, as he was passing by Latrun, he saw them on the high land contiguous to Wadi Ali. This Latron or Latrun, six hours distant from Jerusalem, lies about equidistant from both ends of the route. An hour's distance to the south-east is the village Amwas,[1] which the monkish legends make to be the Emmaus of Luke xxiv. 13-35. The earlier pilgrims speak of this place as a "Castellum Emmaus," and speak, moreover, of the church and the city of the Maccabees as there. The tradition may be one which dates back as far as to the time of the Maccabees. Fabri, who visited Emmaus in 1483, found there a hospice, bearing the names both of Luke and Cleophas, and thanked God that it was permitted to him, a poor pilgrim, who had come so long a distance, at last to tread in the unquestionable footsteps of his Saviour, and to kiss the ground on which He had trodden. At the time that Jerusalem was destroyed, this place also perished, but was restored subsequently by the Romans, and called Nicopolis. The *Bordeaux Itinerary*, written in 333, places Nicopolis[2] in its true position, twenty miles from Jerusalem, and ten from Lydda. It cites no other place as answering to the Emmaus of Luke. Very few people were living there at that time, although large buildings were standing: these may have been destroyed at the time of the subsequent Saracen invasion. As the distance of this Amwâs from Jerusalem, by way of Latrûn, was about seven hours, and as Luke makes the Emmaus mentioned by him sixty stadia or furlongs from the capital, *i.e.* about three hours, the two accounts do not appear to agree; but Rödiger[3] has solved the difficulty by remarking, that in some of the manuscripts the reading is a hundred and sixty furlongs,—an emendation which removes the difficulty, this being not far from the distance of Amwas from Jerusalem.[4]

---

[1] Comp. Robinson, *Bib. Research.* ii. pp. 232, 255.

[2] *Itin. Hieros.* ed. Parthey, p. 283.

[3] Rödiger, *Rev.* in *Allg. Lit. Z.* 1842, No. 72, p. 576.

[4] Thomson objects forcibly to this, that if we read a hundred and sixty

And up to the present time there has been discovered no Emmaus farther east which presents stronger claims to be considered the one mentioned in the Gospels, than the one of which we are now speaking. Dr Barth,[1] who was able to see the village of Amwas from the more elevated Kubab, visited the place, and discovered important fortifications there, which justified the name which it once bore—Castellum Emmaus. Von Prokesch[2] asserts that there is no site between Jerusalem and Ramleh which presents more advantages for a strong military position than does that of the ancient Nicopolis, which cannot be reached except after five hours of hard climbing on horseback. It does not seem open to criticism, to look with Quatremère for the site of the mountain Modin, the home of Mattathias Maccabæus, where, in the neighbourhood of Latrûn, that lofty pillar of the Maccabees family was erected by Simon, which has been discernible by travellers until a recent period. Robinson[3] and other travellers have been unable to discover whence the name of this place, "Castellum s. domus boni Latronis," which has been current since the sixteenth century, was derived. Quatremère thinks that the place had originally an Arabic name, but that subsequently to the Mohammedan invasion the place became a refuge of robbers, and at length received an appellation which was perverted into *domus boni latronis*, and the equivalent *maison du bon laron* of the Franks.[4] Richardson says that in his time the place was a real robbers' nest, as indeed the fact that so many pilgrims passed through this neighbourhood

---

furlongs, and admit that the present Amwas was the ancient Emmaus, it would have been impossible for the disciples to have returned the same evening to Jerusalem, as we are told (Luke xxiv. 33) that they did. Josephus states (*Wars*, vii. 6, 6) that there was a village Emmaus sixty furlongs from the city. Robinson and Thomson coincide in the conjecture that this is the Emmaus of Luke, and that its site was at the modern village of Kuryet el Enab, the supposed Kirjath-jearim of the ancient Scriptures.—ED.

[1] Dr Barth, MS. 1841.
[2] Prokesch, *Reise*, p. 39.
[3] Robinson, *Bib. Research.* ii. p. 228, Note 3.
[4] Quatremère, in Macrizi, *Hist. des Sultans Maml.* T. i. p. 256.

would naturally make it and others of these inaccessible hills near by.

Von Wildenbruch[1] has shown that the position of Amwas and Latrûn is wrongly given on Robinson's map, the former really lying north of the road, the latter south of it. Wolcott has made the same correction. Amwas is a half-hour's distance from Latrun. At the latter place the real entrance is first made into the mountain-land of Judæa, through a pass about five hundred feet wide, shut in between limestone cliffs, which continually become higher and more bare. The way after this passes through Saris, a village little known, and lying on the high land.[2] It was long avoided on account of the robbers who settled there; but within recent years it has changed its character, has been rebuilt, and converted into a fine-looking village. The soil in the neighbourhood is under good cultivation, and yields excellent returns of olives and grapes.[3] The road then traverses the village of Karyet el Enab, which also used to be a perfect nest of robbers. It lost its bad name only so recently as 1847.

Prokesch, who visited[4] Karyet el Enab, speaks of finding spacious and well-built houses there, but noticed particularly a church of the Knights Templar, built with a triple nave, but used at the time of his visit as a storehouse for salt and a cattle-pen. Von Wildenbruch confirms the story of the original excellence of the architecture, and states, moreover, that it is still in a tolerably good state of preservation. The nave is thirty paces long and twenty-four wide: four square pillars on each side sustain the side naves with their roofs: the altar at the east side is lighted by a large window. Throughout the main walls of the church frescoes can be seen: they were well executed at the first, but only the blue has held well. Above the altar, niches and mosaics are set in the arched roof. Beneath the church are arches in a good

---

[1] Von Wildenbruch, *Reise-routen in Syrien*, in *Monatsb. der Berliner Ges. für Geog.* 1843, Pt. i. p. 229, and iv. p. 251.

[2] W. Turner, *Journal*, ii. p. 281.

[3] Dr Barth, MS. 1847.

[4] Von Prokesch, *Reise*, p. 41.

state of preservation. The French traveller L. de Mas Latrie[1] calls this the Church of St Jeremiah, and alludes particularly to the generally good preservation of all excepting the altar and of the pavement. A high castle, which flanks the city on the side of Bethlehem and Jaffa, is called the Pisano Castle.[2] The name of St Jeremiah, applied to the church, is confirmed by Sieber, who at the time of his visit in 1818 heard this appellation alone applied to it.

At the next village, Kulonieh (Colonia), whose site has already been spoken of, Wilson saw a fallen church, probably one which Prokesch speaks of seeing in ruins. Up to this point, where is gained the first view looking back upon the western plain, and the dark line of the sea, there are seen more or less striking marks of agriculture; but from that point on there are very few. Scattered groups of goats[3] may indeed be descried here and there, but the amount of pasturage which they find is very meagre. The hair of these creatures is fine, and black; the horns bent back, and striped with red; and the whole appearance of the animal is very striking.

From Kulonieh, around which a little agriculture is carried on, the road enters the deep Beit Hanina, and then winds towards the south-east, up from one layer of the white limestone to another, till the village of Lifta is reached. After this the traveller remains on the high plateau of Judæa. Towards the south-east there is caught a glimpse of the distant Mount of Olives, and at length of the western wall of Jerusalem, with the minarets, domes, and towers of the city. The view is the least imposing of any one which is gained in approaching this city, whither the pilgrims of all nations throng.

[1] L. de Mas Latrie, *Lettre du Cairo*, 17 Dec. 1845, in *Archives des Missions Scienti,. et Lit. Paris*, 1850, p. 166.
[2] Sieber, *Reise*, p. 33.
[3] Von Prokesch, *Reise*, p. 124.

DISCURSION II.

THE NORTHERN ROUTE FROM LYDDA—THE GREAT CARAVAN ROAD BY WAY OF THE PASS OF BETH-HORON AND EL-JIB (GIBEON)—THE BRANCH ROAD BY WAY OF WADI SULEIMAN.

In the great northern road from Lud to Jerusalem, by way of the pass Beit Ur, the ancient Beth-horon, Robinson must be our guide.

Lud or Lod,[1] an ancient Benjamite city (1 Chron. ix. 12), which was rebuilt after the captivity (Neh. xi. 35), has, in spite of all the changes of fortune, preserved its ancient name (Lydda), slightly modified, until the present day. By the Romans, Greeks, and Christian bishops it was called Diospolis. In its present condition it is a large village of small houses. There are still to be seen the ruins of a church once celebrated as that of St George. The western end has been converted into a mosque, whose high minaret indicates from a distance the situation of Lydda. The eastern walls, including the altar, are still standing. The intermediate parts have been destroyed, although several pillars yet stand together, with a high pointed arch, south of the nave. The width of the church is seventy-eight feet. It is supposed to have been the church built in Diospolis over the grave of St George, who is said to have been born there. The Bordeaux pilgrim[2] makes no mention of any burial-place in Lydda; but Antoninus Martyr, two centuries subsequently, speaks of the grave of St George as there. It was in Lydda that the gospel was early preached with great acceptance, and there, too, that the Apostle Paul restored the sick Æneas. It was from this place that he went to Joppa, near by, working miraculous cures there also. Subsequently this place, which had been raised to the rank of a provincial capital, came into possession of the Emperor Vespasian, and became a prominent centre of Jewish learning. One of the first episcopates in Palestine was that of Lydda or Diospolis, and the signa-

---

[1] Robinson, *Bib. Research.* ii. p. 244.
[2] *Itin. Burdig.* ed. Parthey, p. 283.

tures of its bishops are found dating as late as the year 518. The church is said to have been restored by King Richard of England.[1]

The wadi on which Lydda lies does not send its waters westward to Jaffa, but north to el-Aujeh, which enters the sea two hours north of that port.

The great caravan road runs[2] from Lydda direct to Jimzu, the ancient Gimzo (2 Chron. xxviii. 18), which the Philistines once took, and which has preserved its name down to the present day. The plain around Jimzu is very fruitful, and the harvest begins as early as the 9th of June. It is at this place that the southern route through Wadi Suleiman begins. This takes a direct course past Bersilya to el-Jib, where it unites with the northern road and passes through Beit Hanina to Jerusalem. The two roads lie but a short distance from each other, and are connected by means of short paths. Robinson took advantage of one of these to visit the upper and lower pass of Beit Ur or Beth-horon. The first one reached is lower Beit Ur. Massive stones still indicate the site of the ancient city,[3] whose erection was accomplished by Sherah, a daughter of Ephraim, and a grand-daughter of Jacob. The lower Beit Ur is separated from the foot of the adjacent mountain by a deep wadi which comes down from Ram Allah, and enters Wadi Suleiman farther south. Crossing this there is a long steep pass, which at first is very rough and rocky, but which subsequently assumes in many places the form of a stairway cut in the rocks, evidently of very ancient origin. On the way up are found foundation walls composed of large stones, and probably once belonging to a castle erected to guard the pass. After a full hour's climbing the highest point is reached, where stands the upper village of Beit Ur. This is on the extreme border of the mountain, and discloses a valley both on the north and south. Farther east, towards the high plain of

---

[1] See Robinson, *Bib. Research.* ii. p. 144 et sq. for the history of Lydda; and von Raumer, *Pal.* p. 190.

[2] Robinson, *Bib. Research.* ii. p. 248.

[3] *Ibid.* p. 249 et sq.

el-Jib, the ground rises very gradually towards the rocky mountain-land. The village is small, but shows traces of former buildings and an ancient reservoir. Both of these places are mentioned in Joshua as frontier cities; the upper Beth-horon being mentioned in Josh. xvi. 5 as on the borders of Ephraim, and the lower in Josh. xviii. 13 as on the limits of Benjamin. It is very plain from Josh. x. 10, 11, that after the battle with the five kings of the Amorites at Gibeon, the Hebrew leader pursued them to Beth-horon, and into the valley of Ajalon and Makkedah, where their rout was completed by a fierce storm of hail. A pass so valuable as Beth-horon was not neglected by Solomon, who, according to 2 Chron. viii. 5, and 1 Kings ix. 17, built the upper and the lower towns, and surrounded them with gates and walls and bars. The places escaped capture at the time of the Roman invasion; but in Eusebius' age they had become insignificant villages, and were only brought out into the light again by the labours of Clarke,[1] Nicolayson, and Robinson. The name has continued current in the mouths of the people for 3000 years. At the time of the Crusades, the place was called Bethar, Betheron, or Betelon, a name which is not to be confounded with Batrin in northern Gebail.[2]

The inhabitants of Beit Ur were fully occupied on the 10th of June, the time of Robinson's visit, with their harvest. From the upper town Lydda and Ramlah can be seen. Jaffa is invisible, however; but in the neighbourhood, the deep, green, and fruitful plain Merj Ibn Omeir is in full view, stretching away as far as Ekron.

From el-Gib, the ancient Gibeon, the road runs past the wall known as Bir es Ozeiz, through Beit Hanina, and over the Scopus, past the Graves of the Judges, and so to Jerusalem.

---

[1] E. D. Clarke, *Travels*, iv. pp. 424–427.
[2] Sebast. Pauli, *Codice diplom.* i. p. 420.

DISCURSION III.

THE NORTH-WEST ROUTE FROM JERUSALEM OVER THE MOUNTAINS OF EPHRAIM TO KEFR SABA, THE ANCIENT ANTIPATRIS—FROM BIREH AND JIFNA TO TIBNEH, ON WADI BELAT—THE BURIAL-PLACE OF JOSHUA—PAST MEJDEL YABA, RAS EL AIN, TO KEFR SABA.

The thoroughly traversed road from Jerusalem to Jaffa, the ancient harbour of the Jewish capital, where king Hiram of Tyre landed his cargoes of cedar-wood from Lebanon, which were intended for the erection of the future temple, has in modern times, as in the remotest period, drawn to itself almost all the commerce of the interior. In the intermediate period of Herod the Great, however, the more northerly harbour of Stratonis Turris, lying between Jaffa and Dora, and upon the river Choreus, became the chief port of the land, and through the efforts of the Emperor Augustus it was converted into a city of considerable splendour, receiving in honour of him the name of Cæsarea. During the following three centuries, while the country was under the sway of Rome, this port was the seat of the Syrian government; and after the destruction of Jerusalem by Vespasian it became the most important city of Palestine, and took the name Cæsarea Palestinæ, in contradistinction to Cæsarea Philippi, on the upper Jordan. It retained its importance for four centuries, became a chief episcopal residence, and at the time of Justinian was the seat of the primate. It was in this period that a direct commerce was instituted between Jerusalem, or Ælia Capitolina Hadriana, as it was then called, and Cæsarea, of which not even the Roman itineraries give any account: for in the fullest of them all, the *Itinerary* of Antoninus and the Peutinger Tables, there is no trace of it; and the still existing remains of military roads constructed in the time of the Roman supremacy, still continue the most exact and definite memorials of that time.[1]

In the Acts of the Apostles (xxiii. 23-35), it is stated that Paul, after being charged by the Jews of Jerusalem with

[1] Comp. *Itin.* ed. Parthey, pp. 276, 283 ; von Raumer, *Pal.* p. 131, Note 95 ; Robinson, *Bib. Research.* ii. p. 242, Note 7.

an offence worthy of death, and brought to judgment before Claudius Lysias the chief captain, was sent by him with an escort of four hundred and seventy men to Felix, who was then governor, and lived at Cæsarea, to be tried there. In vers. 31, 32, 33, we are told that "the soldiers, as it was commanded them, took Paul, and brought him by night to Antipatris. On the morrow they left the horsemen to go with him, and returned to the castle: who, when they came to Cæsarea, and delivered the epistle to the governor, presented Paul also before him." It is plain from this that there existed a military road at that time which could be traversed in one night (to avoid the heat of the day, as is still the custom in the Orient) as far as to Antipatris, twenty-six Roman miles south of Cæsarea Palestinæ. From Antipatris the prisoner was taken on the next day, passing from the high land to the coast. It is singular that a road so important as this should have fallen into oblivion so complete, that even the name of Antipatris should have been lost for centuries; that the whole district around it should have become a complete *terra incognita* to geographers; that even the maps of Robinson and Kiepert should display a mere blank in this quarter; and that Berghaus, in his otherwise admirable map, should be obliged to mark this region with groups of unauthorized mountains bearing the general name of Mountains of Ephraim. Even the excellent chart of Jacotin, which has long been the standard as to all that relates to the western coast, is here very defective. All had to be supplemented by the results of Dr Eli Smith's[1] journey through the hitherto unvisited mountain-land of Ephraim and its western slope. This journey was taken in April 1843 in conjunction with the missionary Calhoun, the object held distinctly in view being to trace the course of the Apostle Paul to Antipatris. This place, it was even then suspected, was identical with Kefr Saba, as Josephus says the name of the place known in Roman authors as Antipatris was previously Capharsaba.[2]

[1] Eli Smith, *Visit to Antipatris*, Letter 10th May 1843; Robinson, *Bib. Sacra*, 1843, pp. 478–498.
[2] Reland, *Pal.* p. 455.

ROUTE TO ANTIPATRIS. 245

First day's march.—From Jerusalem by way of Bireh, Jifna, Tibne (the Timnath of Joshua, and his grave), to Mejdel Yaba (thirty English miles).

Bireh, the ancient Beeroth, was reached in two hours from Jerusalem. From that point the regular Nablus road runs northward, passing Jifna, the ancient Gophnah. Of this route I shall have occasion to speak in subsequent pages.

The little river Balua, which runs north-west to the Mediterranean, was now dry. The country east of the great watershed as far as to Bethel was excellently tilled, and produced good returns of grapes and olives. Jifna is a little more than an hour and a half's distance from Bireh. In the place itself Smith found ninety Christians. Here the road begins to bear towards the north-west, and after passing a high hill Beir Zeit is reached,[1] where traces of an ancient Roman road are found. The road runs through a productive and well-tilled country, soon reaching the western margin of the high land, whence there is a view extending down to the plain. The intermediate region does not consist so much of ridges or lines of hills as it does of isolated ones, whose connection is not readily seen, the wadis which separate them not being discernible. This is the case through the entire descent to the coast plain. It was plain, however, that the path which Dr Smith took ran on a watershed between two wadis which are the great drainers of the whole region. The northern one is Wadi Belat, which begins at Jifna; the other bears the name Wadi Ain Tuleib. The first of these is the present boundary between the province of Jerusalem and that of Nablus.

The whole of the way down Smith found to be thoroughly comfortable, the first portion seeming to have a natural pavement of its own. Soon, however, marked traces of the old Roman military road are reached. It was nowhere in a perfect state, and the Americans were compelled to travel by the side of it, and not directly over it. Still Dr Smith asserts that he never discovered more continuous sections of a Roman road than here. This discovery confirmed him in his belief

[1] E. Smith, *Visit to Antipatris*, p. 480.

that he was on the old road running from Gophnah to Antipatris, and the one, moreover, which Paul must have taken on his way from Jerusalem to Cæsarea.

The first important locality which offered much interest in an antiquarian point of view, was a gentle hill crowned with ruins. These are extensive enough to indicate the former existence here of an important city. On the opposite side of the wadi there is a much higher eminence, on whose northern slope there are several excavations bearing much resemblance to what are called the Tombs of the Kings, near Jerusalem. The front of every one had a portico, supported by two pillars, the whole cut out of solid rock. There was no time to open one of these tombs and examine its contents, which is much to be regretted, as there do not seem to be others of a similar character in all the rest of Palestine. The locality bears the name Tibneh—the same appellation which characterized the home of Samson's wife on the confines of Judah and Dan, and north-west of Bethlehem. There were many places which bore the name of Timnath, and this must have been one in the territory of Ephraim.[1]

Josephus, in speaking of the different toparchies of Judæa, names the following: Jerusalem, Gophna, Acrabatta, Thamna, Lydda, and Ammaus. He writes Thamna for Timnath; and in another passage the name comes yet more prominently forward, when he speaks of the cities whose inhabitants were sold into slavery by the Roman Prætor Cassius: they were Gophna, Emmaus, Lydda, and Thamna. The name occurs also in 1 Macc. ix. 50 as that of one of the cities which Bachides fortified. In describing the advance of Vespasian from Antipatris, where he spent two days, Josephus says that he first ravaged the toparchy of Thamna, and then marched by way of Lydda, Jamnia, and Emmaus to Jerusalem. All these data show that the present Tibneh is no other than the ancient Timnath or Thamna, which is spoken of in connection with Emmaus, Beth-horon, Gophna, Pharathan, and Antipatris, near which it lies. The province of Thamna must have bordered on

[1] Robinson, *Bib. Research.* i. p. 17, Note.

the plain, since Vespasian was able to ravage it while on his march.

In the book of Judges (ii. 8, 9), it is stated that when Joshua the son of Nun had died at the age of a hundred and ten years, he was buried " on the border of his inheritance in Timnath-heres, in the Mount of Ephraim, on the north side of the hill Gaash." These words are borrowed from the close of the book of Joshua (xxiv. 28, 29). In xix. 49, 50, it is stated that after the territory had been distributed to the different tribes, Joshua received Timnath-serah, in Mount Ephraim, where he built a city and spent[1] the remnant of his days. These two places are probably identical, as Reland has shown. The site of Gaash and of the Timnath which Joshua built in the mountains of Ephraim had become so completely forgotten at the time of Eusebius and Jerome, that Paula was told that the grave of Joshua was at the more southern Thamna, the one near Beth-shemesh, and in the low land west of Judah, *i.e.* at Tibneh, in the territory of Dan. There is very little doubt that Dr Smith[2] has discovered in the Tibneh, on the road from Jifna to Kefr Saba, the home and the burial-place of Joshua. The mountain on which it lies cannot, therefore, be any other than the Gaash of the Old Testament, although no name seemed to be given to it by the people of the neighbourhood. Still it must be confessed that more investigation is required before we can decide on the antiquity of the tombs found there. Robinson[3] doubts the identity of Timnath-heres and Thamna, the capital of the toparchy of the same name, although he does not doubt that Tibneh and Thamna are the same.

The first day's journey ended at the village of Mejdel Yaba, thirty miles from Jerusalem.

Second day's journey.—From Mejdel Yaba, by way of Ras el Ain and Wadi Aujeh to Kefr Saba, the supposed Antipatris.

[1] Keil, *Comment. zu Josua*, p. 357; comp. v. Raumer, *Pal.* pp. 148, 149.
[2] E. Smith, *Visit to Antipatris*, p. 485.
[3] Robinson, Note in *Bib. Sacra*, 1843, p. 496.

The village of Mejdel Yaba lies upon the top of a hill north of the Belat valley, and contiguous to the plain of Sharon on the west. The house of the sheikh of the place, standing on the ruins of an ancient stronghold, was found to be solidly built, but had been much injured in the course of quarrels with the sheikhs of the neighbouring villages. Dr Smith inferred that Mejdel must formerly have been a place of no inconsiderable pretensions; and the word itself indicates tower or stronghold.

Instead of going directly north from this place to Kefr Saba, Smith[1] and his companion bore to the right, and visited Ras el Ain. It took but a few moments to descend from the hill to the plain, through which the road then ran for a distance of forty minutes. Ras el Ain is a hill which rises from the level expanse, and whose top is almost entirely covered by a square structure, having at a distance the appearance of an old khan like that at Ramleh: loopholes and towers at the corners show, however, that it was a fort. At the western foot of this hill, in a small morass full of sedge and rushes, begins the Wadi Aujeh. Here is one of the largest springs, or series of springs, which Dr Smith had ever seen; in fact, so large as to supply almost all the water which runs through the wadi. While all the other valleys in the neighbourhood were dry, there was so much water here that it could be forded only in certain places, and the channel was as broad as that of the Jordan at Jericho. Its water has a bluish colour, the rate of flowing is very leisurely, and yet the rate of fall is sufficiently great to allow it to drive several mills. It enters the sea a short distance north of Jaffa; and at its mouth, von Wildenbruch[2] tells us that, although at the time when he passed through this region all the other watercourses were dry, the depth of the water in this was so great as to make it impossible to cross it.

The road ran then northward through the middle of the plain. On the east side of this, the mountains of Samaria rise

---

[1] E. Smith, as quoted above, p. 491.

[2] Von Wildenbruch, *Reiseroute in Syrien*, in *Monatsb. d. Geogr. Ges. in Berlin*, Pt. i. p. 232.

gradually, and form the edge of the celebrated plain of Sharon. Towards the north-west there is a row of low wooded hills, which separate the level tract from the sea. The soil is a black loam, almost everywhere tilled, and of so great fertility, that it might easily become the granary of the whole country. Fields of wheat and barley reach away farther than the eye can pierce, with here and there a tract of cotton interspersed.

Forty minutes brought the travellers to Jiljûlieh.[1] This place lies upon a very low and broken range of hills, extending westward from the mountains on the east quite to the middle of the plain. The village is now a small one, but was plainly at one time of considerable importance: its population is entirely Mohammedan. On the south side there is a khan, a structure of some antiquity, and which formed one of a series of caravanserais extending along the whole of the old route from Gaza to Damascus, by way of the coast, the plain of Esdraelon, and Scythopolis.

In the middle of the deserted khan is a great round well, and at the entrance to the court of the building are the ruins of a minaret. This place Smith conjectured to be identical with the Gilgal mentioned[2] in Josh. xii. 23, where dwelt the "king of the nations," and in whose neighbourhood were Carmel and Dor. The *Onomasticon*, indeed, speaks of a Gilgal six miles farther north (Villa nomine Galgulis ab Antipatride in sexto milliario contra septentrionem), while this one lies about two miles south-east from Kefr Saba. On this ground von Raumer proposed, before Smith had made his careful examination of the region, to read "contra meridiem." There is, it seems, a village of Kilkilia lying north-east of Kefr Saba, but the orthography of the name hardly makes it possible to suppose that it corresponds to the place mentioned in the *Onomasticon*.

The next place of antiquarian interest met by Smith and Calhoun was Kefr Saba. The village, like all the others of this plain, is composed of houses built entirely of earth, there

---

[1] Von Wildenbruch, p. 492.

[2] Keil, *Comment. zu Josua*, p. 287; Robinson, *Bib. Research.* ii. p. 243; v. Raumer, *Pal.* p. 139, Note.

being but a single one of stone. This seemed to have once been a mosque, although it was entirely destitute of a minaret. No other traces of antiquity were observable. The only thing which bore the marks of care in the making, was a well east of the village: it was surrounded with a wall of hewn stones, and was fifty-seven feet deep to the water. The village stands upon a slight elevation near the range of hills along the sea-coast, and separated from them by an intermediate plain. Only one wadi could be seen, a small one, running in the direction of the Aujeh. The land in the neighbourhood seemed to be in an excellent state of tillage, and to be very fertile.

As Josephus states that the name borne by Antipatris, before it had received the designation given it by Herod, was Χαβαρζαβά, there is the whole strength of the argument derivable from the identity of the words,[1] that the modern village of Kefr Saba occupies the site of Antipatris. Still there is no certainty that, in the course of time, the place designated may have been lost, while the designation has remained. In some places the old name has been transferred to a spot in the neighbourhood; as, for example, that of Hebron has been carried from the mountain-land down to the valley. At Jericho or er-Riha there has been a change; and Sarepta, which lay on the sea-coast, is now found in the name Sarafend, which is applied to a place on the hills removed a considerable distance from the sea.

The extremely accurate and observing Dr Smith conjectured, even while descending the hill of Mejdel Yaba, that there were the natural conditions of the ancient city of Antipatris, the position being a very commanding one. It was there that he emerged from the mountain land, and entered the more secure region of the plain. The distance, too, from Jerusalem (thirty miles) seemed great enough. The whole ride to Kefr Saba, eight miles farther, would have been a very trying one for a single night, unless it were protracted into the next morning. And this, too, would make correct

[1] Reland, *Pal.* p. 690; von Raumer, *Pal.* p. 131; Robinson, *Bib. Research.* i. p. 401, Note, and in *Bib. Sacra*, 1843, p. 497.

and plain what Jerome says regarding the position of Galgulis, six Roman miles north of Antipatris. The statement of the *Bordeaux Itinerary*, that Lydda was ten Roman miles from Antipatris, would also coincide with this theory. The fact that Alexander Jannæus undertook to cut off the march of Antiochus from Syria to Gaza, by excavating a trench a hundred and fifty stadia long, and extending from Antipatris to Joppa, makes it the more certain that Kefr Saba cannot be the site of the Roman city. To construct a trench of any strategic value from this place to the sea, would be a work of the greatest difficulty, the distance being seven and a half hours; and besides, the work would be useless, on this ground, that the enemy would pass around the eastern end of it, there being no necessity to cross it at all. It is quite otherwise if Mejdel Yaba be considered the site of Antipatris; for in that case the trench would only need to be carried for a distance of two English miles to the Ras el Ain, where it would connect with the Wadi Aujeh, itself one of the most formidable natural barriers to an advancing army.

Yet there are other places in Josephus which militate against this view; as, for example, the one in which he states that Herod built the city called Caphar-Saba in the plain. In another passage he says that this plain was the finest part of his dominions,—an assertion that could by no means be made of Mejdel. Moreover, according to Josephus, Antipatris was surrounded with water, and with fine forest trees. There are only two brooklets in the neighbourhood of Kefr Saba, and these were dry in April, although in the winter-time they might perhaps answer to Josephus' description. There is a well there, moreover, fifty-seven feet deep; while all the drinking water which is used at Mejdel must be brought from Ras el Ain, two miles away. The great argument, however, is the identity of the name—not absolutely conclusive, indeed, but of great weight. In view of all the circumstances of the case, Smith concludes that Kefr Saba best answers to the location of Antipatris and the Caphar-Saba of Josephus.[1] It is a remarkable fact, however, that no traces of buildings

[1] Comp. v. Raumer, *Pal.* p. 130, Note 95.

are to be found, since the city was built with Herod's customary magnificence, if not with even greater splendour than he ordinarily displayed.

It may be remarked, that subsequently to Dr Smith's exploration of this neighbourhood, von Wildenbruch[1] passed through it, but was unable to find any trace of a place bearing the name of Kefr Saba. Nearly all the other particulars given by Smith respecting this region were fully confirmed, however.[2]

[1] Von Wildenbruch, *Reiseroute in Syrien*, in *Monatsb. d. Geogr. Gesell. in Berlin*, i. p. 233, with Plate v.

[2] Robinson, in his *Later Researches*, pp. 138, 139, sees no reason for doubting that Kefr Saba marks the site of the ancient Antipatris. Among the recent authorities who have pronounced upon this question, Van der Velde's voice is emphatically in favour of Mejdel. See Mem. to his Map, p. 285. As the region is not yet thoroughly explored, it may be that some light may still be thrown upon the difficulty.—ED.

CHAPTER III.

THE COAST PLAIN, FROM THE PLAIN OF PHILISTIA TO THE CARMEL RIDGE.

SEPHELA AND SAROM, WITH THEIR CITIES AND MAIN HIGHWAYS—JOPPA OR JAFFA—RAMLEH AND THE PLAIN OF SHARON—THE EASTERN OR MOUNTAIN ROAD AS FAR AS THE PLAIN OF ESDRAELON—THE WESTERN OR COAST ROAD TO CÆSAREA, AND THE PROMONTORY OF CARMEL.

WE have already followed the coast road from Gaza past the cities of Askelon, Ashdod, Jabneh, and Ekron, as far north as Nahr Rubin. We now continue our course northward over the coast plain of Sharon, and along the side of the Syrian mountain ridge as far north as the Carmel range, which forms the natural barrier between Samaria and Galilee.

The coast plain of Sephel was entirely in the possession of the Philistines as far north as the Nahr Rubin, which ran between Akir (Ekron) and Yebna. North of it lay Phœnician cities, to which the Philistines seem to have laid no claim; all their efforts at conquest having been made on the eastern or Judæan side. There is not a trace of an expedition having gone northward. The nearest Phœnician cities were Joppa and Dor. These came subsequently under the sway of the Israelites; but even there they appear to have been for a long time under Phœnician influence.

DISCURSION I.

JOPPA AND RAMLEH.

1. *Joppa, the Port of Jerusalem.*

Joppa, lying on the meridian of 32° 2′ north lat.,[1] and six hours north of Jamnia—the present Jebnah—was the

[1] Niebuhr, *Reise*, Pt. iii. p. 41.

most important commercial place on the whole coast. It is a city of great antiquity, and is characterized by Pliny in the following words: "Joppe Phœnicum antiquior terrarum inundatione, ut ferunt. Insidet collem præjacente saxo, in quo vinculorum Andromedæ vestigia ostendunt. Colitur illic fabulosa Derceto; inde Apollonia;" and then he adds: "Cæsarea ... finis Palestinæ ... deinde Phœnice,"—an indication that he wished to discriminate between the Phœnician city and Phœnicia itself. According to his account, the place was standing before the Deluge. Pliny, as well as Strabo, doubles the "p" in the name, although it should probably be written with but one,—Jope meaning, in the Phœnician language, high place.[1] Strabo, who never was in Palestine, describes the situation of the place truthfully, except in respect to the height at which it lies. He remarks that Jerusalem could be seen from it.

In the book of Joshua the place is known under the name of Japho (Josh. xix. 46), the 'Ιάφα of the later Maccabees, whence came the orthography Jaffa,[2] Jafa, which, at the time of the apportioning of the territory of Dan, in which it lay, was rejected. There is no existing trace of evidence that this important commercial city was in the earliest times in the possession of the Philistines or the Israelites; but the legends anciently current there relating to a former deluge of the place, to Andromeda, and to the visit of a huge marine monster, appear to indicate that at a very early period the Phœnicians had established business relations with the place. The assertion of Hiram the king of Tyre, that he "will cut wood out of Lebanon, and bring it in floats by sea to Joppa," whence it was to be carried up to Jerusalem, shows the closeness of the connection between the Tyrians and this port. This connection existed until the time of Cyrus; for the roads of Joppa were the anchoring-place for the Phœnician ships which brought the cedar-wood for the rebuilding of the temple (Ezra iii. 7). The prophet Jonah embarked in a ship

[1] Movers, *Die Phönizier*, Pt. ii. p. 177; Hitzig, *Die Philistäer*, pp. 131-134.
[2] Keil, *Comment. zu Josua*, p. 356.

at Joppa, wishing to escape from the command of the Lord which had called him to Nineveh, and to go to the Tyrian colony of Tartessus or Tarshish. As this ship belonged to men who did not believe in Jehovah, it is plain that it was manned by Carthaginians. Indeed, the narrative presupposes that Joppa was a Phœnician harbour. It was subsequently used as a port for transporting the corn of Judæa to Tyre and Sidon, which was paid for by gold sent in government ships. Only the presupposition of a very ancient connection of this place with Jerusalem can explain the allusion of Herodotus to Kadytis, as if Jerusalem itself were meant.[1] And, indeed, a seafaring nation like the Phœnicians had good reason to establish a colony here, where, as Strabo remarks, the coast-line turns from the west more to the north, where two good springs were found, and where were the best harbours, poor though they were, that were found upon this whole coast. Still it is impossible to conjecture, from its present condition, what it may have been at an earlier period.[2] It only passed into the possession of the Israelites at the time of the Maccabees, since which period its destinies have been connected with those of Palestine. The older form of the name, Jope or Joppe, has been used interchangeably with the later and more general form, Jafa or Jaffa, down to the present time.

The present Jaffa has experienced many changes: it has often been destroyed, and again rebuilt. Through all the centuries since Christianity was introduced into Palestine, it has been a prominent resort of pilgrims. It was the first city that was taken by Godfrey of Bouillon, who gave it to the Church of the Holy Sepulchre,[3] together with many other places taken subsequently,—all of which again reverted to the Mohammedans in the time of Saladin. When Rauwolf[4] visited Joppa in 1573, he found it wholly destroyed, not a house being left standing; and it was hard for him to believe

---

[1] Krafft, *Topogr.* p. 143.

[2] Niebuhr, *Reise,* iii. p. 42.

[3] Sebastiano Pauli, *Codice diplomatico, etc., de Regno di Gerusalemme,* i. p. 441.

[4] Rauwolffen, *Reyss, etc.,* p. 15.

that there was the site of a city which had once been large and prosperous.

Yet the favourable position of Joppa, as the only harbour of all central Palestine, could not allow it to remain long in ruins. At Niebuhr's time (1766) there were between four and five hundred houses, and several mosques. A marsh lying in the neighbourhood had been drained and converted into gardens, making the atmosphere more healthful than it had been before. The houses on the sea-shore were built of limestone. The most of them lay on an eminence, at least a hundred and fifty feet high. A couple of unimportant towers served to guard the town. The harbour, always a dangerous one, had been so filled up with rubbish, that ships were obliged to anchor half an hour's distance away, and still were not always secure, even in the summer-time. It seemed to Niebuhr [1] as if the shore had very much changed; and an old man assured him that he could remember the time that small craft could come close to the shore, and anchor where it was then perfectly dry. A stone quay [2] of great antiquity is to be seen, from which a flight of steps descends to the strand. The water is now twenty paces from the extreme end of this quay. Dr Barth, who has recently visited Jaffa, remarks that the contracted basin forming the harbour of this place secures very little safety to ships, and that the men who attend to the unloading are compelled to wade in the water up to their shoulders, to get the goods to the land. The Egyptian vessels which come hither never find it easy [3] to discharge their cargoes; and even their passengers must sometimes be carried ashore on the shoulders of men. The commerce of the place is therefore now very unimportant. Lusignan speaks [4] more fully about the harbour privileges of Jaffa, and shows that its very want of safety was what converted it into the nest of sea pirates which both Josephus and

---

[1] Niebuhr, *Reise*, iii. p. 42.
[2] O. v. Richter, *Wallfahrten*, p. 10.
[3] Sieber, *Reise*, p. 17.
[4] S. Lusignan, *Letters*, ii. p. 79; Wilson, *Lands of the Bible*, ii. p. 257; Buckingham, *Trav. in Pal.* i. p. 229.

Strabo declare that it was. He says that the port extended from north to south, and lay close to the main town; that it had two approaches by sea, the northern one of which was the widest but the most dangerous, because it was obstructed by sand-banks; the second, at the west, was very narrow, and had a channel of only about ten feet deep. Browne [1] says that there are extensive coral formations near Jaffa. The northern part of the basin, in which vessels anchor, is only about twenty paces wide; the southern not even so broad as that, but a little safer, since it is in part sheltered from the wind by high rocks, and by the wall of the town. The very slight depth of six, and at most ten feet, is the result of the filling up which can be seen in some places going on. If the harbour were artificially deepened, it would be easy to find room for fifteen or sixteen vessels, of a hundred and fifty tons each, to lie at anchor. When Vespasian had passed through Syria with his army, and had taken possession of Tiberias, he sent his general, Cestius, to Jaffa, to exterminate the pirates of that port, who were ravaging the whole coast. The pirate fleet, closely driven together in the little harbour, became the victim of a vehement storm from the north,—the melansboreas of Josephus, and the black boreas of even the present day. Its violence was so great as to permit none of the vessels to encounter the waves which rolled on from the west, and the pirates all perished. The statements in the defective passage of Strabo (xvi. 759) seem to have related to this occurrence. The rocks, which even in the time of Jerome bore the name of the "Place of Deliverance," are the same where the fettered Andromeda is said to have been released from Perseus,—a proof of the existence of ancient Assyrian mythology on this coast. In the Acts of the Apostles Joppa is mentioned in well-known connection with some of the incidents in the life of Peter (Acts ix. x. xi.).

It is easy to see what perils must have been encountered during the long centuries of the middle ages by the thousands of vessels which landed at Jaffa with their tens of thousands

[1] W. G. Browne, *Reise in Darfur und Syrien*, trans. by Sprengel, p. 353.

of crusaders and pilgrims; and how formidable those perils must continue to be down to the present day.

The houses and walls of the present Jaffa, which lies on a long ridge rising in terraces, are by no means insignificant; yet when one enters the town he finds a confused mass of modern houses built in that unattractive manner which characterizes the towns of the East. Russegger[1] conjectured that the population was between two and three thousand. The view from the top of the hill is enchanting. On one side is the emerald-green sea, on the other the fertile plain of Sharon, seamed with shallow watercourses; in the immediate foreground there are dense groves of fruit trees, in the distance there are the blue hills of Judæa and Ephraim. The combined beauty of all these objects is some compensation for the repulsive aspect of the streets of the city itself. At the east gate there is a fountain of white marble, finely ornamented in the Saracen style,[2] and bearing a gilded Arabic inscription. The best houses of the merchants and the best shops are on the sea side of the town, under the protection of the battery. Here, too, are the dwellings of the European consular agents; here is a Franciscan convent for Catholic pilgrims, and a small Greek and Armenian one. The bazaar is well supplied with goods, and the gardens which surround the city furnish it with a great abundance of fruit. When Browne[3] was here at the close of the last century, the trees which had formerly been the great charm of Jaffa had been converted into firewood by the Mameluke soldiery which had invested the place. Yet they shot up again with such rapidity, that when Russegger was here in 1838 the lemon and orange trees were bending under the weight of their burdens, and the red pomegranates filled the air every evening with fragrance. The irrigation is effected by means of wheels. The figs and oranges of Jaffa are noted for their size and flavour. The water-melons, which thrive on the sandy soil around, are in great repute,[4] and are carried

---

[1] Russegger, *Reise*, iii. p. 118; Richter, *Reise*, p. 10.
[2] Sieber, *Reise*, p. 20.   [3] Browne, *Reise in Darfur und Syrien*, 352.
[4] Eli Smith, in *Bib. Sacra*, p. 495.

in great numbers to Alexandria and Cairo. Through all Syria, too, they have a reputation. A camel-load of them cost in Sieber's time (1818) only a few pence. The vegetables of Jaffa, too, are abundant and cheap: the soil yields as freely as it did centuries ago. The horticulturist Bové,[1] who visited the place in 1832, was surprised at its great fertility. He observed three kinds of figs, apricots, almonds, pomegranates, peaches, oranges, pears and apples, plums, bananas, and grapes. The sugar-cane grows to a height of five or six feet, but no sugar is made from it. The vegetables raised are chiefly tomatoes, maize, and cabbage. All the gardens are hedged with the thorny *cactus opuntia*, which makes the best fences, since the wood grows to the size of small trees. Bové was surprised to see that the agave, or North American aloe, which had thriven so well in the western basin of the Mediterranean, was almost entirely lacking in Palestine and Syria.

Wilson found[2] only a small number of Jews in Jaffa, about twenty-six families in all, comprising a hundred and twenty souls, making an insignificant portion of the entire population of the place, which he estimates at five thousand. The Sephardim Jews had mostly come within recent years from North Africa, and plied the occupations of tradesmen, silk-workers, and carpenters. The silks of Jaffa have for a long time had a good reputation in the East. The Armenian Convent in Jaffa,[3] once the hospital of the French, is shown as the place where Bonaparte caused his wounded men to be poisoned, in order to prevent their falling into the hands of the victorious Turks; and a quarter of an hour's distance from the city is the camp where he ordered the inhabitants of the place to be hewn down in cold blood, after the town had fallen into his hands.

[1] Bové, *Bulletin*, quoted above, iii. p. 381.
[2] Wilson, *Lands*, etc., ii. p. 258; Kinnear, *Cairo*, etc., p. 214.
[3] Irby and Mangles, *Trav.* p. 184; Buckingham, *Pal.* i. p. 249.

2. Ramleh or Ramula.

South-south-east from Jaffa, past the fruitful fields of Jesor,[1] past Kubab, Beit Dejan (Beit-dagon), and Surafend (Sariphæa), at a distance of three hours, lies Ramleh, on the great commercial thoroughfare from Egypt to Damascus, and at the junction of the main road from Jerusalem to Jaffa. The situation is one, therefore, which would always make the place of no little importance. The whole of the road from Jaffa thither is sandy, yet the ground is rolling, and under good cultivation. The towns and villages on every side are generally built on knolls, and display some traces of antiquity. One of these is Surafend, which occupies the site of the ancient Sariphæa, which, according to Reland, was the centre of an important bishopric.

Like Gaza and Jaffa, Ramleh is surrounded by fruitful[2] gardens and the finest oranges and other fruits. Browne speaks of seeing olive trees here, together with carobs and sycamores. Sieber saw palm trees at Ramleh, but says that the climate is too cold for them to bring their fruit to maturity. The tufted tops are finer in appearance, he asserts, than even those of Egypt.[3] The town with its white-washed houses lies on the eastern side of a low and broad hill rising out of the sandy but fruitful plain. The streets all sink towards the east. There are stone houses neatly built in some of the streets; scattered among them are several mosques, which probably occupy the site of earlier churches; and here, too, is found one of the largest Latin convents[4] in all Palestine, surrounded by high walls. This was founded as a hospice in 1420 by Philip of Burgundy, but since the eighteenth century it has been used as a home for Franciscan pilgrims. Its church is very small. The convent pays to the governor, in return for his military protection, a hundred piastres and three yards of cloth.

---

[1] Robinson, *Bib. Research.* ii. p. 229 et sq.; Wilson, *Lands, etc.*, ii. pp. 259-263.

[2] Robinson, *Bib. Research.* ii. p. 242.

[3] Sieber, *Reise*, p. 27.   [4] Prokesch, *Reise*, p. 87.

The place has a population of three thousand inhabitants, including a thousand Armenians and Greeks. The water is supplied from good cisterns. About ten minutes west of the city, on the highest spot in the neighbourhood, and among the ruins of a large square wall which seems to have once belonged to a stately khan, rises the celebrated tower which far and wide announces the situation of Ramleh.

The former town of Ramleh, or Ramula as it was called, appears, judging from the extent of the ruins and from the number of cisterns, to have been a place covering much more ground than does the present one. One of the structures still partially remaining, north of the city, has, according to Prokesch, twenty-four arches still standing, and is ascribed, as so many other buildings in Palestine, to the Empress Helena. Scholtz[1] says that this ruin is of remarkable size and strength, it being thirty-three feet long and thirty feet wide. In the neighbourhood of the Frank Convent there are very large cisterns, giving an ample supply of water for the city.

At the time of Jerome this place was pointed out to Paula, while she was journeying in the neighbourhood of Lydda, as the site of Arimathæa, the residence of Joseph, the rich friend of the Saviour, who begged the body of Jesus, and wished to lay it in his own tomb (Matt. xxvii. 57; John xix. 38). Whether this Ramleh or Ramula existed in the New Testament times, or whether it was built subsequently, it is now very difficult to determine, in consequence of the number of places known by this name. It may only be said here, that Reland as well as Robinson[2] doubt the location of Arimathæa there, and suspect that the Toparchia Thamnitica lay farther north. The latter could discover no evidence that it was south of Diospolis (Lydda), since this was the capital of a toparchy. In spite of all the learning which von Raumer[3] has devoted to the identification of Arimathæa, he has been unable to solve the question with any certainty.

---

[1] Scholtz, *Reise*, p. 149.

[2] Robinson, *Bib. Research.* ii. pp. 234, 239 et sq.; *Bib. Sacra*, 1843, Note to p. 565.

[3] Von Raumer, *Pal.* p. 197, and in App. p. 405.

The name Ramleh, *i.e.* the sandy, is first met in the year 870 in the account of Bernardus, *de Loc. Sanctis;* and this name Ramleh, which Rödiger[1] decides cannot be derived from Ramah, appears to confirm the comparatively modern origin of the place. According to him, the name is not at all ancient, but a genuine modern name, which denotes the sandy nature of the soil; while the Hebrew Ramah indicates a "high place," and is as little applicable to the situation of Ramleh as it is to any other place in the whole plain along the coast. William of Tyre[2] expresses his doubts as to the antiquity of the place (called by him Ramula), and states that at the time it was taken by the Christians, the Mohammedan population fled for safety to Askelon, and the city was converted into a stronghold.

The Arabian authors, Abulfeda[3] among them, assert that this Ramleh is no ancient city, but was built by Abd el Melek, who built a palace there after Lydda had fallen into decline; yet it is possible to believe with Clarke, that only the restoration of a former city was meant by those writers. At all events, since the time when they wrote, Ramleh has become a large and important place. According to Edrisi[4] and Abulfeda, it was the joint capital of Palestine, and a great centre of trade. It was in that flourishing time that the important structures whose ruins are now standing were erected. At the era of the Crusades, Ramleh had a citadel and twelve gates; with important bazaars at those which led to Jaffa, Jerusalem, Ascalon, and Nablus.

Robinson[5] has given the history of Ramleh down to its decline. According to his view, the ordinary belief that the high tower of this place, with the ruins lying around it, was a church of the time of the Crusades, is entirely unfounded; as well as the supposition that it was a church founded by the Empress Helena. He asserts that it is of Saracen

---

[1] Rödiger, in Rev. already quoted, p. 576.
[2] *Gesta Dei*, ii.; Will. Tyr. *Hist.* x. 17, fol. 785.
[3] Abulfeda, *Tab. Syriæ*, ed. Koehler, p. 79.
[4] Edrisi, in Jaubert, p. 339; Mejr ed Din, in *Fundgr.* ii. p. 135.
[5] Robinson, *Bib. Research.* ii. p. 235 et sq.

architecture, and before the year 1555 no one ever called it a Christian edifice. Scholtz[1] says that three hundred paces west of Ramleh (which he calls Rama) lie the ruins of the great building known as Jamea Elabidh, or the Church of the Forty Martyrs. It was six hundred paces in length, and was erected at the time of the Crusades by the Knights Templar. There are still to be seen the churches above and below ground, with their nine pillars and two naves, the subterranean dwellings, magazines, and cisterns, and the external walls with the cells. In later times the Arabs converted it into three mosques, as appears from the inscription: the largest lies in the northern, the other two in the southern part of the square building, between which parts there were two chapels. The wall of the lofty minaret, to whose top a hundred and twenty-five steps lead, is inferior in strength and beauty to the one which was built by the Christians. The authorities had in vain tried to use the ancient masonry, but could not overcome the difficulties that lay in the way of working it.

Robinson[2] alludes to the great skill in the workmanship, to the massiveness of the subterranean masonry, and to the rooms, which he considered storehouses for wares. The tower, he says, stands at present all alone: it is Saracenic in its architecture, is square in form, and is composed of well-hewn stones; the windows are various in form, and all of them arched. The corners of the gate are supported by long slender pillars, the sides broken by several storeys, and all become narrower as they ascend. There is a staircase of a hundred and twenty steps inside, leading to the gallery, which is at least that number of feet from the ground. Robinson compares the extensive view from the top with that from the Milan Cathedral over the plain of Lombardy. At the east are the steep mountains of Judæa, at the west the blue waves of the Mediterranean. Southward the view extends over the fine Philistine territory, and northward the plain of Sharon stretches away as far as the eye can reach,

[1] Scholtz, *Reise*, p. 148.
[2] Robinson, *Bib. Research.* ii. p. 230; Prokesch, *Reise*, pp. 37-39.

covered with its green carpet of millet, or dotted with corn and cotton[1] fields. Directly beneath are the flourishing gardens and groves of Ramleh. It is only a short distance to Lydda at the north, and to Akir at the south with its pleasing minarets. The whole landscape is a beautiful one, particularly in the light of the setting sun.

The French traveller and antiquarian De Mas Latrie,[2] who has recently inspected the tower, declares it to be a masterpiece of workmanship, dating from the time of the Knights Templar when in their power and prime: he thinks it was intended to serve as a hospice for the pilgrims who visited the Holy Land, and that, at the same time, it was meant to serve as a fortress in case of assault. This, he thinks, is proved by the small windows, which could also serve the purpose of port-holes.

Rödiger,[3] on the contrary, judges from an inscription on the tower, that it is a work of Saracen origin. Von Wildenbruch[4] regards it as a work of the Templars, and thinks that the inscription was added subsequently. A copy of this has been translated by Professor Larsow, and runs as follows:—

"This tower was begun by the Sultan Abul Fetach Mohammed, son of Sultan Said Malek el Mansur Saif eddonia Wa ed din (the Sword of the World of Faith), our Lord of the Salechite Hasem, Prince of Faithful: may God prolong his days, and extend his victorious banners. And ended was this tower in the middle of the month Shâbûn, in the year 718 of the Hegira" [*i.e.* A.D. 1318].

The Sultan Abul Fetach Mohammed mentioned in this inscription is of the dynasty of the Baharidic Mamelukes of Egypt, whose power was extended over Syria from A.D. 1293 to 1341. It is probably the same one who repaired the aqueduct at Jerusalem. Thus it seems that not all the doubts are solved regarding this remarkable tower, which is

---

[1] Rauwolf, *Reise*, p. 19.
[2] L. de Mas Latrie, in *Archives de Miss. Scientif. et Lit.* p. 106.
[3] Rödiger, in Review quoted above, p. 575.
[4] Von Wildenbruch, in *Monatsb. d. geogr. Ges. in Berlin*, i. p. 239.

interesting not only for its antiquarian character, but also for the extensive prospect which it affords.

DISCURSION II.

THE PLAIN OF SHARON, AND THE ROADS WHICH TRAVERSE IT: THE GREAT DAMASCUS HIGHWAY OVER MOUNT CARMEL TO THE PLAIN OF ESDRAELON.

From the three central points which have now been named, there stretches away to the north as far as the Carmel ridge, and between the mountains of Ephraim and the Mediterranean, the great maritime lowland of Palestine, the largest part of which is usually known under the appellation of the plain of Sharon.

The portion of the coast-plain extending southward from Eleutheropolis to Gaza, Gerar, and Beersheba, was known as Darom or Daromas: the name Sephela was given at the time of Jerome to all the coast-plain from Eleutheropolis north and west as far as Askelon and Jerusalem, and even as far as to Jabneh and Akir, one of the most fertile and best tilled[1] portions of the whole country. Saron or Saronas was the name given to the tract extending from Joppa to the Carmel ridge. The latter separates it from the plain of Esdraelon. In a more limited sense, only the immediate neighbourhood of Lydda and Joppa was called Sharon—a tract which was and still is of almost unexampled fertility. It was in Sharon that David found pasturage for his flocks (1 Chron. xxviii. 29).

Isaiah praises the glory of the Lebanon, and the beauty of Carmel and Sharon. The Song of Solomon extols the rose of Sharon and the lily of the valleys. The old beauty of the place has continued down to the present day; but the plain has become a solitude, and a soil rich enough to supply all Palestine with food is in great part untilled. On this account an effort has been made of late to colonize the district with Germans,[2] there being not only the best of land

---

[1] Sieber, *Reise*, p. 19.

[2] *Missionsblatt des Rhein-Westph. Vereins für Israel*, July 1850, No. 7.

and an abundance of water, but also many walls and arches which would afford a temporary shelter. Besides this advantage, there is not an absolute deficiency in wood, although in all probability it was once more plentiful than at the present time. In fact, a part of the plain once bore the name $Δρυμὸς$, *i.e.* the forest; and this portion,[1] as well as Mount Carmel, fell into the hands of Vespasian at the time when he vanquished the pirates of the coast. Even at the present day there is no lack of oak in the northern portion of the plain, although, as it approaches Jaffa, it begins to degenerate. Otto von Richter says,[2] that on the best tilled and most populous side of the plain the view is charming: in the spring the ground is covered with roses, lilies, tulips, narcissus, anemones, and other flowers. The vegetation of the plain is determined in some measure by more or less elevated strata of sandstone on which it grows: the places which lie deepest produce sesamum and cotton; the knolls around all the villages are covered with olive groves, while the whole surface of the ground is a mass of green. The numerous stone houses which are to be seen on every little elevation of ground, although the most of them are in ruins, give to the whole an animated look, which is missed in Egypt, where the huts are built of mud.

Among the streams which cross the plain of Sharon, the only one which seems to be of any importance is the Nahr Aujeh. This, however, has no permanent character. It commences at the Ras el Ain, where there are very profuse springs, and, flowing westward, enters the sea two hours north of Jaffa. At the mouth it has such width and breadth, that von Wildenbruch[3] was unable to cross it. Its sources were discovered by Dr Eli Smith,[4] as I have mentioned on a previous page.

There is a second stream, given in most of the maps as the Nahr Arsuf, deriving its appellation from the village of

[1] Reland, *Pal.* pp. 188, 370.
[2] O. v. Richter, *Wallfahrten*, p. 13.
[3] Von Wildenbruch, in *Monatsb. d. geog. Ges. in Berlin*, i. p. 232.
[4] E. Smith, in *Bib. Sacra*, p. 495.

Arsuf, north of Jaffa. Von Wildenbruch,[1] who thoroughly explored this region, was, however, unable to find any trace whatever of it. Between Nahr Aujeh and Abu Zabura no river flows into the sea. There are indeed marshes and little lakes extending as far as Arsuf (near el-Burj, according to von Wildenbruch's notes), yet they form no connected stream.

The present ruined village of Arsuf was, at the time of the Crusades,[2] a tolerably strong military position, and was considered to be the site of Antipatris, although bearing even then the name by which it is now called. No remains of ancient masonry are, however, according to Buckingham,[3] to be seen there.

The entire row of hills which run west of the plain of Sharon, and parallel with the coast, are evidently composed of drift sand which has been thrown up, and are entirely different from the soil of the plain itself. Hasselquist[4] was one of the first to notice the distinction between the composition of the two. The soil of the plain proper is a red sand, which probably gave currency to an old myth, that in the country of the Hebrews near Joppa there is a fountain whose waters are as red as blood. The explanation given for this colour was, that after Perseus had slain the sea-monster which watched Andromeda, he washed off the blood which stained his person at this spring. Buckingham[5] claims to have discovered the fountain itself, about a half-hour's distance north-east of Jaffa; but he says that the water is clear, cool, and refreshing. Not far east of Lydda, the peculiar characteristics of the plain of Sharon disappear, and the uniform horizontal strata of limestone and chalk appear,[6] which characterize the whole mountain region from Gaza to Samaria. These often assume the form of an amphitheatre, and ascend by a series of steps, favouring the terrace culture

---

[1] Von Wildenbruch, as above, p. 232; v. Prokesch, *Reise*, p. 36.
[2] Will. Tyr. *Hist. in Gesta Dei*, ii. ix. 19, fol. 774.
[3] Buckingham, *Travels in Pal.* i. p. 219.
[4] Hasselquist, *Reise in Pal.* ed. by Linné, p. 141.
[5] Buckingham, *Travels*, i. p. 251.
[6] Sieber, *Reise*, p. 31.

of ancient times. It is from one of these steps to another that the mule now pursues his perilous way.

The whole plain north of Lydda appears to rest upon a loose tertiary sandstone. It is very light and fertile; and where it is not given over to thorns and thistles, it affords excellent pasturage.[1]

DISCURSION III.

THE EASTERN OR MOUNTAIN ROAD THROUGH THE PLAIN OF SHARON: THE GREAT CARAVAN ROAD FROM LYDDA OVER THE CARMEL RANGE TO THE PLAIN OF ESDRAELON.

The great caravan road from Egypt, which runs along the eastern border of the plain of Sharon, and which forms, in contrast with the western route, a real mountain road, passes over Ras el Ain, and through Kefr Seba, Gilgoul, Kulensawe, Kakun, and Kannir. It then bears towards the north-east, leaving the direct northern route, and passes over the Carmel ridge, traversing el-Lejjun (Legio or Megiddo), and then entering the great Esdraelon plain, to continue its course towards Tiberias and Damascus. The fullest account of this route we owe to von Wildenbruch,[2] although it is only too brief.

From Ramleh to Lydda it is three-quarters of an hour: a Wadi Nahr Musrara (perhaps identical with Nahr Betra) is here crossed by a bridge, near which are some lions hewn out of the stone, and bearing an inscription which states that Mohammed Sultan was the builder of the bridge. From this point to a Gothic bridge is an hour's distance: north-west of this lies Jehudie. Two hours and a half more bring one to Remthieh. From there it is four hours to Ras el Ain. Right of that, and past the hills of Bir Adas, is Gilgoul, five hours away. Thence it is eight hours to Kulensawe, in the immediate neighbourhood of which is a fort similar to that at Ras el Ain.

[1] Wilson, *Lands of the Bible*, ii. p. 253; Russegger, *Reise*, iii. p. 118.
[2] Von Wildenbruch, *Reiseroute in Syrien*, p. 233, and Tab. v. No. 6.

From this point Kakun seems to be about nine and three-quarters hours. From Kakun to an old Roman cross road it is two hours, and thence to the uninhabited and ruined village of Bedouss it is two and three-quarters hours. This whole way, which runs through a hitherto unexplored tract, is everywhere covered with the remains of former towns and villages. From Kakun to a Roman aqueduct it is three and three-quarters hours. Here is the opening of the pass running north-west through the Carmel ridge. The oaks which adorn the northern part of the plain of Sharon now cease. Six and a quarter hours from Kakun bring one to the watershed of the range, and a view is gained of Tabor and the hills of Galilee. After seven and a half hours, most of it on a well-preserved Roman road, the Khan el Legoun is reached. It lies on the great pile of ruins marking the old Campus Legionis (Megiddo), which commanded the entrance to the plain of Esdraelon.

DISCURSION IV.

THE WESTERN OR COAST ROUTE THROUGH THE PLAIN OF SHARON: KAISARIYEH, THE ANCIENT CÆSAREA PALESTINÆ—CÆSAREA MARITIMA, ORIGINALLY STRATONIS TURRIS, SUBSEQUENTLY CÆSAREA STRATONIS.

From Cæsarea southward over the Kudeira and Nahr Abu Zabura to Muchalid is a distance of three and a half hours; from that point it is six hours to Jaffa, the road passing Arsuf and el-Aujeh. This was the time taken by Wildenbruch.[1] Dr Smith[2] travelled in company with ladies, and more leisurely, taking thirteen hours for the ride.

In the ancient history of Palestine there is no mention of Cæsarea, despite its subsequent form, and even its name indicates its comparatively modern origin. Although lying between Joppa on the south and Dor[3] on the north, and the only harbour of any importance between them, it was only at

[1] Von Wildenbruch, i. p. 232; v. Prokesch, *Reise*, pp. 28-35.
[2] Robinson, *Bib. Research.* ii. Note xl. p. 528.
[3] Movers, *Phönizier*, ii. p. 176.

the time of Herod that this port assumed prominence. In the Old Testament there is no allusion to the place, although its northern neighbour Dor is mentioned twice in Joshua (once as Naphath Dor) as a part of the subjugated territory. At Solomon's time Dor belonged to Israel, for there lived one of the twelve men whose duty it was to supply the king's household with provisions (1 Kings iv. 11). It is true that Dor and Naphath Dor are not precisely identical, the former being the port, and the latter the city connected with it, but lying inland a little way from the sea. Dor, like the other cities of the coast already alluded to, would seem to have been in the possession or under the control of the Israelites. At the time of Darius of Persia, Scylax of Karyanda calls the city of Dor, between Carmel and Joppa, a city of the Sidonians; and Steph. Byz. terms it a little city, inhabited by Phœnicians, while Naphath Dor belonged to the Jews. The port seems to have been indispensable to the former, as Joppa was the port next south of Tyre. But both of these harbours, Joppa and Dor, were of insignificant pretensions; and unquestionably in consequence of this, Herod formed the plan of building a new one between them. He devoted twelve years to this magnificent undertaking, and gave to the place thus created the name Cæsarea Palestinæ,[1] in honour of the Emperor Augustus Cæsar. The briefer appellation Palestinæ came subsequently into vogue, to discriminate between it and Cæsarea Philippi, at the head waters of the Jordan. The Cæsarea on the sea also bore the name of Cæsarea Maritima. Strabo mentions it under the mere appellation of Strato's Tower, and speaks of a landing-place there. Ptolemy subsequently alludes to it as Cæsarea Stratonis; and Pliny shows that this must have been identical with Cæsarea, by the words, "Stratonis turris, eadem Cæsarea, ab Herode rege condita; nunc Colonia prima Flavia, a Vespasiano Imperatore deducta;" and adds to this, that here was the northern extremity of Palestine, what was contiguous to it being Phœnicia. This language may have borne relation to the Phœnician Dora, which lay between Cæsarea and Carmel.

[1] Reland, *Pal.* pp. 27, 670.

Josephus gives the most detailed description of the great undertaking of Herod alluded to above. It might be taken as exaggerated, did it not correspond to the general magnificence of the other known works of Herod, and to the extent of the ruins seen at the present day, and described by Prokesch,[1] Wilson, and Barth. Herod, says Josephus, found at the deserted place called Strato's Tower (Strato seems to have been a Greek wanderer) an admirable position for a harbour, which should shelter vessels against the violent south-west winds, which were so injurious to the shipping along that coast. There was then not a single good harbour for vessels to betake themselves to in time of storm. He not only built a magnificent city, but excavated and walled in a harbour larger than the Piræus at Athens. The whole of the building materials had to be brought from a great distance at large expense, there being none suitable there. Herod was able, says Josephus, to conquer the obstacles of nature, and to unite massiveness with taste and beauty. The immense hewn stones, many of them fifty feet long, nine feet wide, and as many high, were sunk fifteen yards in the water, and so up to the surface. In this way a dam two hundred feet broad was built out into the water. On this mole a wall rose two hundred feet above the surface of the water, the lower hundred of which were intended to encounter the dashing of the waves. Dr Barth found the remains of these in such quantities as to admit of but one entrance into the harbour. The wall was fortified by means of towers, the largest of which, a structure of great size, received the name of the Drusus Tower. There were several arches for people to land upon, and a broad quay ran around the harbour, making a beautiful walk. The entrance was on the north side, the wind being the lightest in that quarter. Three colossal figures, supported by columns, stood on each side of the channel. At the left there was a massive tower which served as a breakwater, and opposite it were two more massive aggregations

---

[1] Von Prokesch, *Reise*, pp. 28-34; Wilson, *Lands of the Bible*, ii. pp. 250-253; Dr Barth, MS., *The Christian in Palestine*, p. 230, Tab. xlv.

of masonry, which were firmly bound together. Along the whole seaboard Herod built a row of elegant houses of white hammered stone. These stood equidistant from each other; and on a height near the sea there was a temple of remarkable size and beauty, which navigators could see from a great distance. On this temple stood the colossal statue of Cæsar Augustus, constructed after the model of the Jupiter in Olympia, and one of Roma, taken from the Juno at Argos. The foundation of the city was traversed by numerous arches and long passages, so that the city in flood time could be thoroughly cleansed and purified. For the citizens of Cæsarea Herod built a theatre, and behind this an amphitheatre, intended to accommodate a large number of spectators. After the completion of these great works, he sent his two sons Alexander and Aristobulus to Rome, to offer their services to the emperor, and to pay him their homage.

At the time of the apostles, this port of Cæsarea became a very important centre in diffusing the gospel. Philip, we are told in Acts viii. 20, preached in all the cities along the shore, from Ashdod or Azotus to Cæsarea: Peter went from Joppa to the latter city, where he met the pious Cornelius, who was at the head of the Italian band (Acts x. 1, xi. 1). Here Herod came to the dreadful end recounted in Acts xii. 19-24. It was to Cæsarea that Paul was brought, after the attempt had been made to put him to death in Jerusalem; and thence it was that he was sent to Tarsus (Acts ix. 30). On his return from Greece to Palestine he found a church in Cæsarea, and after greeting its members he went thence to Antioch (Acts xviii. 22). Coming back from Tyre and Ptolemais to Cæsarea, he tarried several days in the house of Philip the evangelist (Acts xxi. 8), and then went up to Jerusalem to the feast, although it was told him that he would be bound there, and delivered into the hands of the Gentiles. Persecuted there to the death by the people and the high priest Ananias, he gave himself up as a Roman to Claudius Lysias, and was escorted to Cæsarea to be tried. Here he was heard in the judgment-hall of Herod, as he pleaded his innocence, and showed himself

to be not worthy of punishment. Yet in Cæsarea he was
detained two years as a prisoner; and after Festus had
been made governor, he had a hearing before king Herod,
his wife Bernice, and Festus. Although found not worthy
of death, he was placed with other prisoners on board a
ship, and sent from this port of Cæsarea, by way of Sidon,
Cyprus, and Lycia, to Rome, to be tried there. In consequence of the establishing of a Roman colony there by
Vespasian, and in consequence, too, of the speedy downfall
of Jerusalem, Cæsarea became a place of great importance.
The Christian church was strong there also; and under
the Christian emperors it became the Metropolis Palæstinæ
Primæ. The church had a very strong support there in the
celebrated scholar Bishop Eusebius. Church councils were
held there in the years 198 and 553; and for a long time
it had the supremacy over Jerusalem. Regarding the seven
years' siege by Omar, and the subjection of the city by
the caliphs, the historians give very diverse[1] accounts; yet
the two hundred thousand pieces of gold which the defence-
less citizens offered to pay after the Roman troops had
cowardly deserted them, show how wealthy the city must
have been. The booty, too, taken by Godfrey[2] of Bouillon
was very great, when, supported by Pisanese and Genoese,
he took the city in 1101, after a fifteen days' siege. It had
always been the landing-place of the crusaders, and now
became the capital of the archbishopric which was established. Fearful was the slaughter of all the Moslems in their
mosque, formerly a Christian church, which stood upon the
same height where Herod's temple had stood. In it they
found that celebrated six-sided green glass jar which was said
to have come down from the Saviour's time, and to have been
used at the Last Supper. So high did it stand in honour, that
the Genoese preferred to take it rather than gold as their
share of the spoils, and it may now be seen in the Church of
St Laurence, Genoa. The wondering crowd is told that it

[1] Gibbon, *History of the Decline and Fall*, etc., Ger. ed. Pt. xiv.
p. 331; Weil, *Ges. der Chalifen*, i. pp. 80–83.
[2] Wilken, *Ges. der Kreuz.* ii. p. 102.

was a present from the queen of Sheba to Solomon, and that it formed a part of the ornaments of the temple on Mount Moriah.[1]

During the Crusades, Cæsarea was twice destroyed by the Moslems, and twice rebuilt by the Christians. In 1251 it was fortified with extraordinary pains by Louis IX. For fourteen years it endured the assaults of the bold Sultan Bibar, who took it in 1265, and so thoroughly demolished it, that, according to the possibly somewhat hyperbolical expression of Makrizi,[2] not one stone was left upon another. About the middle of the twelfth century, Edrisi speaks of it as a large city, having suburbs and a fortification. Benjamin of Tudela alludes to it as a well-built port, and thought it to be the Gath of the Philistines. In the time of his visit, there were only ten Jews and two hundred Samaritans there. Abulfeda,[3] on the other hand, who well knew its former greatness, saw but a single ruin there. Wilson, who wandered over the widely scattered masses of masonry which once adorned this capital, found only a single herdsman and two Jewish families inhabiting the place. At D'Arvieux's[4] time some fishermen lived there, who were very skilful in plying their trade, but who suffered many losses by the invasion of corsairs. Irby and Mangles[5] could devote but a few hours to studying the ruins of Cæsarea, for a thorough examination of which Dr Barth asserts that several days would hardly suffice. They held the ancient moats and the city walls to be of Saracenic origin. At the extreme end of the promontory they saw what seemed to them the ruins of a Roman temple, supported by colossal granite pillars. They noticed also a little cove, on whose northern side was a landing-place, still exhibiting arches that once sustained warehouses. Near this the traces of an aqueduct could be seen, though no water now flows near, but that of the disagreeably tasting Nahr Zerka. In

---

[1] *Guide de Gênes*, p. 134, with Sketch, Tab. *il Catino*.
[2] Edrisi, in Jaubert, i. p. 348; *Benj. Tud. Itin.* ed. Asher, i. p. 65.
[3] Abulfedæ *Tabul. Syr.* ed. Koehler, p. 80.
[4] D'Arvieux, *Nachricht.* ii. p. 13.
[5] Irby and Mangles, *Trav.* p. 189.

addition to the Saracen walls, they found a marble column, bearing a Latin inscription, and the name Septim. Severus. This Mr Banks declared to be a Roman milestone.

The fullest existing description of the ruins of Cæsarea we owe to von Prokesch, although very much remains to be done by future antiquarians. The results of Dr Barth's careful inquiries there have not yet been given to the world.

Perpendicularly to the western coast-line on which the ancient Cæsarea was built, a mass of rock extends for about four hundred paces direct into the sea; and a little cove on the north and the south side of this serves to protect the tiny vessels which navigate the neighbouring waters. Whether the Tower of Strato stood upon the extremity of this mass of rock, is a fact now hard to learn. There are walls and gates still standing, according to von Prokesch; and yet, although they are large enough to afford comfortable shelter, the fear of the Beduins prevents any from resorting to them for protection. In wandering over the ruins it is necessary to use great precaution, for fear of slipping through the weeds and grass, which have grown up profusely, into arches, holes, fountains, and other pitfalls.

The right angle in which the city was built measures five hundred and forty paces from north to south, and three hundred and fifty from east to west. In the eastern wall ten towers are to be counted, in the northern three, and at the north-west corner there is a kind of fortification. On the west side, along the sea, there are three towers. At the south-west corner of the place the rock which sustains the citadel projects into the sea; and at the extremity of it may be seen the remains of a tower, which Barth thinks to have been the one which in ancient times bore the name of Strato. Strong breakwaters have been carried from the citadel out into the sea. Left of the rocky reef is the southern cove, which serves as a harbour.

Both of these havens were artificially widened to a breadth of two hundred paces, and protected with walls. The land side has even now a moat thirty-six feet wide, guarded with towers: they seem to date from the time of the crusaders. The walls of the city itself, according to von Prokesch, are from twenty to thirty feet high, and six feet thick. The towers stand at unequal distances from each other, varying from fifty to ninety feet. The city appears to have had three gates, two of which are still standing. That on the east side is so ruined, that a man can ride directly over it. A fourth gate, which probably leads to the middle part of the northern haven, has wholly disappeared. There is still a lofty tower discernible in the square castle. It is separated from the city by a passage a hundred and twenty-five feet long and twenty-five broad, which connects the two coves. A great number of grey and red granite pillars, which were once brought from Egypt by Herod, are now to be seen in a shattered state. In a like ruined state are the dam on the north side of the northern harbour, and that which ran out into the sea south-westward. These were composed very largely of granite pillars. At the northern base of the castle, and lying in the water of the northern harbour, there is a granite pedestal, more than six feet wide. Before the gate of the castle there is a cistern and a deep shaft: two archways of the gate have remained. From the top of the tower an extensive view is obtained seaward and landward.

The ruins in the interior of the city are for the most part of brick, and have very little that is striking. At the northwest corner, hard by the wall, stands a subterranean church; and among other ecclesiastical ruins, is one of very massive walls, which perhaps was the residence of the archbishop, a man of such authority that twenty bishops were subject to him. Near the southern gate there are to be seen the remains of a stadium; nothing being left of it, however, but a few granite pillars and hewn stones, on one of which the name Fibianus Candidus can be made out. There are traces of walls and towers around the southern port, which was probably a suburb.

Dr Barth holds the castle, a massive work, sixty feet square, to be a citadel of the middle ages, and says that its vastness makes it even now an imposing object. He remarks, also, that the structures of the ancient and the more modern times are so mingled together, that it is a very difficult task to discriminate between them. Yet it is plain that the works executed by Herod were on the grandest scale. Barth recognised in four large pillars the remains of the metropolitan church which was erected on the site of the ancient Roman temple. At the north-western part of the city he discovered the remains of a building, which must once have been of great magnificence, although but a few colossal stones and pillars remain to attest it. He found the whole circumference of the city, from the sea round to the sea again, to be 3600 paces. He examined the aqueduct carefully, which once conveyed water into the city from Nahr Zerin. It is to be wished that he would communicate a full account of his observations to the world.[1]

DISCURSION V.

COAST ROUTE FROM CÆSAREA TO CARMEL, BY WAY OF DANDORA (TANTURA, DOR, DORA) AND ATHLIT (CASTELLUM PEREGRINORUM).

1. *Road to Dandora (Tantura), the ancient Dor, the Naphath Dor of Solomon, and the seat of a Sidonian fishing colony. The purple mussel, and its fishery.*

From the ruins of Cæsarea there still run northward the traces of an aqueduct along a bight in the coast; and the traveller accompanies this along the shore, passing the numerous sand-dunes, till he reaches the remains of a castle

---

[1] I translate the above words in Berlin, just after the death of the illustrious Barth, whom it is a sincere pleasure to have known and loved as a friend. Almost his last conversation was with me regarding the introduction of his great teacher Ritter's writings to the English-speaking world of Great Britain and America; and his last gift to the library of the Geographical Society of Berlin was a copy of Ritter's *Lectures on Physical Geography*, published a year since, by the house of Blackwood and Sons. His papers are in the hands of the distinguished geographer Kiepert; and it is to be hoped that his careful examinations of the whole Mediterranean basin, including the results gained in Syria, will be given in full to the world.—ED.

and other buildings, and comes to the Nahr es Serka[1] with its ruined Roman bridge. The water of this stream is not deep, but bad. Here a ridge of hills commences, which reaches away towards the east, and in a few hours joins the Carmel range. After two hours from Cæsarea the Nahr el Belka is reached,—a name which, according to von Wildenbruch,[2] is not known in the interior. On its lower windings, and only twenty minutes distant from it, is the place which is generally written Tantura on the maps, but which Barth[3] heard plainly pronounced Dandora by the natives. Irby and Mangles visited the place, and found ruins there which did not seem to them to have much interest; and D'Arvieux, who calls it Tartoura, says that it is merely a market-town to which the Beduin bring their plunder to sell it to the natives, who give in return rice and linen, and other articles from Egypt.

The present town is built up around a stately structure erected in the middle ages. The ancient city, which can hardly fail to occupy the site of the Dor of the Canaanites and Phœnicians, and whose name is still seen in the modern form Dor or Dora, lies with its not unimportant ruins[4] some minutes north of a little dirty pond, surrounded by a swamp.

North of a small bight or cove, and on a projecting rock, is a castle dating from the middle ages, but resting upon foundations which are ancient. The south side serves for quarries, but on the north side may be seen the remains of a stately structure, whose long walls betoken its former great size. The true city once lay on a low range of hills northward to another bight, which is entirely surrounded by ledges of rocks, in the midst of which there is a quay paved with mussel shells, which has given rise to the conjecture that there has been here a gradual elevation of the shore. The whole ridge is covered with ruins, among which are several

---

[1] Prokesch, *Reise*, p. 27.
[2] Von Wildenbruch, in *Berliner Monatsch. der geog. Ges.* i. p. 232.
[3] Dr Barth, MS.; D'Arvieux, p. 11; Irby and Mangles, *Trav.* p. 190.
[4] Barth, MS.; v. Prokesch, *Reise*, p. 27; Wilson, *Lands, etc.*, ii. p. 249.

fragments of pillars, including a very fine Ionic capital, very much worn by time, and made out of the same kind of stone which was used in the buildings of the city. The city extended as far as the well-watered plains on the east, in which there are the most fertile corn-fields. This was the case, at any rate, in the later Roman time, when Gabinius restored it and provided it with a harbour. Quarries and several tombs are now found in the neighbourhood of the ruins of the city.

Dor was one of the Canaanite cities which did not yield to Joshua, but whose inhabitants subsequently were so far overcome as to be included in the half-tribe of Manasseh, and compelled to pay tribute. Under Solomon it became an official post, and later was a powerful stronghold, which endured many a siege under the Maccabees. Polybius speaks of it as an important post, which witnessed valiant service during the wars between Ptolemy and Antiochus, but was assailed by the last in vain. It seems to have become a more important place after being restored by Gabinius than it had been before; for Jerome speaks of it as a very powerful city, whose ruins excited the admiration of the pilgrim Paula. Other authors mention Dor, however, as only an insignificant place; yet when it is spoken of in this light, the great inland city is not meant, but the smaller one on the sea which belonged to the Phœnicians, of which Scylax and Stephen of Byzantium and Hecatæus of Miletus speak, and concerning whose rise Claudius[1] Julius has given a remarkable sketch in his work on Phœnicia. This Dor, he says, lay in the neighbourhood of Cæsarea, and was a small place inhabited by Phœnicians, who, in consequence of the abundance of purple mussels found there, had built them huts and entrenched their town for security. But when they had become very successful in their fishery, they enlarged their quarters, quarried the rock, made a harbour, and gave the place the name of Dor, borrowing it from the Greek tongue. Some said, however, that Dorus, a son of Neptune, was the builder of the town.

[1] Steph. Byz. ed. Meinecke, p. 255, s.v. Δῶρος.

On the Peutinger Tables Thora is put down as eight miles from Cæsarea. Jerome gives it as nine. When Wilson was here in 1843, he found the place a little collection of huts.

Whether the purple mussels are taken to any extent on that coast now, is exceeding improbable. The earliest trace of them in that region is preserved in the blessing pronounced by Moses on Zebulon and Issachar (Deut. xxxiii. 19), "For they shall suck of the abundance of the seas, and of treasures hid in the sand ;" an enigmatical expression, signifying nothing else than that the gain was to accrue from the purple mussel and from the manufacture[1] of glass. Dora, however, was a place belonging to Manasseh, but in the territory of Issachar, and upon the boundary between Israel and Phœnicia. But Zebulon was to lie upon the sea, as we learn from the blessing of Jacob (Gen. xlix. 13), and was to extend to Sidon, which almost makes it certain that Scylax was correct in his statement that Dor was a city of the Sidonians. That Moses, while in Egypt, was acquainted with the costly purple dye which was brought from the Phœnician coast, and it may be from Dor itself, the nearest port of that country, is made evident by the allusion to blue, purple, and scarlet which were used in adorning the tabernacle which was set up in the wilderness. The purple mussels now seen on that coast agree fully with those described in all ancient accounts; and that they are now discovered, is confirmed by the statements of repeated travellers. Von Prokesch saw them[2] in great quantities, and Seetzen speaks of noticing two varieties, the Murex trunculus Linn. and the Helix janthina, which yielded the celebrated purple of the ancients. Olivier found the janthina in great abundance upon the Syrian coast from Tyre to Alexandria, and held it to be the same which yielded[3] the most of the purple, although he found almost all the buccenites to yield a red colour, particularly those of the finest species. The living

---

[1] Movers, *Die Phönizier*, ii. pp. 210, 176.

[2] Von Prokesch, *Reise*, p. 34; Buckingham, *Trav. in Pal.* i. p. 196.

[3] Seetzen, in *Mon. Corresp.* 1808, p. 445; Olivier, *Voy. in Orient*, T. ii. p. 251.

janthina is greenish or white, but it becomes red and then purple when it is exposed to the air, and after it dies it is violet. The little creature yields the most of their income to the fishermen of Dora; but an immense quantity of them must be taken, in order to extract the purple which is contained in a little pouch not as large as a pea. Hence arose in ancient times the great costliness of the purple which the Phœnicians prepared,—an article which only Persian kings and Roman emperors, senators, and the wealthiest could buy. The much cheaper cochineal has in modern times entirely destroyed the demand for the Tyrian purple. The Helix janthina of Lamarck,[1] or Janthina fragilis, thought by Lessow to be identical with the buccinum of Pliny, is considered by that naturalist as identical with the ancient Phœnician dye. Lamarck disagrees with this, however, and thinks that the Purpura patula was the real dye. Still it must be confessed that a variety of views[2] prevails regarding buccinum and murex. The art of making purple was known not only on the Phœnician coast, but also on that of North America, particularly in the Minor Syrtis, on the island of Meninx.[3]

2. *Route from Dandora to Athlit, the Castle of Pilgrims, Castellum Peregrinorum, Castello Pelligrino, Petra incisa, and to the southern base of the promontory of Carmel.*

From Dandora[4] northward to Athlit it is a two hours' journey along the coast, according to von Wildenbruch, and thence three hours to the convent on Carmel.

Three-quarters of an hour from the ruins of the ancient Dor, the traveller reaches the village of Surfend, the Sura-

---

[1] Leunitz, *Synopsis*, Pt. i. pp. 382, 390.

[2] Winkelmann, *Geschichte der Kunst der Alterthums.* iii. p. 9; Voss, *Comment. zu Virgils Landbau*, iv. 373, p. 855; Mongez, *Mem. de l'Institut Hist. et B. L.* T. iv. p. 259; Lessow on Tyrian purple, in Jameson, *N. Edin. Phil. Journal*, 1828, Apr. to Sept. p. 403; Wilde, *Narrative of a Voyage along the Shores of the Mediterranean*, ii. p. 151, and Appendix.

[3] Dr Barth, *Wanderungen durch das Punische und Kyrenaische Küstenland*, p. 261, Note 643, p. 378, after Abu Obeid Bekri, in *Notic. et Extr. de la Bible du Roi*, T. xii. p. 480.

[4] Von Wildenbruch, quoted above, p. 232.

fend of the maps, up to which point a low range of hills accompanies the road on the east, in the plain adjoining which a few palms are left standing. In the village, in which some twenty shepherd families live, von Prokesch spent a night in the mosque. On the heights running north there may be seen here and there the remains of walls, which perhaps correspond to the places mentioned by Strabo as once lying between Carmel and Strato's Tower, but of which even in his time nothing but the names existed: these he cites as Sycominonpolis, Bucolonpolis, and Crocodilopolis. Yet Sycomina is alluded to by Jerome as identical with Ephe, the present Haifa, at the northern[1] base of Carmel. Regarding the remarkable Crocodilopolis, which is mentioned by Reichard as hard by the coast, south of Cæsarea, but which is put by Berghaus farther in the interior, Pococke[2] has propounded the theory that it was an Egyptian colony, by the members of which the crocodile was held in honour. The creature is believed by him to have propagated itself in a swamp, which takes its name from the animal. He states that he has heard of crocodiles of five or six feet in length being brought to Acre. Later travellers make no allusions to it.[3]

Twenty minutes before reaching Athlit, von Prokesch[4] noticed a gap in the range of hills, and farther on several ancient wells by the roadside. He also saw magazines for grain hewn in the sides of the rock, with openings two to four feet in width.

Irby and Mangles[5] took up their quarters for a night in the village of Athlit, which lies elevated above the sea on a kind of peninsula, in the midst of ruins. The place is of small extent, is similar in appearance to a citadel, and is surrounded by ancient walls, outside of which are others

[1] Von Raumer, *Pal.* pp. 140, 141.
[2] Pococke, *Beschr.* ii. pp. 75, 76.
[3] Thomson, however (*Land and Book*, ii. p. 244), makes it very probable, though not certain, that crocodiles are found there at the present day.—ED.
[4] Von Prokesch, *Reise*, p. 25.
[5] Irby and Mangles, *Trav.* pp. 191-193.

still. The southern extremity of these runs down as far as to the sea-coast: there are two gates upon the east, one on the south side. There are several places where there are flights of stairs to ascend. One of the external walls runs near to the citadel, while the other appears to have surrounded the ancient town or city; at present, however, the enclosure is a mere desolate space. On the south side of the little peninsula or promontory there is a small cove, which was probably at one period the harbour of the place.

Within the citadel there is a massive structure in the form of a hexagon, or decagon according to some, the half of which is still standing. From beneath the exterior cornices spring in bold relief several animals' heads, together with one of a man joined to the body of a lion. The external walls of this structure have double lines of arches in Gothic style: the lower row is broader than the upper, the architecture light and elegant.

No ancient name of this place is known, although the convenience of the little harbour close by, the numerous quarries and tombs in the neighbourhood, and the fruitful plains by the side, make the conjecture a safe one, that here was once a considerable centre of population before the time of the Crusades, when the citadel began to bear the name of Castrum Peregrinorum, from which is derived the modern name of Castello Pelligrino.

Dr Barth, who has examined these ruins with more care than any one, was so much struck with the traces of great antiquity there, and with the extent of the fortifications, as to believe that a place of such strategic importance in its relation to the highway between Tyre, Carmel, Philistia, and Egypt, could not have been unused at the time when Canaan and Phœnicia were in their period of pride and power. But no ancient name is known, unless the one given it by the Arabs—Athlit, according to E. Smith, or Athalit, as Wilken[1] gives it—be a contraction of the primitive designation.[2]

Oliverius, Scholasticus of the church at Cologne, and

[1] Wilken, *Ges. der Kreuz.* Pt. vi. pp. 158, 311, vii. p. 772.
[2] Robinson, *Bib. Research.* ii. Note xl. p. 528.

apostolical legate, who in 1231 incited many pilgrims and crusaders to visit the Holy Land, gives a detailed description of the situation of this pilgrims' castle,[1] and says that in ancient times it was called Districtum, a name perhaps derived from the manner in which the sea limits the course of the road (*propter viam strictam*), and from the narrow pass which perhaps gave the name *Petra incisa* to the place. Raymond of Toulouse is said to have laid the foundation of a castle near Tortosa, to which he gave the name Castellum Peregrinorum,[2] which might easily be confounded with this. This *Petra incisa* would seem to have been first built as a protection for pilgrims on their way to Jerusalem, and to have subsequently fallen into ruin till about the time when the kingdom of Jerusalem passed away, when it was restored in 1218 by the Knights Templar as the chief seat of their order, and remained their last stronghold even after Jerusalem had fallen into the possession of the unbelievers. Previously to these events the neighbourhood of Petra was rendered very dangerous by the robbers who infested it, and in 1103 King Baldwin was mortally wounded by them. At the time when the excavations were made for the walls, Oliverius Scholasticus states that several gold coins were found bearing a stamp which the discoverers did not recognise, but which served to defray in part the expenses of the undertaking. Jac. de Vitry, *Epis.* i. *ad Honor.* iii. p. 289, says that the amount of wealth discovered there was incredibly great. Unquestionably these were Greek or Roman coins, or possibly Phœnician ones, and their discovery confirms the probability that this fine strategic position was early improved. The place which seemed to be so blessed of God in consequence of this discovery of gold, received the name *Castrum filii Dei*. The fortress was so admirably built and protected, that the Saracens, in spite of the most determined efforts to take the place, protracted for eight days and nights, were unable to make any impression upon it, and were compelled to raise the siege. The crusaders and pilgrims who had

---

[1] Wilken, *Gesch.* as above, vi. p. 99.
[2] Will. Tyrens. *Hist. Hieros.* lib. x. p. 26, fol. 791.

sought a last refuge there after the rest of Palestine had passed from their hands, gradually withdrew to their homes; and in the year 1291, this castle, in conjunction with Tortosa (Dor), had been deserted by all the Christians,[1] and they were destroyed by the Sultan Melek el Ashraf, into whose hands they had fallen. Ottokar of Horneck, in his account of Suders (Sidon), and the misfortunes of this citadel, calls it Chast Pilgrim, and says that "*there* once stood a fine city."

And, in truth, the extensive ruins still seen there confirm this expression; for within them D'Arvieux saw the remains of a massive and even elegant church, which both Barth[2] and Wilson have declared bomb-proof. Pococke, too, has alluded to the beautiful character of this architecture even at the time of his visit. The greater part of the walls were then used as quarries, and the finest material was applied to the building of Acco; yet granite columns were strewed around, and Barth, who examined with great care the harbour, the thickness and length of the immense walls, the gates and towers, was so amazed at the magnitude of this citadel, that at the close of his manuscript journal he states that the remains of this Castellum Peregrinorum are among the best which exist, for the light which they throw upon the study of fortification in the middle ages, as well as in the Roman period, from which the walls appear to date. A topographical survey of these ruins would be an excellent substitute for the general description of them, which is all that we thus far possess. The only place in the older records of travel which seems to relate to this site is in the Bordeaux pilgrim's *Itinerary*, where the Mutatio Certha[3] finis Syriæ et Palestinæ, viii. mill. southward from Carmel, seems to coincide with the position of Athlit.

As one goes northward from this Castello Pellegrini, amid whose remains nothing is now to be seen excepting a few poor miserable huts with their inhabitants, a line of ruins can be traced for a distance of ten minutes beyond the walls,

---

[1] Wilken, *Gesch.* vii. pp. 766-793; Robinson, *Bib. Research.* p. 469.
[2] D'Arvieux, as cited, ii. p. 10; comp. Wilson and Pococke, ii. p. 83.
[3] *Itiner. Hieros.* ed. Parthey, p. 276, ad. 585.

to a place where a road is cut through the natural wall of rock. This separates the fruitful eastern plain from the western one, which is much covered with sand-dunes extending northward from Athlit, and seemingly the work of many centuries. That the peninsula of Athlit was once an island, is expressly stated by Adrichomius (before 1585), who says: "Castrum Peregrinorum quondam in insula in corde maris sita, dicta Petra incisa."[1] Pococke thinks that the insular form was first given by the fosse which was carried across the peninsula for purposes of fortification,[2] but that it was subsequently filled up by the sand.[3]

There are traces still discernible of an ancient road once running farther northward: upon the high land the remains of a watch-tower can be seen in one place: then follows a beautiful although contracted plain near the base of Carmel, separating it from a spur of rock which shuts the view of the sea from the traveller. The road then runs northward to a fountain lying about an hour's distance from Athlit. The rocky spur has some openings through which the eye occasionally catches glimpses of the sea. There is no special object of interest after this till the high promontory of Carmel is reached.[4]

---

[1] Barth, *Reise*, MS.; Wilson, *Lands, etc.*, ii. p. 247; Pococke, *Trav.* ii. p. 83.

[2] Scholtz, *Reise*, p. 151.

[3] Von Raumer, *Pal.* p. 138.

[4] Comp. Buckingham, *Trav. in Pal.* i. pp. 190-193; Wilson, *Lands, etc.*, ii. p. 248.

SAMARIA, THE CENTRAL PART OF PALESTINE.

CHAPTER IV.

SUBSEQUENTLY to the restoration of the temple of Jerusalem under Ezra and Nehemiah, after the return from the Babylonian captivity, the name Samaria, derived probably from the city so called, began to become the stated appellation of the district which had been settled by strangers while the Jews were in Babylon, and which therefore from that time forms a very marked contrast to Judæa. For this district not only took no part in the erection of the new temple, but its rulers made decided opposition to it (Neh. ii. 19, iii. 34, iv. 2; Ezra iv. 10), and erected for themselves a temple of similar character on Gerizim. At the time of Hosea and Hezekiah the inhabitants of a large part of the kingdom of Israel, through repeated invasions of the Assyrian kings, particularly Shalmaneser's, who pillaged Samaria also, were carried away captive into Assyria and the neighbourhood of the Tigris, settling at Khabur. In the meantime, the Assyrian colonists who established themselves in the Israelite territory, took, according to Josephus, the name of Samaritans from the main city. This became the first centre of this new population, which at a later period, however, removed to the ancient Shechem, the present Nablus.

The tribes of Ephraim, half Manasseh, Issachar, and Naphtali, were led into captivity from the region around Gennesareth and a portion of Galilee, and their place was occupied by immigrants from Babel, Cutha, Hamath, and other places. These brought their idolatrous worship with

them, but entered into close relation with the remnant of
Israel, and so became a very heterogeneous population, taking
the name of Cuthites from the former home of a portion,
or Samaritans from the home of another part. This strange
blending of populations was an incredible spectacle to the
Jews who had been carried into captivity. When that
portion of the Israelite territory was ravaged by wild beasts,
the new settlers looked upon this as a punishment by the
Divinity of the country, whom they had not known how to
propitiate. They begged, therefore, of the Assyrian king a
priest, and had their request granted. The monarch ordered,
as we learn from 2 Kings xvii. 27–41, that a Jewish priest
should be sent to them to teach them the manner in which
they might propitiate the Divinity of the country. This func-
tionary came, and lived for some time in Bethel, imparting the
doctrines which prevailed about God among the Hebrews.
Yet we learn from the book of Kings that this was all in
vain; each community of the Samaritans made its own God,
and set it up in the houses or on the high places: for while
they had a certain fear of Jehovah, they served their own
idols as their fathers had done. Thus it remained down to
the latest time. Separated completely from the Jews as they
were, the Hebrew and the Assyrian elements of the Samaritan
nation began to blend, and to become homogeneous. They
were heathen indeed, with the intermixture of a Jewish
element; for although we do not hear of any subsequent
immigration of Jews, yet we find the Samaritans asking for
Jewish priests. This shows conclusively[1] that they were a
mixed race, and that they cannot be considered a true heathen
people, although regarding this the opinions of commentators
were for a long time at variance. The Samaritan woman
at the well (John iv. 12) confesses to a common lineage
with the Jews when she asks, "Art Thou greater than our
father Abraham, who gave us the well?" by which words
Hengstenberg's view that the Samaritans were only heathen
is completely overthrown. Josephus gives the name Mannasses

[1] A. Knobel, *zur Ges. der Samaritaner*, in *Giessener Denkschriften*,
i. p. 130.

to the Jewish priest who was located in Samaria, and who afterwards married the daughter of the Assyrian governor, for which act his brother Jaddus, the high priest at Jerusalem, deprived him of his office. It was through the Jewish influence, which in religious things always was paramount, and which may have been constantly strengthened by the addition of Jewish refugees, that the capital was removed to the hallowed site of Shechem, although the people carried the name Samaritan with them, and always retained it.

When Zerubbabel and Joshua were beginning to rebuild the temple of Jehovah on Moriah, at Jerusalem, their opponents came from Samaria (Ezra iv. 10), and said to them (2-4), "Let us build with you, for we seek your God as ye do." But Zerubbabel and the chief Jews answered them, "Ye have nothing to do with us to build a house unto our God; but we ourselves together will build unto the Lord God of Israel, as king Cyrus the king of Persia hath commanded us. Then the people of the land [Samaria] weakened the hands of the people of Judah, and troubled them in building." Upon this the Samaritans accused the Jews of rebellious designs against the Persian government. Under the foreign yoke of the Persians, the Seleucidæ, and the Romans, the division between the two peoples must have been continually growing greater, inasmuch as political was joined with religious hatred.[1] The subsequent civil arrangement of the districts of Palestine—Southern, Middle, and Northern—must have contributed its share to perpetuate this national hatred. "To the Samaritans and the Philistines, as well as to the stupid Shechemite populace, I am a hearty enemy," says Jesus Sirach (l. 28). The Samaritans were put under the ban by the Jews, and Jesus himself calls the Samaritan in this sense a stranger (ἀλλογένης), Luke xvii. 18. At another time His disciples wondered that He talked with a Samaritan; and the woman at the well was equally surprised at His asking a draught of water from her, for, said she, "the Jews have no dealings with the Samaritans."

Josephus says of them, that for political reasons the Sama-

[1] Reland, *Pal.* p. 180; von Raumer, *Pal.* pp. 127-131.

ritans gave themselves out as Jews when it favoured their interests; as, for example, at the time of Alexander the Great, who showed great kindness and consideration to the Jews. On the other hand, they concealed their connection with the Hebrew race when it seemed expedient to do so; and in a letter which they sent to Antiochus Epiphanes, whom they addressed as God, they called themselves Sidonians, and besought that they might be permitted to give the name of Jupiter Hellenius to their temple on Gerizim. At a subsequent period we find the Samaritans contending before the Egyptian king Ptolemy Philometor, that it was not the temple in Jerusalem, but that on Gerizim, which had been built in accordance with the law of Moses. They claimed the latter to be the true temple, because, according to their assertion, it was there that the twelve memorial stones which had been taken out of the Jordan had been set up; and the words of the Samaritan woman (John iv. 20) hint at the same. Reland remarks that it was the people alone who were held by the Jews as unclean, but not their land, their water, nor their mountains; and on this account the Galilæans could take their course to Jerusalem through the heart of Samaria without polluting themselves. He states, however, that there were certain places where Samaritans were forbidden to live; for example, the country around Tiberias, Nazareth, Diocæsarea, and Capernaum. In other places, however, they were permitted to locate themselves, and in some they acquired great influence. Silvestre de Sacy[1] derives the name of the people not from the city of Samaria, because this is denied by the fathers, but from Shomer, pl. Shomerim, *i.e.* "to guard." He terms them therefore the watchmen, as did St Epiphanius also, and ascribes to them the function of being the true guardians of the laws of Moses. He finds the same etymology in Eusebius and Jerome: "Rex Chaldæorum ad custodiendam regionem Judæam accolas misit Assyrios, qui emulatores legis Judæi facti, Samaritæ nuncupati sunt, quod latina lingua exprimitur custodes." This

[1] In *Notic. et Extr. de la Bible du Roi*, T. xii. 4-6; *Correspond. des Samaritains de Naplouse;* Wilson, *Lands*, etc., ii. p. 46, Note.

etymology was first accepted from the Samaritans themselves: the Jews have not used the word down to the present time, but have called this people Cutheim and Cuthæi, because the great proportion of the Assyrian colonists appear to have come from the province of Cutha. It is only through the use of the Greek language in the New Testament, and through the diffusion of Josephus' writings, that the name Samaritans became common. If the word had been really derived from Shomeron, Wilson thinks that they would have been called not Shomerim, but Shomeronim. He accordingly adopts the conclusion, that it is far more probable that the name Samaritan is derived from the city Samaria, and the district which they inhabited for so long a time.

Regarding the province of Samaria, which is generally omitted by the Jews when they speak of the districts of Palestine, but which is included in Josephus' list of the toparchies Samaria, Galilee, and Peræa, I have spoken in a previous part of this work. I have there not only given a general sketch of the district and its boundaries, but I have also alluded in various places to points on the boundary between it and Benjamin or Judæa. The limits on the north are the Carmel range and the plain of Esdraelon, on the east the desert of the Ghor, on the west the Mediterranean coast. Yet, definite as this seems to be, and contracted as are the limits of the province thus enclosed, it is impossible to trace the boundary line with absolute precision.[1] We cannot tell, for example, whether Antipatris, Athlit, Dor, Bethshean, Hepha, and Jezreel, are or are not to be reckoned among the cities of Samaria.

The present southern boundary between the province of Jerusalem in the south, and the province of Nablus in the north, begins with Wadi Belat, and the villages lying partly on its northern and partly on its southern side; and it was exactly in this region that the ancient *Via Romana* deviated from the great northern road running northward to Damascus, and ran north-westward towards Antipatris and the sea. The present southern boundary appears to coincide with the ancient

[1] Reland; v. Raumer, *Pal.* pp. 128, 150.

one, which separated it from Dan and Benjamin, and passed not far from Bethel (Beitin), Gophna (Jifna), Ophra (Taiyibeh), Ain Si'a, Bir es Zeit, and through Sinjil. The territory of Ephraim was not very different from that which was subsequently called Samaria, but parts of Manasseh and Issachar belong to it also. Ginæa, the present Genin, at the entrance of the plain of Jezreel, was the frontier city of Samaria on the north.

If now we follow the few travellers through this region who are able to throw light upon it—the only way of becoming acquainted with it, since the accounts of antiquity and of the middle ages are very meagre concerning it—we shall discover that only a very limited region has been explored, and that outside of that the whole district is *terra incognita*. I have already referred to the researches into the region towards the Jordan, effected by Robinson, Barth, Berggren, and Schultz, and relating to Rimmon, Taiyibeh, Sinjil, Seilun (Shiloh), Turmus Aja, and Karijut, and the ancient Acribitene. I have also alluded to the researches made west of the main road by Dr Eli Smith and others in search of Antipatris. It only remains for me to speak of the great highway running northward from Jerusalem to Nablus and Samaria, for no other one has ever been taken by the countless tourists who have traversed the country between these two cities. All who have added to our knowledge of the country have done so by either turning aside here and there from the main route, or by crossing the country in an exactly opposite direction.

The heights of the main geographical features of Samaria are as follows, to which I add that of two or three others for purposes of comparison:—

Jerusalem, 2349 Paris feet, von Wildenbruch; 2472, von Schubert.
Ain Yebrud, north of Bethel, 2208, von Wildenbruch.
Sinjil, near Turmus Aja, 2520, von Schubert.
Nablus, 1568, von Wildenbruch; 1751, von Schubert.
Samaria, 926, von Schubert.

Gerizim,[1] 2398, Schubert.
Genin, 514, Schubert.
Esdraelon, 438, Schubert.
Convent on Carmel, 582, Schubert.
Peak of Mount St James, 1500, Schubert.

DISCURSION I.

THE NABLUS ROAD FROM BEITIN (BETHEL) BY WAY OF JEFNA (GOPHNA), SINJIL, SEILUN (SHILOH), THROUGH THE PLAIN OF MUKHNA TO NABLUS (NEAPOLIS, SHECHEM).

We have already examined the route running northward from Jerusalem to Bireh and Beitin (Beeroth and Bethel), and which forms the commencement of the main highway to Nablus. It was in Bethel, the so-called house of God, that Abraham pitched his tent, and that Jacob erected an altar. It was thither that the road ran up from Gilgal, over which the great prophet passed every year to render judgment at Bethel. Only an hour and a half west[2] of Bireh, Dr Eli Smith discovered the station Jefna, lying four hours north from Jerusalem, and directly in the route which he took in his search for the ancient Antipatris. Robinson had, however, passed over the same route at a still earlier date.

It was in this divergent road westward, which offered, according to Maundrell,[3] a great contrast to the bare and repulsive Judæan hills farther south, that traces of a *Via Romana* were discovered, which continued to be observed by Dr Smith at frequent intervals all the way to Cæsarea. At Jefna a good piece of this was in a fine state of preservation. This village was conjectured by Robinson to be the ancient Gophna, mentioned by Josephus and Ptolemy. The name does not occur in the Scriptures; but Josephus asserts that Titus, while on his march from Cæsarea to Jerusalem, passed through Gophna. The fertility of the valley accounts for the fact that it was confounded in the *Onomasticon* of Jerome

---

[1] D. Steinheil, *Resultate, aus v. Schubert's Reise, in Gel. Auz. d. bayersch. Akad.* 1840, No. 47, pp. 382, 383.

[2] Robinson, *Bib. Research.* ii. 261.   [3] Maundrell, *Journey,* 64.

with the Vale of Eshcol,—an error, however, into which Eusebius does not fall.

The village of Jefna contains, in addition to a spring of living water, some ruins of not insignificant appearance, among which the most striking are those of the Church of St George, and a fortress. Robinson thought that there was some ground for considering the place as occupying the site of the scriptural Ophni mentioned in Josh. xviii. 24, it being in the neighbourhood of Bethel, Ophra, and other well-known cities.

The next place visited by Robinson lay north-east of Jefna, and bore the name Ain Sinai.[1] Here commences a well-watered valley which extends north-westward and then westward to the Mediterranean, and bore the name Wadi Belat. On the western side of this lies Atara, perhaps the ancient Atharoth on the borders of Ephraim. Robinson did not visit it, however, for he lost his way; and after passing through the village of Jiljilia, he came back into the main Nablus road. Others—as, for instance, Wolcott in 1842, and Wilson in 1843, who kept on the main road—passed Ain Yebrud,[2] Jibea, Ain el Haramiyeh, and reached Sinjil. Ain Yebrud has a very fine position, surrounded on all sides by fruitful valleys and hills, those on the west affording an unobstructed view.

Going north-north-east from Yebrud, a half-hour brings one to an eminence from which the ruins of a fort can be seen, bearing the name el-Burj Azzil. Farther down in the valley, on whose eastern side ruins are soon passed, ten minutes bring one through the deep Wadi el Jib, lying along the northern base of the ridge. On the west side of this wadi Wilson discovered a place bearing the name of Jibea, from which the wadi probably derives its own designation. Robinson's map gives the place according to the date assigned it by Maundrell; he himself did not visit the place. He thinks, however,[3] that it is the Geba of Eusebius and Jerome, which

---

[1] Robinson, *Bib. Research.* ii. p. 264.

[2] Wolcott, *Excursion* in *Bib. Sacra*, 1843, vol. i. No. 1, p. 71; Wilson, *Lands*, etc., ii. pp. 40, 290.

[3] Robinson, *Bib. Research.* ii. p. 265, Note.

lay five Roman miles north of Gophna, on the road to Nablus. Yet Robinson thinks that they wrongly confuse it with the Gebim of Isa. x. 31. With this Wilson concurs, although he does not hold it to be the Gibeah of Phinehas on the mountains of Ephraim (Josh. xxiv. 33), where his father, Eleazar the son of Aaron, died and was buried, mention being made of a mountain there. Yet Jibea lies very high certainly, and it is worthy of the attention of future travellers; for here was probably[1] that sacred city of Benjamin, in which the high priest was buried. The tradition of the Jews and Samaritans transfers that spot to the neighbourhood of Shechem, for which there is no historical basis. From Wadi el Jib, Wolcott went northward over the mountains, and reached a great water basin called Ain Haramiyeh, lying in a narrow and beautiful valley. The bevelled stones seen there seemed to indicate the existence of a castle at that spot. A half-hour from that point the deep valley coming from the east is left, a village known as et-Tell is passed, and the watershed is reached, from which the road runs through Wadi Sinjil, east of the village of that name, to Seilun with its lovely valleys close by, the former resting-place of the tabernacle. This was the Shiloh of the Scriptures, which, as Josephus (*Antiq.* v. 1, 19) says, was chosen as the seat of the Jewish worship, on account of its convenience and its attractive character. Robinson and Wilson have examined the character of this place, and its topography, and demonstrated its identity with the ancient Shiloh, as described in Judg. xxi. 19, as "a place which is on the north side of Bethel, on the east side of the highway that goeth up from Bethel to Shechem, and on the south side of Lebonah." Jerome, however, was unable in his day to discover any remains of the ancient city: "Silo tabernaculum et arca Domini fuit, vix altaris fundamenta monstrantur."

Robinson passed by way of Jiljilia and Sinjil to Seilun.

The large village of Jiljilia lies on the western edge of the mountainous tract, and from it there is an extensive view westward over the low coast plain, and also eastward as far

[1] Keil, *Comment. zu Josua*, p. 410.

as the mountains of Gilead on the farther side of Jordan. Here was the place, too, where Robinson first saw the lofty height of Hermon. Close by the village on the north side begins a broad valley extending east and west, which unites with the Wadi el Belat. Farther north Wadi Lubban can be descried, which comes down from the main road to Nablus, and also enters Wadi Belat. All these wadis come together not far from Ras el Ain, and form the bed of the Nahr Aujeh. Jiljilia, which probably is the modern name of a certain Hebrew Gilgal, of which, however, no memorials remain, cannot be the eminent Gilgal of the Bible, which was on the Jordan. It is thought by Keil,[1] however, to be mentioned in Deut. xi. 30, as lying in the neighbourhood of Gerizim and Ebal, where Joshua had pitched his camp when the Gibeonites came out to meet him. The site of the modern village answers well to these conditions. The present inhabitants appear to be a very timid folk, probably because they live off from the main highway, and seldom see travellers, or it may be because they held the strangers to be emissaries of Ibrahim Pacha.

The direct road from Jiljilia to Nablus is said to run through deep valleys, and to be a very difficult one. In order, therefore, to turn back to the main highway, the road to Sinjil[2] was taken, and the village reached in about an hour, lying on the high border of a wadi, perhaps two hundred feet above the level at the bottom. This high locality extends eastward to the broad plateau, on one of whose elevations lies the village of Turmus Aya. The main road runs by Sinjil, ten minutes distant from the village, and passes by the Khan el Lubban on its way to Nablus.

The distances on this route are the following: from el-Bireh to Beitin (Bethel), forty-five minutes; to Ain Yebrud, one hour forty-five minutes; to Ain el Haramiyeh, one hour thirty minutes; to the valley below Sinjil, one hour; to Khan el Lubban, one hour ten minutes;—altogether, six hours and ten minutes.

[1] Keil, *Comment. zu Josua*, pp. 148, 160.
[2] Robinson, *Bib. Research.* ii. p. 265.

Sinjil, where Robinson spent a night, has a population of two hundred and six taxable men, and eight hundred souls. A hundred of the men were compelled to bear arms.

On the fourteenth of June he left the place in order to examine the neighbouring village of Seilun.[1] He heard much about this place, but nothing which indicated that the country people suspected its historical interest. The result of his inquiries confirmed his conjectures, and led him to a discovery which must be regarded as one of the most important which he ever made. Even von Schubert,[2] who passed through Sinjil only a year before, and ascertained the barometrical altitude of the place to be 2520 Paris feet above the sea, passed by Seilun without suspecting its historical interest. In the neighbourhood of Sinjil he saw excellent fig plantations; on the limestone sides of the mountains he saw roses in bloom, which he recognised as belonging to the variety known as the *Rosa sempervivens*. During the nights in the middle of April there was a heavy dew. A half-hour from Sinjil, after passing through the valley, and then ascending the high land at the north, Robinson came to the fine plain, on a slight eminence on which lies Turmus Aya, with its surrounding margin of millet and wheat fields. A half-hour in the same direction are the ruins of Seilun, which, though encircled with hills, look down on the plain at the south. Five minutes from the place[3] are the relics of an ancient tower or a church, with four thick walls; the ruin is twenty-eight feet square, and within are three overturned pillars with broken Corinthian capitals. Above the entrance is an amphora carved between two garlands, and at the side is a wall thrown up evidently for defence. The chief ruins of the place lie on a knoll, which is separated by a deep wadi from a higher mountain at the north, and is well guarded against attack. Between the ruins of modern houses lie great stones and fragments of pillars. Under a stately oak at the southern extremity stands a little mosque. At the distance of a

---

[1] Robinson, *Bib. Research*. ii. 266; Bartlett, *Walks, etc.*, pp. 247-249.
[2] Von Schubert, *Reise*, iii. p. 129.
[3] *The Christian in Palestine*, p. 123, Tab. xxx. xxxi.

quarter of an hour a fine spring issues from the ground, which forms a well eight or ten feet deep, where many of the neighbouring shepherds water their flocks. In the narrow valley where the spring is found, Robinson noticed several opened tombs.

It was to this place under its ancient name of Shiloh that Joshua went up from Gilgal; and here it was that the tabernacle was erected, and the division of the country made to the several tribes. Here Samuel spent his boyhood in the service of the Lord, and here it was that he was called to be a prophet of the Lord, recognised as such from Dan to Beersheba (1 Sam. iii. 20, 21): it was at this place, too, that many of his greatest deeds were done. It was in Shiloh that a feast was made to the Lord every year, at which the daughters of Shiloh danced; and it was at one of these feasts that the Benjamites made an invasion upon them, as the Romans did upon the Sabines, and carried them away to make them their wives; for, as we read in Judg. xxi. 24, "at that time there was no king in Israel, and every man did that which was right in his own eyes." After the Philistines had carried the ark of the covenant away from Shiloh into their own country, the place was deserted of the Lord, laid under a curse (Jer. vii. 12, 14), and never named after the exile. Jerome can scarcely have known where its site was, and at the time of the Crusades it was utterly unknown: according to the statements of the monks, it was at Neby Samwil. A certain Bonifacius a Raguso, a guardian of the Holy Sepulchre, is the only one in the sixteenth century who appears to have been aware[1] of the real location of Shiloh.

Wilson also visited these ruins[2] of Seilun, which he reached in a walk of forty-five minutes from Khan Lebban. He found them more extensive than he had expected, and adds some particulars to Robinson's account. They lie, he says, on rising ground, surrounded by yet more elevated land, however. Among the shattered pillars and the ruins of comparatively modern buildings, he discovered an old arched structure,

[1] Quaresmius, *Elucidatio Terræ Sanctæ*, ii. lib. vii. fol. 798.
[2] Wilson, *Lands of the Bible*, ii. pp. 292-297.

which his guides called Mazarah, with two columns in the middle, and a space like that within a mosque: before the entrance there is a great scindian oak. Two bow-shots away from these ruins are still others, among them a pyramidal-shaped structure, which was called Jama es Sittim, the Mosque of the Sixty. The peculiar shape was owing to the pillars: the enclosed square was about twenty yards by fourteen. The whole seemed to Wilson to be very ancient. Over the entrance he noticed a carved jug, which reminded him of the manna jug on the ancient Jewish coins, such as those of Simeon the Just, for example: around the jug there were garlands and branches, in the style of those on Helena's grave. He also saw some inscriptions, which were so much effaced as to be illegible. Several pillars and Corinthian capitals were lying around. Wilson prepared a small sketch of the neighbourhood of Shiloh,—a name which Josephus[1] gives with many different spellings, but which cannot fail to designate the place which was so sacred among the ancient Hebrews.

From Seilun the road winds down through a deep valley, in which lie the ruins of Khan el Lubban: near this is a fine spring, and north-west of it, on a high slope, is the village of Lubban. The wadi continues westward through a narrow seam in the mountains. Robinson found several graves lying north of the village just named. This place seemed to him to correspond to the ancient Lebonah referred to in Judg. xxi. 19 as lying between Bethel and Shechem. Olshausen doubts this, however, while he admits the identity of the two names. Maundrell,[2] as early as 1697, conjectured that Leban, as he wrote the word, occupies the site of the ancient Lebonah.

From the fine basin of Lubban, which affords a view westward through a gap in the mountains, Robinson[3] went south-eastward through a narrow gorge, which widens towards the north into an open plain, on which stand the village of

---

[1] Winer, *Bibl. Realw.* ii. p. 459, art. *Silo*.
[2] Maundrell, *Journey*, p. 62.
[3] Robinson, *Bib. Research.* ii. p. 272; Wolcott, *Excur.* in *Bib. Sacra*, 1843, p. 73.

Sawich, and the khan of the same name. These lie upon the watershed, on the north of which begins another wadi, whose name Robinson could not ascertain, but which Wolcott found to bear the name Wadi Yetma. It runs parallel to Wadi Lubban, and enters the Nahr Aujeh. At the right, between olive and fig trees, there are two villages, Kubelan and Yitma, whose names may be found upon Robinson's map. Going northward from that point, Robinson discovered in the neighbourhood of Yitma the ground walls of a tower, whence the mountains of Samaria could be descried, and the extensive plain of Mukhna, on whose northern border lies the city of Nablus, occupying the site of the ancient Shechem.[1]

The many-peaked mountains of Nablus are seen from this place in all their beauty; and Gerizim, or Grisim, as it is now called, adorned with the wely on its highest point, crowns the view on the north. On the north-east is the entrance into the valley of Nablus. North of this entrance, and on the farther side of Gerizim and the valley, are the steep sides of Ebal. The long plain of Mukhna extends along the eastern base of the mountain range, its waving lines being discernible as far as to Nablus: on its eastern side it is bordered by gentle but attractive hills.

The steep descent from the ruins to the plain, which here forms a sharp angle, passes by a cistern; and the plain leads on the west to a narrow wadi, probably the Wadi esh Shaar of Wolcott. This, like the Wadi Lubban already mentioned, runs westward, and enters the Nahr Aujeh. It passes between the villages of Kuza and Ain Abus, which lie upon its two sides. The slopes surrounding the southern extremity of the Mukhna plain are beautified with cistus roses; the drier heights are overgrown with *poterium spinosum;* while the depressions and vales, from half an hour to three-quarters of an hour broad, are transformed into the finest fields of wheat and millet.

Robinson's route led through the valley, winding around the foot of the mountain, passing the height on which the village of Hawara lies, where the eastern declivity is steeper,

[1] Robinson, *Bib. Research.* ii. p. 273; Schubert, *Reise,* iii. p. 136.

and the plain broader. Farther on he passed the village of Kefr Kulin, which lies on the border of Mount Gerizim. The dwellers in the villages there appeared to be very much intimidated by the terrors of the Egyptian sovereign. The path winds along the base of Gerizim, and then leaves the broader plain, and enters the narrow valley running westward between it and the more northern mountain of Ebal, passing the ruins of the village of Belat. In the midst of this narrow valley stands a small white building in the form of a wely, called Joseph's tomb; and nearer the foot of Gerizim the people point out the ancient well of Jacob. Opposite, on the hills lying towards the north-east, there are three villages, —Azmut, Deir el Hatab, and Salim. From Jacob's well the path runs on through the narrow valley to another more copious well with reservoirs adjacent to it, but without trees: it continues then through an olive grove to the city of Nablus. On the north side of the town an uncommonly fertile valley extends westward, forming a noble field of vegetables, well watered, and forming a magic picture, with which there is nothing in Palestine to compare (Bartlett calls it the unparalleled valley of Nablus). Under an immense mulberry tree, and by the side of a murmuring brook, Robinson pitched his tent. The Jew, Mordecai of Bombay, who accompanied Wilson on his journey through Palestine, and who could not reconcile the boasted excellence of the country of his forefathers with the country as it now is, he having been reared in the tropical Indies, confessed that here was the true "land of promise," which "flowed with milk and honey."[1]

Wilson, who pursued the same route along the west side of the plain el-Makhneh,[2] names not only the village of Hawara, but also other villages, such as Baulin and Kafr Kallin, from which point he diverged on a side path in order to have a better view of Gerizim. When he at length entered the narrow Valley of Nablus, the steep face of Ebal which encountered him, which is usually so sterile, seemed to him to be overgrown with the Indian fig tree.

[1] Wilson, *Lands of the Bible*, ii. p. 45; Bartlett, *Walks*, p. 250.
[2] Wilson, *Lands, etc.*, ii. pp. 43, 45.

DISCUSSION II.

THE CITY OF NABULUS OR NABLUS, THE ANCIENT NEAPOLIS, THE ROMAN FLAVIA NEAPOLIS—SHECHEM AT THE TIME OF JACOB—MABORTHA, THE PASS—GERIZIM AND EBAL, THE MOUNTAINS OF BLESSING AND OF CURSING—THE CUTHITES, OR SAMARITANS—THE WELL OF JACOB AND THE GRAVE OF JOSEPH.

The city of Nablus,[1] or, according to Abulfeda's orthography, more strictly Nabulus, the Neapolis of the Romans, lies along the north-eastern base of Mount Gerizim a half-hour west of the great plain of Mukhna, and in a valley between Gerizim and Ebal, and extending westward for a considerable distance. The houses of the place are high and well built, the material being stone: there are cupolas upon the roofs, as at Jerusalem. The valley between the mountains runs south-east and north-west, and has a width of about 1600 feet. It forms a true saddle, the city of Nablus lying on the watershed. From it the springs on the east side run to the Jordan, while those on the west side send their discharges to the Mediterranean.

Before the time of Robinson this peculiarity had been unnoticed, and the reason undetected why Nablus should be the medium of commerce between the Jordan valley and the Mediterranean. The extensive bazaars of the city show even now the magnitude of the trade between Damascus and the places on the coast. North and south of the town rise the sides of Ebal and Gerizim, mostly sterile and bare, their vegetation being mainly confined to a few olive trees. They ascend to a height eight hundred feet above the city, which itself lies fifteen hundred feet above the level of the sea. The wall of Ebal, on the north side of the city, is full of ancient burial-places; on the south side of the town, along the base of Gerizim, there are springs and trees.

Wolcott[2] investigated the three wells which supply the city with water. The Nahr Kuriyum he found to flow as a

[1] Robinson, *Bib. Research.* ii. p. 275; v. Schubert, *Reise*, iii. p. 142.
[2] Wolcott, *Excursion*, in *Bib. Sacra*, 1843, p. 73.

strong stream through the upper part of the town, it first coming to the light under a dome-shaped structure, where there is a flight of steps leading down to the spring. Ras el Ain, the second source, issues from a gorge a hundred rods south of the western extremity of the city, and sends a supply of water through an aqueduct to the town. Directly below this, and within the city, is the third spring, Ain el Asal. Buckingham [1] speaks of a fourth, but Wolcott could find no trace of it.

The name given by the people to the mountain of Ebal, on the north side of the city, was not ascertained by Robinson. Wolcott, however, calls it Sitti Salamiyeh,—a designation which he does not claim to have had from unquestionable authority. Gerizim is still called, as it was in former times, et-Tur: even in the life of Sultan Saladin it is designated as Tourum, and only the Samaritans retain the old scriptural name as given in Deut. xi. 29. When Joshua, after the destruction of Ai, was following up his victory, we are told in the book bearing his name (viii. 30), that he built an altar upon Ebal, as the Lord had commanded him, using whole stones, which had never been touched with iron, *i.e.* which were up to that time inviolate. There he offered burnt-offerings and thank-offerings; and after covering the altar with plaster, he inscribed upon it all the words of the law, in accordance with the commandment recorded in Deut. xxvii. 2. In Josh. viii. 33, 34, we read: "And all Israel, and their elders, and officers, and their judges, stood on this side the ark, and on that side, before the priests the Levites, which bare the ark of the covenant of the Lord, as well the stranger, as he that was born among them; half of them over against Mount Gerizim, and half of them over against Mount Ebal; as Moses, the servant of the Lord, had commanded before, that they should bless the people of Israel. And afterwards he read all the words of the law, the blessings and cursings, according to all that is written in the book of the law." This solemn transaction was the fulfilling of the command of God, and the public commemoration of the promise that those

[1] Buckingham, *Trav. in Pal.* ii. pp. 421–474.

who kept the law should be blessed, and that those who disobeyed it should be cursed (Deut. xi. 26-28). The order of the ceremony had all been prescribed[1] in advance. The command had been given that the ark should stand still in the valley of Shechem, and that six of the tribes should take their places on Gerizim, and pronounce the blessings which should follow obedience; and six upon Ebal, and pronounce the curses which should follow disobedience; or rather, should listen to the blessings and curses, and seal them with an audible Amen. There have been various reasons assigned and inquiries made why the curses should have been pronounced from Ebal, the mount on which the altar was built, and not from Gerizim. But there is a very natural and simple explanation of the fact, notwithstanding what Schubert says, that Gerizim is better adapted for an altar than Ebal. The Levites who guarded the ark of the covenant were always compelled to stand with their faces turned towards the rising of the sun; and in this case, as we are expressly told by Josephus (*Antiq.* iv. 8, 44), they took their usual position. At their right hand, which was always the place of honour, was Gerizim—the natural location, therefore, for the blessing to be pronounced; while at their left, which was always the subordinate place, was Ebal. Moreover, there was nothing unnatural or contradictory in the fact that an altar was erected on Ebal. The curse had nothing to do with the mountain on which it was pronounced, but only with the transgressors of the law; and the altar was an impressive memorial of the fact that Israel had nothing to fear from the curse, so long as it should live in covenant relations with Jehovah.[2]

The Samaritan copies of the Pentateuch deviate from the Jewish text in this, that they do not locate this altar upon Ebal, but upon Gerizim, the mountain esteemed hallowed. Their priests have charged the Jewish scholars with corrupting the original in this respect; an accusation which was supposed by Kennicott and other earlier writers to be well founded,

[1] Keil, *Comment. zu Josua*, p. 153.
[2] *Ibid.* pp. 153-155.

but which has been disproved by the more modern critics.[1] Maundrell,[2] as recently as 1697, was the first traveller to set the public right respecting the character and appearance of these two mountains. He cites the opinions of the Samaritan priests of that period—from whom, however, he extorted the confession, that not a trace of an altar could be found on Gerizim—to support his opinion that the original one was built on the side of that mountain. The opposite side of Ebal, Maundrell found to be not less favoured by nature for the erection of an altar than was that of Gerizim: the wall of Ebal seemed, perhaps, a little more barren than that of Gerizim opposite, in consequence of the direct rays of the sun falling upon it. Ebal was not ascended by any European till the visit of Bartlett,[3] whose adventure was not unattended by danger of being robbed by the wild inhabitants of the district. His visit was so hasty, therefore, that the scientific gain from it was very slight. He ascended the mountain from the western base, passed by a small wely, reached the summit, and rode over a rough tract a mile across, without encountering a single human face. There were traces, indeed, of former habitations, but none of any importance. The view was satisfactory enough, however, to recompense him for his toil, the trans-Jordan district being visible; Gerizim, with its ruins, at the south; the fair vale of Nablus between, extending itself, by a slight slope, to the Mediterranean; and the sea itself beyond all. On his return he was met by some reapers, who threatened to attack him, but from whom he happily escaped without suffering violence.

Through a gorge south-west of the city Robinson passed on his way up Mount Gerizim, an eminence which is steeper and more difficult to ascend than Ebal, but not inaccessible. The summit was reached at the end of twenty minutes, and was seen to be no peak, but a tract of table-land, extending west and south-west. A walk of twenty minutes more brought him to a wely, standing on a small elevation on the eastern

---

[1] Keil, *Comment. zu Josua*, p. 150, Note 7.
[2] Maundrell, *Journ.* Mar. 24, p. 59. Comp. Robinson on the same.
[3] Bartlett, *Walks*, etc., p. 251.

edge of the mountain, and serving the Samaritans as a kind of temple, to which they go up four times in the year in order to hold divine worship. This seems to occupy the highest point of all, and from it an extensive prospect is gained. From it not only can many villages be discerned,[1] but the summit of Hermon is also visible. Wolcott did[2] a great service to chartographical science in taking numerous angles from this point.

Robinson was shown[3] the place where the Samaritans sacrifice on Afseh, *i.e.* the feast of the passover, seven lambs as an offering for sin, believing that bloody offerings are more acceptable to God than the fruits of the earth, because "in blood there is life." The other occasions when the Samaritans ascend the mountain are on Whitsuntide, the feast of tabernacles, and the great day of atonement. The sacrifice is made upon a pile of rough stones, near which is a pit in which the offering is roasted: this must be eaten with bread and bitter herbs, according to their law. The Turks, in the wanton exercise of authority, often forbid the Samaritans making these religious excursions.[4] Wilson, who ascended by the same path with Robinson, speaks of passing a place called the Church of Adam, where the legend says that his first daughter Mokada was born.

Beyond the place where the sacrifices are offered,[5] lie the ruins of an immense structure of hewn stones, as if once a massive and strong fortification. It consists of two portions lying quite apart, each extending two hundred and fifty feet from east to west, and two hundred feet from north to south. The stones are bevelled, and are taken from quarries in the neighbourhood: the walls are nine feet thick. In the northern portion there is a Mohammedan wely and a place of burial.

---

[1] *The Christian in Palestine*, p. 95, Plate 23, and p. 121.

[2] Wolcott, as already cited, pp. 73, 74.

[3] Wilson, *Lands*, etc., ii. p. 66.

[4] Stanley gives, in *Hist. of the Jew. Ch.* i. 119, a vivid picture of the celebration of the passover on Gerizim, with the literal usages which must have marked it in the most ancient times.—ED.

[5] Robinson, *Bib. Research.* ii. p. 277.

The Samaritans merely call this place "the Castle," and connect no sacred associations with it: Robinson regarded it as a structure put up by the Emperor Justinian. Wilson heard it called es-Luz and Bethel. Beneath the walls of the castle, the guides of the last-named traveller asserted that the twelve stones lie which were brought by the Israelites from the Jordan. The Samaritans believe that they will lie there till the Messiah, alluded to in John iv. 25, shall come. Benjamin of Tudela[1] asserts that upon these stones the Samaritan temple at Gerizim was built. South of the pile of ruins, Wilson's guide drew off his shoes, pleading that it was holy ground, and that he was forbidden to tread it except with bare feet. A few steps farther west lies a naked patch, which is said to have been the place where the tabernacle stood. The guide had never heard anything about the existence of a temple. Yet in the neighbourhood there were traces of ruins, which seemed as if they might have once formed part of a temple. The dimensions appear to have been about sixty feet from north to south, and forty-five from east to west. After the destruction of the first temple erected on Gerizim, however, which stood about three hundred years, and was razed by John Hyrcanus, it seems not to have been rebuilt, although Mount Gerizim was for a long time placed on the Roman coins of Neapolis,[2] probably from the fact that worship was still continued at an altar on the mountain. According to Photius Damascius,[3] a temple in honour of Jupiter was erected on Gerizim. This place served the same purpose to the Samaritan that the kaaba does to the Arab: it was the object to which he turned while offering his prayer. Near by the place is pointed out where, at the command of Jehovah, Abraham intended to sacrifice Isaac: the mountain hence bore the name of Moriah, and the burial of the dead was prohibited upon it, and could take place only at its base.

Wilson, to whom the same place was shown, considered it

---

[1] Benjamin von Tudela, ed. Asher, i. p. 66.
[2] Robinson, *Bib. Research.* ii. p. 292.
[3] Von Raumer, p. 145, Note 131.

the site of a temple, but saw no traces of masonry—nothing but an excavation in the rock sloping gently westward towards a small tank. In the neighbourhood there is a spring, near which the Samaritans believed that their expected Saviour would make his appearance.

South of the spot here alluded to, Robinson discovered extensive ruins, which seemed to indicate the former existence there of a city: there are also the traces of numerous cisterns, all of them destitute of water.

The view from Gerizim is very extensive, and is of an entirely different character from that presented in the neighbourhood of Jerusalem: all here is fresher and greener. From Sinjil northward the hills are less high and steep, the valleys are attractive and fertile, and assume the form of plains and basins. Of these, Mukhna is the largest. Yet, notwithstanding the extent of the view, it is by no means so interesting, historically speaking, as that from the Mount of Olives. Hermon is not discernible, it being shut out from sight by the intermediate Ebal. North-east of the Mukhna plain Salim can be seen, which used formerly to be identified[1] with Shechem. This latter place is the very ancient city which, as early as the time of Joshua (Josh. xx. 7), was considered sacred, like Kedesh and Hebron. It was considered the middle point of the whole country.

The locality[2] known as Shechem, and the city bearing that name, are mentioned as early as the time of the patriarchs (Gen. xii. 6, xxxiii. 18, xxxv. 1). Abraham went thither while the Canaanites were still in possession of the land, and pitched his tent there, and then went on to Bethel and Hebron. From the place last named the sons of Jacob went to pasture their father's cattle at Shechem; and in that neighbourhood they caught their brother Joseph, and sold him into the hand of the Midianites (Gen. xxxvii. 12, 14, 28).

Through the erection of Abraham's altar at Shechem the

---

[1] Wilson, *Lands of the Bible*, ii. p. 72; v. Raumer, *Pal.* p. 145, Note. Comp. Gross, Anmerkung, in *Zeitsch. der d. Morgen. Ges.* iii. p. 55.

[2] Reland, *Pal.* pp. 1004-1010; Robinson, *Bib. Research.* ii. p. 280. Comp. v. Raumer, *Pal.* p. 144.

place was consecrated to the worship of Jehovah; and in consequence, after the entrance of the children of Israel into the country, the bones of Joseph were deposited at Shechem, in the parcel of land which Jacob purchased of the children of Hamor, the father of Shechem, for a hundred pieces of silver (Josh. xxiv. 32).

As a city of the Levites, at a time when there could be no mention of Jerusalem, it became the central point of union to all the tribes: during the epoch of the judges it was conquered by Abimelech, burned, and utterly destroyed (Judg. ix.). Rebuilt at a subsequent period, it was the place where Rehoboam consulted with the leaders of the people, and where he uttered his threat, to be afterwards so bitterly paid for, that whereas his father chastised them with whips, he would chastise them with scorpions (1 Kings xii. 14, 15). During the exile Shechem is mentioned (Jer. xli. 5); and after the exile, notwithstanding the fact that Samaria had been the previous capital[1] of the country (Neh. iii. 34; Ezra iv. 10), at the building of the new temple on Gerizim, Shechem, which was hard by, was made by Manasseh, probably before the time of Alexander the Great, the chief centre of the Samaritan worship; after which time its inhabitants were an especial object of Jewish scorn. According to Josephus, John Hyrcanus destroyed this temple on Gerizim 129 years before Christ, and after it had stood about 200 years. With this event the prophecy recorded in Amos vi. 1 was fulfilled: "Woe to them that are at ease in Zion, and trust in the mountain of Samaria, which are named chief of the nations, to whom the house of Israel came." The woman of Samaria makes no allusion in her conversation with Jesus to a temple on Gerizim, but simply says, "Our fathers worshipped in this mountain, and ye say that in Jerusalem is the place where men ought to worship." After the lifetime of the Saviour,[2] the site of Shechem became the place where the disciples laboured and formed churches (John iv. 39; Acts

---

[1] A. Knobel, *zur Ges. der Samaritaner*, in *Giessener Denksch.* 1847, i. p. 168.
[2] Winer, *Bib. Realw.* ii. p. 454.

viii. 5-25, ix. 31), the town taking the name of Neapolis in the writings of Josephus, Pliny, and Ptolemy, and being called on the Roman coins Flavia Neapolis. It would appear that the name is derived from Flavius Vespasianus, who restored the place on the site of the ancient Shechem.

Josephus says that Neapolis was called Mabortha by the natives; and Pliny, too, who died in the year A.D. 79, uses this expression (*Hist. N.* v. 13), "Neapolis, quod antea Mamortha dicebatur." Many explanations have been given of this term; but they are all so unsatisfactory, that Robinson was able to come to no satisfactory conclusion regarding them. The only one which seems to me to rest upon a satisfactory basis is the one which makes the word Mabortha a true Aramæan form, signifying "Pass," it being so strictly in harmony with the depression which is found between Gerizim and Ebal. The form Sychar,[1] used instead of Shechem, and which became current in the first centuries, is shown by Jerome to be an incorrect form. The name Agazaren, which is met in the *Bordeaux Itinerary*, is probably an abbreviation of Gerizim.

It took forty minutes for Robinson to descend from the summit of the mountain to the Samaritan synagogue in the south-western part of the city, and on a slight spur extending from the base of Gerizim. It is a well-built structure of large size, and comfortably fitted up.

The city proper has only two long streets, reminding Richter[2] of his pleasant Heidelberg home; and all the more from the fact that the city, which is surrounded by green gardens, ascends, terrace-like, the side of Gerizim for a little way. The main street[3] runs from east to west, and is furnished with many shops and storehouses: the bazaar is an extensive one, while the workshops of the artisans are in the most retired parts of the city. At the time when von Schubert passed through Nablus, a part of the town lay in

---

[1] Robinson. *Bib. Research.* ii. p. 291, Note 1; and Winer, *Bib. Realw.* ii. p. 455, Note 1.

[2] O. v. Richter, *Reise*, p. 56.

[3] Von Schubert, *Reise*, iii. p. 142.

ruins, the effect of the earthquake which had been recently experienced. The beautiful and well-watered gardens of the city, fantastically ornamented by the white minarets which peer above them, produced excellent oranges, lemons, pomegranates, apricots, which all flourish particularly well on the shaded side of Gerizim, while on the more exposed face of Ebal nothing grows but olive trees. The gardens are ornamented with little Turkish arbours, which are often shaded by fragrant orange trees, and surrounded with honeysuckle and clover.

Robinson estimated the number of Mohammedans[1] in Nablus at the time of his visit—1838—at 8000 souls. Besides these, there were about a hundred and twenty Greek Christians, who were subject to taxes, implying a population of at least five hundred souls,[2] reckoning women and children. The Samaritans proper number only about thirty men who are liable to pay taxes, and the whole Samaritan population would hardly surpass a hundred and fifty souls. There are, in fact, about as many Jews in Nablus as there are Samaritans. The province had its own governor. There seemed to be only one rich man among the Samaritans; all the others appeared to be in merely moderate circumstances. They had not the Jewish physiognomy; they were strict observers of the Sabbath; they recognised the well near by as Jacob's well, but designated it usually as the well of the Samaritan woman; they also asserted that the Mohammedan wely in the neighbourhood was the grave of Joseph.

Wilson, who was here in 1843, visited[3] the little Jewish synagogue, and found the number of that communion very small, numbering but twenty families and sixty souls. The rabbi asserted that many of the Jews at Jerusalem would find their way thither and make it their home, if the rabbi

---

[1] Robinson, *Bib. Research.* ii. p. 286.

[2] Consul Rosen states the population of Shechem in 1860 to be about 5000; of whom 500 are Greek Christians, 150 are Samaritans, and a few are Jews. See his article in *Zeitsch. der Deutsch. Morgenland. Gesel.* for 1860.—ED.

[3] Wilson, *Lands of the Bible*, ii. p. 62.

at Jerusalem would allow it. Only two of the Jews at Nablus were shopkeepers, one was a goldsmith, and all the others were poor people. The rabbi spoke very depreciatingly of the Samaritans, and asked why Wilson as a traveller put up with them, and not with his people. Wilson answered that Jews are to be found everywhere, but Samaritans only at Nablus. He then invited the Jew to come the next day, and call upon him in the Samaritan quarter. On the morrow the rabbi came, as he was invited; but when the Samaritan priest espied him, he asked, " Who has asked this creature to come to my house?" showing that there is no abatement in the old hatred.[1] Yet it ought to be remarked that, on the occasion of Wilson's second visit, the same Samaritan priest was very courteous to some Jews, who came in their need to the Englishman to get some pecuniary assistance. On the occasion of this visit, Wilson and his companion Mr Graham, a missionary at Damascus, obtained permission—hitherto refused to all Europeans—to visit an ancient church which had been transformed into a mosque. The only notable features within were a few columns of red granite, and a window profusely ornamented.[2]

Robinson met the priest of the Samaritans at the synagogue, and found him a man of about sixty years, clothed in a tunic of red silk, and wearing a turban. His companions had red turbans. Their ordinary language was Arabic. Their reception of him was very polite: they answered all questions, and were particularly desirous to hear about America. Their prayer-books and commentaries were written in Hebrew. They possessed the first volume of the London Polyglott, and confessed the accuracy of the Pentateuch there: they complained much of the changes made in the original by the Jews, and claimed that their own copy was much purer.

They drew off their shoes on entering the synagogue.

[1] Mr (now Dr) Graham tells me that on a recent visit at Nablus, made in Jan. 1865, he visited both the Jewish and Samaritan priests, and that the signs of hatred were more manifest than ever.—ED.

[2] Wilson, *Lands, etc.*, ii. p. 297.

The building is small, and is simply arched over. It has an alcove, behind whose curtain[1] lay their religious writings, on which they laid great worth. One of these manuscripts, a roll of parchment done up in silk with great care, is said to date back to the time of Abisua, the son of Phineas, the son of Eleazar, and to have an antiquity of 3460 years: all the others are of much more recent origin. The priest alone attended to the copying of them. Wilson, who wished to purchase one of the ancient Samaritan manuscripts, was unable to do so. He learned that when the quarterly visit to Gerizim is made, extracts from the books of Moses are read, but that the other books are repeated only in the Jewish synagogues. Wilson and Graham have given an account of these manuscripts in the *Lands of the Bible*. At the time of Wilson's second visit, he found the floor of the synagogue covered with mats, and saw three marble slabs covered with inscriptions which were only seventy years old. The place devoted to prayer is so situated that the worshippers look directly out upon Gerizim.

When Wilson entered[2] Nablus he asked at the gate for the Samaritani, but the Arabs did not understand the word: no more did they the corresponding Hebrew term Shomeronim. But when he uttered the word Samarah they comprehended directly, and a young man took him at once to the Samaritan quarter. On the way he encountered a priest with a white turban and a white beard. "I am," he said, "Salamah Ibn Tobiah, the correspondent of the French scholar Baron de Sacy." He is probably the man who called himself Salamah Cahen[3] in the letters which passed between Paris and Nablus in 1808, 1820, and 1826. This man was delighted to find that Wilson, who came from India, brought him letters from Samaritans living in Bombay. That, he

---

[1] *The Christian in Palestine*, p. 107, Plate 24.
[2] Wilson, *Lands of the Bible*, ii. pp. 47-63.
[3] *Corresp. des Samaritains de Naplous*, 1808, in S. de Sacy, *Notic. et Extr. des MS. de la Bibl. du Roi*, T. xii. pp. 1-235; Daunou, in *Journ. d. Savans*, 1833, pp. 108-112; S. de Sacy, *über den gegenwärtigen Zustand der Samaritaner*.

said, was what he had long wished. He conducted his guest through a garden at the west extremity of the city, then through a dark passage, and to a staircase leading up to his dwelling over the synagogue; and said to his guest, Here you are to feel yourself at home. Upon Wilson expressing doubts whether the letters which he brought him from Bombay were from true Samaritans, the priest repeated at once the articles of his faith. They ran as follows:—

Allah Wahid (God is one).

Musa Nabiyah (Moses is His prophet).

Et Torah hi el Kutab (the Torah is the book of the law).

Karizim el Kiblah (Gerizim is the Kiblah).

Yakun vom el-keiamat wa ed-deinunat (there will be a resurrection at the last day).

In establishing these articles, he made many references to the Scriptures. In displaying the curiosities of the place, he showed his guest a very finely written copy of the Samaritan Pentateuch executed on paper. He read it with a peculiar intonation, quite different from that of the Jews. At breakfast he used a rare service of silver. Wilson[1] gave his son, an accomplished man of thirty years, a copy of the Arabic New Testament; and read with him and his father the fourth chapter of John, in order to draw from him his views of the Messiah. Wilson gives the conversation in his work, and shows on what grounds the Samaritans refuse to accept Christ as the Saviour who should come.

On the following day Wilson sat down to a finely served dinner; and in the evening he met an assemblage of men, women, and children, numbering forty-five in all, and completely filling the room. They were all eager to see and to converse with the stranger.

Wilson says that their appearance and deportment were striking, but not disagreeable, and that they resemble in many respects the Kathis of India. The most of them had a strong family resemblance. There was nothing Jewish in their physiognomy; their faces were much rounder than are those of the Jews. All the men wore red turbans, while the

[1] Wilson, as already cited, p. 51.

priests had a white one, and their hair done up behind their ears. Almost all wore striped woollen stuffs. Some of the children were very fair, and had the fine fresh colour of Europeans. The family of the priest claimed to descend from the tribe of Levi, all the others from Ephraim and Manasseh. The names which they bear are, according to Wilson, the same which occur in the period subsequent to Solomon, although with some Arabic modification.

They were unacquainted with any other Samaritan community than that at Nablus; that in Egypt, which was existing in 493, and had a synagogue there as well as in Rome, became extinct about 260 years ago. A century ago there were, indeed, according to Edrisi, scattered individuals of the Samaritans in Askelon, Gaza,[1] Joppa, and Damascus, but none now are to be found. Those living at Nablus never leave their own city, because elsewhere they find too many obstacles in the way of living and worshipping after their fashion: they cannot have any communion with Jews or Mohammedans in these things, and must always repeat their prayers before and after their meals. When Wilson told them that the people in Bombay who pretended to be Samaritans not only worshipped Jehovah, but also snakes and gods of wood and stone, they cried out in horror, *They* can be no Samaritans; *they* cannot accept Gerizim as their Kiblah.[2]

The conversation regarding their doctrines, festivals, and other peculiarities, lasted till late into the night. Wilson's report[3] of the conference establishes fully the accounts which had been brought to us by their own communications, with which they deserve to be compared. They practise circumcision and monogamy, and their prayers they regard as mere thank-offerings to Jehovah. They are strict observers of the Sabbath, making no fire, and doing no cooking on that day. They celebrate the first day in the year: the new moon begins the month with them, and they hallow its advent by prayer before and afterward. They are no husbandmen, but

---

[1] Edrisi, ed. Jaubert, i. p. 339.
[2] *On the Sect of the Beni Israel in Bombay*, in Wilson, ii. pp. 667-677.
[3] Wilson, *Lands, etc.*, ii. pp. 65-68.

merchants, copyists, weavers, tailors, and the like. Wilson was rejoiced at his second visit to accomplish what he failed to do on the occasion of his first—namely, to purchase[1] some of their manuscripts. In his volumes are to be found interesting statements regarding the literature of the Samaritans,[2] and some transcripts of their writings.

The eminent traveller Della Valle seems to have been the first European who purchased Samaritan manuscripts. This was in 1616. He brought them back to his own country, and thereby awakened much interest in the remnant of this long-forgotten and extremely interesting nation, which, though so small, connects the present with an epoch so ancient. The scholars of Europe immediately began to examine the contents of these writings, and their studies were shared with some of the most distinguished of eastern scholars. It was found that differences existed not only between the text, but also the contents of their Scriptures, and those which the Jews hold. Among those who took an active interest in these writings were the Scaligers; Job Ludolf, distinguished for his acquaintance with Ethiopian history; Maundrell,[3] who was prompted to make further personal investigations regarding the Samaritans; and Reland. Robert Huntington, chaplain to the English agency in Aleppo, visited the Samaritans in 1671, and amazed them by his familiarity with the language of their sacred books. They could not restrain the belief, in consequence, that there must be colonies of their countrymen in Europe; and they proposed to open a correspondence with them, hoping in this way to somewhat alleviate their poverty. This gave rise to communications, not with European Samaritans, but with European scholars, particularly with the Abbé Gregoire, who proposed to them the most eager questions regarding their numbers, dwellings, habits, customs, faith, their synagogue, their relation to the Caraites and other Jews, their sacrifices on Gerizim, their literature, etc. The most important of their answers, parti-

[1] Wilson, *Lands, etc.*, ii. p. 297.   [2] *Ibid.* pp. 687-701.
[3] Maundrell, *Journey*, Oxford, p. 60; H. Reland, in *Dissert.* vii. *de Samaritanis*, in his miscellaneous writings.

cularly those sent by Salamah Cahen, have been published by Silvestre de Sacy, and form a valuable body of testimony regarding this peculiar people, to our knowledge of whose character and peculiarities Robinson[1] and Wilson have since added so much. The history of Nablus and the changes in its population are fully[2] detailed by Robinson.

There are still two localities on the east side of Nablus which require description, Jacob's well and Joseph's grave, regarding whose genuineness various questions and doubts have been raised. The Samaritans regard them both as exactly what their names define them to be, although Joseph's grave is now surmounted with a mere Mohammedan wely. Christians sometimes give the name Well of the Samaritan Woman to the one to which the name of Jacob is almost uniformly applied.

Robinson found[3] this a half-hour's distance east of the city, entirely dry at the time of his visit, but bearing traces of antiquity. It was not only deep, but was said to have water in it at other seasons of the year. As it was evening, he was unable to take any measurement of its depth. That the original well was deep, is made abundantly clear by the passage in John iv., where the fact is distinctly stated.

Maundrell[4] long ago paid especial attention to this well, which has so great interest not to the Samaritans alone, but also to Jews and to Christians. To the objection that it lies too far from the city to be the well from which the woman drew water, he answers that the abundant traces of walls in the neighbourhood prove decisively that the ancient Shechem extended far toward the east, and that it was only when Neapolis was built that it began to reach westward. Over the well, Maundrell goes on to say, there once stood a great church, built by the Empress Helena, but of which there remained at the time of his visit only the slightest traces.

---

[1] Comp. E. D. Clarke, *Trav.* iv. pp. 272-280; A. Knobel, *zur Ges. der Samaritaner*, p. 129.

[2] *The Christian in Palestine*, p. 119, Plate 27.

[3] Robinson, *Bib. Research.* ii. p. 291.

[4] Maundrell, *Journ.* pp. 62, 63.

The well was arched over, and a flight of steps led down to the water. The excavation was a hundred and five feet deep, and three paces in diameter. The water was fifteen feet deep at the time of his visit. Maundrell's description appears, however, to be much overdrawn. As the well is found at the eastern extremity of the Valley of Shechem, it is not at all improbable that the plain formed a part of the land which Jacob gave to his son Joseph, and in which his remains were placed after being brought up from Egypt (Gen. xlviii. 22 ; John iv. 5 ; Josh. xxiv. 32).

In earlier times there stood, according to Bonifacius de Ragusio, in the year 1555, an altar in the vault over the well, and at this altar mass used to be celebrated once every year. The tradition regarding the well of Jacob and the grave of Joseph (which lies a little north of the well) goes back to the time of Eusebius, although he speaks only of the tomb. The *Bordeaux Itinerary*,[1] however, speaks in A.D. 333 of both the tomb and the well. He says that the latter was the one where the Saviour talked with the Samaritan woman, and that plane trees had grown up around it. Eusebius makes no mention of a church there; but Jerome alludes to it, stating that his pupil, the Roman pilgrim Paula, visited it in 404: it was therefore not built by Paula, as so many are falsely said to have been. All subsequent pilgrims speak of the church down to the time of the Crusades, when it appears to have been destroyed, as Brocardus finds it in ruins in 1283.

Although there is no proof of the identity of the present well and the ancient one known by the name of Jacob, yet its situation in relation to the city, on whose eastern side there still runs the great highway to Galilee, which unquestionably Jesus took, is such as to make it in the highest degree probable that the place is the one which has become so hallowed to all readers of holy writ.

Wilson speaks of finding the traces of a church[2] which must once have stood over the well. The mouth of the latter

[1] *Itin. Burdig.* ed. Parthey, p. 276.
[2] Wilson, *Lands of the Bible*, ii. p. 54.

was covered with two great stones, which he removed with the help of the Arabs. The opening he likens to one which was once covered by an arch. The diameter was only about two feet; the shaft was dark and deep. Three years before, the missionary Bonar had dropped his Bible down into the well; an Arab descended with a light, and found it dry. After emerging, the poor fellow seemed to be quite exhausted; but he soon rallied sufficiently to ask for bakhsheesh, and a sovereign completely restored him. The well was ascertained to be seventy-five feet deep, and had the appearance of great antiquity. The tomb of Joseph, which is coupled immediately with the well by the older travellers, lies two to three hundred paces to the north, directly across the valley. The present structure is small, but substantial, roofed over, and having altars at the ends which bear the appellations of Ephraim and Manasseh. On the walls within, Wilson[1] saw the names of many Samaritan and Jewish pilgrims, and also some writing in a character which was thought to be of Egyptian origin. Excavations carried on beneath this structure, which Wilson thinks nothing more than a Mohammedan wely, would lead, in his opinion, to valuable results respecting the authenticity of the place as the grave of Joseph. The Jews have lately made some repairs there, but have brought no new facts to light respecting it. Von Schubert describes a Mohammedan ceremony[2] which he witnessed there, but which seems to stand in no relation to the grave.

In modern times,[3] the neighbourhood of Nablus has been considered one of the most dangerous in all Palestine; and travellers going northward choose the road from Jaffa along the sea, in preference to the regular highway over the hills. The city used to be comprised[4] in the pashalic of Damascus; then in that of Acca: but the chief men of the neighbourhood were in reality all-powerful, and their sway was only supplemented by that of the pasha. In consequence

---

[1] Wilson, *l.c.* ii. p. 60.
[2] Von Schubert, *Reise*, iii. pp. 139-142.
[3] Robinson, *Bib. Research.* ii. p. 301.
[4] W. G. Browne, *Travels in Syria*, p. 359.

of this, the people have always been given to disquiet and
uprisings and robbery. Jezzar Pasha never succeeded, with
all his power, in reducing the people of Nablus to subjection; and so formidable were they, that the French never
subdued them, but were themselves routed, Junot with 1500
soldiers being compelled to withdraw from the field. It
was difficult to go safely through the city even with a military escort: from which circumstance so little was known
formerly regarding it. It was only when the strong hand
of Ibrahim Pasha had been laid upon the province, that the
local chieftains were compelled to yield, and the city was
made safe for travellers. Since then it has been visited by
large numbers of Europeans, and its topography and antiquities largely explored.

DISCURSION III.

THE ROAD FROM NABLUS TO SEBASTE, THE ANCIENT SHOMRON OF THE HEBREWS, THE SAMARIA OF THE GREEKS, THE SEBASTE (AUGUSTA) OF THE ROMANS, AND THE USBUSTE OF THE LOCAL POPULATION: THE ANTIQUITIES OF THE PLACE.

On the route from Nablus to the ancient Samaria, the
present Sebastieh, the Sebaste of former times, Robinson[1]
is again our most trustworthy guide, and all subsequent
explorers have acknowledged the fidelity of his descriptions
and the accuracy of his explorations. Leaving Nablus, he
went W.N.W. and N.W., following the upper course of Nahr
Arsuf downward. On his way he met an Egyptian caravan
laden with salt, going to Jenin, and so onward to Damascus.

There is a more direct road to Jenin, skirting the eastern
base of Ebal. Between the eastern and the western routes
there is a third path running northward directly over the
crest of Ebal. It is the most direct route from Sebastieh
to Nazareth and Esdraelon, but is seldom taken: the only
person who has mentioned its existence is Otto von Richter,[2]
who passed over it in 1816.

[1] Robinson, *Bib. Research.* ii. p. 302 et sq.
[2] O. v. Richter, *Wallfahrten*, p. 57.

The road running westward from Nablus passes by several springs, the valley being rich in them. It is to its ample supply of water that it is indebted for the fine gardens and orchards and fields which fill the valley, and which so entirely use the contributions of the numerous springs, that nowhere is there a brook of any importance. It is to the beauty of this region which Hosea refers when he says (ix. 13, following the German translation), "Ephraim, as I saw it, is as fair and as fruitful as Tyre."

When Wilson passed through this valley on the 26th of May,[1] the fields were literally "white to the harvest,"—an expression which seemed all the more appropriate from the fact that the grain is often allowed to hang for a long time ungathered when there is no promise of a cold winter. Two months earlier[2] the fruit-trees of the valley were mostly in bloom; the figs and olives were ripe; swarms of singing birds filled the air—a phenomenon which is exceedingly rare in Palestine.

The ancient Shomron of the Hebrews, Samaria, the present Sebastieh, or the Usbuste of the common people.

The site of this ancient city, says Robinson, is still marked by buildings, which climb the side of the hill on which it stood, and where is a narrow terrace, which runs round it like a girdle. Below this terrace the declivity inclines gradually towards the valley. Higher up there are traces of other terraces, on which it may be that the streets of the ancient city ran.[3]

The present village of Usbuste (926 feet above the sea, according to von Schubert) lies upon the eastern portion of this even girdle: it is modern, the houses being built of fragments of the ancient city. The inhabitants are represented as very restless, and given to uprisings;[4] yet Robinson[5]

---

[1] Wilson, *Lands of the Bible*, ii. p. 300.
[2] *Ibid.* ii. p. 80.
[3] Barth, MS. communication, 1847.
[4] Reland, *Pal.* pp. 979–983.
[5] Robinson, *Bib. Research.* ii. p. 314.

says that he had a cordial reception from them, while other tourists give a very unfavourable report in this respect.[1] The first thing which the traveller meets which surprises him is the ruined Church of John the Baptist, lying on the site where the legend asserts that he was killed and buried. The eastern extremity rises strikingly above the steep edge of the slope, and is seen before the village itself is descried. All the way up the declivity the sight is riveted, says Barth,[2] by the masterly architecture of the church, which, although executed in the middle ages, shows how powerful an influence Roman art exercised even after the era of the Crusades. The interior of the great recess is one of the finest and most ornate examples of the Roman style, and detains the visitor in long-protracted contemplation. Robinson[3] gives a very close description of this church. On the west side there is a narrow vestibule: the walls, which still remain very high, enclose a space, wherein a mosque and another small building are standing. The church has a length of about 153 feet, and a width of about 75 feet. The altar niche, which occupies a great part of the eastern rounded portion of the church, is an imposing specimen of mixed architecture, in which the Greek style predominates: three arches of the windows are uncommonly ornamented; the upper arches in the interior of the church are pointed, as are also the great ones in the nave. The last rest upon pillars, which belong to no special architectural order, the capitals of which, however, are an impure Corinthian. The windows are high and narrow. The whole church has the appearance of a military position, and the pillars on the outside have contributed their share to this effect. Within, Robinson saw some large marble tablets set in a modern wall, on which numerous crosses of the order of St John are wrought in relief. The Mohammedans, however, have done much to injure them. Robinson makes no allusion to a great arch for water, which others have mentioned as on the south side of the church, and

---

[1] Von Raumer, *Pal.* p. 143; Winer, *Bib. Realw.* ii. p. 368.
[2] Barth, *Reise*, MS. 1847; *The Christian in Palestine*, p. 116.
[3] Robinson, *Bib. Research.* ii. p. 365.

which Barth measured. He found the length up to the point where it is in ruins to be 140 feet, and its breadth to be 30 feet.

Tradition inaccurately ascribes this church to Helena: the eastern part may possibly date from the time of the Crusades. The many crosses found in it make it not improbable that it stood in connection with the Latin bishopric which was established here by the Knights of the Order of St John, regarding which we have no authentic historical testimony.

Within the ruins of the church the Arabs point out and hold in great devotion the grave Neby Yehya, *i.e.* of John the Baptist, a small chamber deeply hollowed out in the rock, to which twenty-one steps descend. The legend asserts that here was the place where the Baptist was for a long time a prisoner; but both Josephus and Eusebius transfer the scene of his confinement to Machærus, on the east side of the Dead Sea, and the legend which connects his name with Samaria appears to be of modern origin.[1]

Wilson was prevented visiting this church by the rude conduct of the people of the village. The foundation walls seemed to him of more ancient origin than the upper portion of the structure. Like Robinson, Wilson noticed other ruins on the south side of the place, but was unable to learn their original use and significance; and many of the fragments of rock lying in the valley he conjectured to be mere bits of the mountain which had been detached, and had rolled down the side. In the upper part of the village lay the threshing-floors; and here Robinson saw used for the first time a machine which was dragged by cattle over the grain, and which cut the straw. The whole mountain of Samaria, says Robinson, is fruitful, and cultivated to the top. Not a trace is to be seen of the ancient Shomron, the capital built by Omri, the king of Israel, which suffered the same fate with Shechem, being pillaged by John Hyrcanus. It was restored by Gabinius, however; but nothing remains of it now, with the

---

[1] Robinson, *Bib. Research.* ii. p. 306 et sq.; Wilson, *Lands of the Bible*, ii. pp. 82, 301.

possible exception of a few ruins on the very summit of the hill. The prophecy contained in Micah i. 6 has been literally fulfilled: "I will make Samaria as a heap of the field, and as plantings of a vineyard; and I will pour down the stones thereof into the valley, and I will discover the foundations thereof." The area around the summit is still strewn with limestone pillars, fifteen of which are in an erect and two in a horizontal position. They are almost eight feet in circumference. They are of a very uncertain architectural character, and appear to have once belonged to a heathen temple, although no foundations of such a structure are now to be found. Phocas and Brocardus speak of a church and a convent as existing there in their time, but Robinson was unable to detect any traces of Christian edifices. Wilson says that there are no capitals on the columns: they seemed to him to have once belonged to a quadrangle two hundred and twenty paces long, and eighty-four wide. Robinson speaks in terms of special warmth regarding the noble panoramic view afforded by this position, extending far and wide, and taking in the blue waters of the Mediterranean.

On descending the mountain, which is here and there overgrown with fine groups of olive trees, and which is so noble in its situation that Bartlett[1] compares it with the site of Jerusalem, Robinson discovered on the west side[2] a noble colonnade, which appears to have once run entirely round the mountain as far as to the locality where the modern village is situated. It begins at a heap of ruins, where once stood the tower of a temple, or an arch of triumph it may be, and from which there is an extensive view. It is not impossible that here was the former entrance to Herod's Sebaste. From that point the colonnade runs for a thousand feet toward the E.S.E., and then bears to the left, following the base of the mountain. In the western portion there are sixty limestone pillars now standing in the midst of cultivated fields; farther east there are twenty more standing at various

[1] Bartlett, *Walks, etc., l.c.* p. 255.
[2] Robinson, *Bib. Research.* ii. p. 307; Wilson, *Lands of the Bible*, ii. p. 301; *Christian in Palestine*, p. 110.

distances apart, and many more may be seen prostrate. Robinson was able to trace the fragments as far as to the village. The height of the perfect ones was only moderate, about sixteen feet; the capitals were very unlike in character; the diameter of some was less than two feet, and at the top one foot eight inches. The two parallel rows were fifty feet apart, and their whole extent, reckoning the many breaks where none are now found, was as much as 3000 feet. Robinson thought that they were unquestionably the remains of the colonnade which Herod built to adorn Sebaste Augusta in honour of his imperial patron. Wilson, on his second visit, inspected these ruins much more closely than he did the first time he was there, and agrees entirely with Robinson in his estimate, but adds that the space between the columns was about eight feet, the width of the avenue twenty-two paces of a horse, and the length of the colonnade 1172 such paces. He counted only seventy upright columns on the terrace. Judging from the fragments which he found, he considered that the Ionic order was the one in which they were finished. Wilson supposes that this colonnade formed the border of the tract in whose centre stood the temple erected by Herod. Josephus closes chap. viii. 5 of Book xv. of the *Antiquities* with a description of the structures built by Herod in Samaria. After giving an account of the other buildings erected by Herod at Jerusalem, Askelon, Cæsarea, and elsewhere, he says that the many conspiracies of the Jews against this tyrant were the occasion of his building the fortifications which he did, such as that at Gaba in Galilee, Heshbon in Gilead, and Sebaste in Samaria. To accomplish the last of these undertakings he carried thither a garrison of 6000 men, and set them at work at building a temple, which should serve alike for the purposes of worship, as a place of refuge in time of danger, and as a memorial of his magnificence. He hoped also to make this a central point, from which to extend undisputed authority over the whole population of the neighbourhood. He divided the adjacent land among his colonists; he surrounded the city which he called Sebaste (Augusta) with very strong walls, and availed himself of the steepness of the mountain for the purposes of fortification. The previous size of the city was not large enough to suit his purposes, as he determined to make it the equal of the most celebrated cities. It was twenty stadia in circumference, and the largest portion of the wall which surrounded it was of great strength. In the heart of the place was a sacred space reserved of three and a half stadia in circumference, within which he built a temple of remarkable size and beauty. Every part of the city, too, received its appropriate ornaments. Of all these magnificent works nothing remains but a few broken pillars and some fragments of hammered stone, and the prophecy of Micah has been literally fulfilled.

The history of ancient and modern Samaria may be found fully portrayed in the volumes of Reland[1] and Robinson.[2] The records of the latter are meagre, however, in the first centuries of the Christian era, and during the epoch of the Crusades. We know, however, that a Latin bishopric was established there, it being alluded to once or twice; and even earlier, at the time of the Council of Nicæa, the names of bishops occur at Sebaste, although nothing is known respecting the state of the church there. The New Testament informs us, however, that the gospel was preached by Philip in Samaria, and in its neighbourhood, before he went to Gaza, Ashdod, and Cæsarea, and that it was joyfully received, so that many men and women were baptized (Acts viii. 5-25). The present titular bishop of Sebaste resides in the convent at Jerusalem. I ought not to close without alluding to Wolcott's[3] valuable chartographical contributions respecting Sebastiyeh and its neighbourhood.

[1] Reland, *Pal.* pp. 979-983.
[2] Robinson, *Bib. Research.* ii. p. 309 et sq.
[3] Wolcott, *Excursion*, in *Bib. Sacra*, No. 1, 1843, p. 74, Note 2.

DISCURSION IV.

ROUTE FROM SEBASTE TO THE SOUTHERN ENTRANCE INTO THE PLAIN OF ESDRAELON AT JENIN, TA'ANUK, MEGIDDO, AND THE NORTHERN BORDER OF SAMARIA.

The northern routes from the two chief places in Samaria, Nablus, and Sebaste, usually conduct the traveller to the great Damascus road by way of Jenin. The route then traverses the extensive and celebrated plain of Esdraelon, and passes between Gilboa and Little Hermon, by way of the ancient Jezreel (the modern Zerin), and so on to Beisan, the Roman Scythopolis; or else it skirts Tabor, and runs directly to the Jordan. This portion of the route has been discussed fully in a previous part of the work, and need not be considered further here. But the road between Sebaste and Jenin, which has been described by Robinson, Schubert, Wilson, Barth, and others, has not yet been fully described in this work.

Much less known is the direct route leading northward from Gerizim and Ebal over Sanur to Nazareth; and the district lying west of Sebaste, and between it and the plain of Esdraelon, has been traversed by no European.

Robinson, who still remains our most trustworthy guide in these regions, says: There are two roads which may be taken from Sebaste to the great Damascus road at Jenin: a southern one, the more easy of the two, by way of Beit Imrin, and probably the one which was taken by Wilson[1] in May 1843, and of which he says that it is traversable by vehicles, this being impossible in other parts of Judah and Ephraim. The more northern one of the two, leading over Burka,[2] was the one which Robinson took. This village, which was reached in three-quarters of an hour's constant ascent, is large: it lies upon an elevated terrace, and, like all other Samaritan villages, it is surrounded with olive groves.

Directly north of this beautiful landscape a distant view

---
[1] Wilson, *Lands of the Bible*, ii. p. 502.
[2] Robinson, *Bib. Research.* ii. p. 311.

of the Mediterranean Sea[1] is gained. Northward there are to be seen the most charming plains. A fine broad valley running from east to west divides this irregular region into two parts, the eastern one of which forms an extended oval plain, on whose north-western side lies Sanur, although it is not to be seen. The western one is narrower, less regular in form, and inclines gently towards the Mediterranean.

The ruins of Sanur lie on a rocky height once defended with fortifications, which are now in a state of decay. In the year 1801, when Dr Clarke passed through this region, they bore a resemblance to an ancient Norman fortress, and were occupied by a hospitable chief, who gave him a courteous reception. Clarke called the place Santorri; and as he was unacquainted with the situation of Sebaste, he held the place to be the site of the ancient Samaria, although Maundrell, to whom he refers, was acquainted with the latter. Sanur lies so securely upon the rocky height which sustains it, that Jezzar Pasha besieged it for two months in vain.

Descending from this high land through fine olive groves and well-watered valleys, no long time transpires before the traveller reaches the city of Jenin, lying on the frontier of Samaria and the plain of Esdraelon. Wilson came over this route,[2] and made some halt under the olive trees outside of the city, where shepherdesses were milking their cows; their ornaments were of so striking a character as to recall the words of the Song of Solomon (i. 10): "Thy cheeks are comely with rows of jewels, and thy neck with chains of gold."

Wolcott,[3] who in leaving Sebaste took the road running northward by way of Burka, left Fendekumieh on the right, passed the wadi which is the usual route of travellers, and which runs eastward through Jeba, and went still northward over the rolling land till he reached the village of Ajjah. On a height at the west he saw the village of er-Rameh, which

---

[1] Robinson, *Bib. Research.* ii. p. 312; Wilson, *Lands of the Bible*, ii. p. 83; Barth, MS. *Reise.*

[2] Wilson, *Lands of the Bible*, ii. p. 83.

[3] S. Wolcott, in *Bib. Sacra*, 1843, p. 75.

commands an extensive prospect. Going on thence, first to the N.N.E. and then N.W., he passed through an open tract, with rounded knolls and broad green valleys; then over meadows surrounded by high hills, on which he saw several villages, among them Kubatiyeh. He had then approached tolerably near the ordinary route to Jenin, and shortly after he reached the village of Burkin. From this place Wolcott espied, a half-hour westward, Kerf Kud, whose site apparently corresponds to the Kaparkotnei of Ptolemy,[1] which was his conjectural Capernaum. In the Peutinger Tables, Capacorta is given as an intermediate station between Cæsarea and Scythopolis; but we know nothing further regarding it than that it was twenty-eight Roman miles from the former, and twenty-four from the latter. From Burkin, where Wolcott[2] passed a night, he went on through Wadi Rustuk, and entered a large plain on whose eastern side lies the village of Kefr Addan. Going northward, he passed a small wadi leading to Yamon, and in two hours reached the border of Esdraelon. In three-quarters of an hour more he came to the insignificant village of Taanuk, lying five minutes distant from the road, on the south side of a small hill, crowned by a bit of level land, and first observed by von Schubert.[3] A wely with a sculptured portal, and with the broken capitals of a pillar, convinced him that it was a place of some antiquity. It unquestionably occupies the site of Taanach,[4] the ancient Canaanite city, which is mentioned in Josh. xii. 21 in conjunction with Megiddo, and as one of the thirty-one Canaanite cities which Joshua took and gave to the Israelites. Though lying in the territory of Issachar, these conquered cities were reckoned as the property of Manasseh: the original inhabitants, however, were not driven out, but continued to live in them, and to pay tribute to their conquerors (see Josh. xvii. 11). At the time

---

[1] Reland, *Pal.* p. 460; Ptolem. ed Bertii, fol. 140, p. 161; v. Raumer, *Pal.* p. 402.

[2] Wolcott, p. 77.

[3] Von Schubert, *Reise*, iii. p. 164; Robinson, *Bib. Research.* ii. 306.

[4] Reland, *Pal.* p. 1032; v. Raumer, *Pal.* p. 148; Keil, *Comment. zu Josua*, p. 236.

of Deborah, the kings of the Canaanites strove at the waters of Megiddo, but they gained no advantage from their contest, and were eventually conquered by Barak. At the time of Solomon one of his twelve purveyors lived there, each one of whom had the duty of supplying the palace with provisions for a month. The district from which this one drew his supplies was perhaps the most fruitful in the whole land, and included Megiddo, Jezreel, and Bethshean. The city of Taanach lay, according to Jerome, three or four Roman miles from Legio. Robinson and von Schubert saw the place at a distance; Wolcott was the first to explore it.

From this place the last-named traveller advanced to the little village of Ezbuba, lying a half-hour away; and after ten minutes more he saw Salim with its olive groves and its mosque: it occupies the site which Robinson designates as Lejjun, but which he did not visit, contenting himself with looking at it from Zerin.

After fifty minutes more Wolcott reached the Nahr Lejjun, a stream which was at the time of his visit five or six feet wide, and which drove three or four mills. As he looked along the left bank of the stream he saw at a distance of ten minutes the ruins of a khan, but without a tree or any other object near. Here, Wolcott supposed, is the site of the ancient Legio, and here he closed his day's march, not without vivid thoughts of the great conflict which had once been witnessed in that spot, hard by the waters of Megiddo, at the time of the judges, of Sisera and of Jabin. Down the waters of the Kishon the bodies of Sisera's immense host floated, while Deborah sang her song of triumph. And in later times this same plain became the scene of the meeting of king Josiah of Judah and Pharaoh Necho of Egypt, and of the fall of the former by the hand of a hostile archer (2 Chron. xxxv. 20–25). The death of this king occasioned that long succession of dirges which were sung to his memory in the plain of Megiddo by Jeremiah, and by the successive poets of Israel.

After speaking of the three separate passages through the northern Samarian frontier into the plain of Esdraelon, at

Megiddo, Zerin, and Jenin, I take leave of this attractive land of Samaria, and pass to the discussion of Galilee. I cannot close without alluding to the value of the observations made by Mr Wolcott for a future map of Palestine.[1]

[1] The reader need hardly to be told that Van der Velde, in his elaborate map, has adopted this suggestion.—ED.

GALILEE, THE MOST NORTHERN DISTRICT OF PALESTINE.

CHAPTER V.

INTRODUCTION.

GALILEE, THE LAND OF THE HEATHEN IN THE CANAANITE EPOCH—THE EXTENT OF THE TERRITORIES OF ZEBULON, ISSACHAR, ASHER, AND NAPHTALI AT THE TIME OF JOSHUA—THE LATER PROVINCE AND TOPARCHY OF GALILEE—UPPER AND LOWER GALILEE AT THE PERIOD OF JOSEPHUS.

I HAVE had occasion to allude elsewhere to the primitive application of the name Galilee to the district directly west of the waters of Merom (the present Lake Huleh), afterwards the territory of Naphtali, and of which Kedesh, one of the cities appointed as a refuge for those who accidentally committed manslaughter, was the centre. This limited use of the word was afterwards widened, and applied to the whole country in the neighbourhood of the Sea of Tiberias. The old scorn which rested upon the Galilæans of Joshua's day, in consequence of their close affiliation with their heathen neighbours, was afterwards transferred to all who bore the name, partly because they, more than the people of southern Palestine, had formed more intimate relations with the primitive inhabitants of the country (Judg. i. 27–33), and partly because there, as in Samaria, the Assyrians in the time of Shalmaneser, and other foreigners at later times, had come into the land and mingled and married freely with the Israelites.

I have already spoken in general terms of the fortunes of the northern tribes, and of the results of their relations with the Phœnicians, so far as these results were felt throughout northern Galilee and the tribes which inhabited it. I have also in the same connection considered the physical character of the province of Galilee, as distinguished from Samaria and Judæa.

Nor are the special geographical features of Galilee wholly unknown to the reader of the earlier pages of this work; for the sea of the same name and its fruitful shores have been fully discussed, and the whole upper part of the Jordan valley. The plan which I adopt has rendered it necessary to discuss in preceding pages the physical character of the mountainous country east of the plain of Esdraelon, and the passes of Zer'in (the ancient Jezreel) and of Jenin.

In order to gain a complete picture of the territory of Galilee, it remains to describe the great plain of Esdraelon, lying between Jenin, Nazareth, Megiddo, and the Carmel range, and intersected by the Kishon, and then to pass to the mountain region lying northward of it. This embraces the country lying west of the Jordan valley (which we have already examined in detail), and the western slope of the mountain range as it gradually declines to the Mediterranean. This shore we shall then follow northward till it merges in the coast land of Phœnicia. The *promontorium album*, Sur, the ancient Tyre; the fruitful plain Merj Ayun, lying between el-Huleh and the mountains of Tyre and Sidon; the contiguous southern extremity of the Lebanon range, running down into Belad Besharah; the mountain country of the Druses, through whose wild gorges the Litany breaks with its mad rush;—all these combined form a perfect natural wall between ancient Palestine, or more especially between northern Galilee and the ancient Phœnicia; and just north of this great barrier there rises the yet more lofty barricade formed by the whole Lebanon and anti-Lebanon ranges, with no southern outlet saving the narrow gateway which the Leontes traverses, and the sloping vale where the waters of the Hasbany find their way to form the head waters of the Jordan.

The name Galilee occurs very early in the Scriptures, but its first application is widely different from that which comes subsequently. It is alluded to in Josh. xiii. 2, but the meaning there is not the territory *occupied by* the Philistines, but the country on the *border*[1] *of* the Philistines. The words of the translation, though literally true, are yet applied in a different way from the usual signification of "the borders." The Galilee hinted at in this passage coincides exactly, therefore, with that which is named in Josh. xx. 7, xxi. 32, where Kedesh is spoken of as a city of Galilee, appointed as a place of refuge. Kedesh evidently was a central spot: it was once the residence of Canaanite kings, and was in close connection with Hazor, Merom, Taanach, Jokneam, and Carmel (Josh. xii. 19–23). The scornful name of Cabul, with which Hiram king of Tyre designated the twenty Galilæan cities which were given him by Solomon (1 Kings ix. 13), is not to be confounded with the Cabul which Joshua refers to incidentally (Josh. xix. 27) as one of the cities in the territory of Asher. Its location is not at present known.[2]

The same uncertainty rests upon the site of the twenty cities which Solomon gave to Hiram. They were probably small and unimportant places, since none of them is mentioned by name. Many of them are probably comprised in the list of cities mentioned in the second apportionment of land among the four northern tribes (Josh. xix. 10–40). Did we know the position of the sixty-nine cities referred to in those four groups, we should be able to follow with exactness the boundaries of those tribes, and to compare the territory which they possessed with the domain to which the name of Galilee was applied.

By far the greater number of the places indicated in the topographical lists of Joshua have not been identified, even with all the painstaking research of Keil, the most recent commentator.[3] Valuable as have been the researches of Robinson, Smith, and others in Judæa and the adjacent districts on the south, almost no light has been thrown on the historical

---

[1] Keil, *Comment. zu Josua*, p. 240.
[2] *Ibid.* p. 346.   [3] *Ibid.* pp. 337–353.

topography of northern Galilee, because the country has been little open to explorers; and perhaps it can be conjectured, that the speedy loss from the memory of the invaders of the primitive names of places, may indicate the lack of sufficient greatness, to give the people of the northern districts equal historical interest with those of the south. Most of the names of places were in Joshua's time probably of Canaanitic origin, and had a meaning to those who used them; but after the entire derangement of the political relations of the land, consequent upon the Israelite invasion, the old names wholly perished, leaving no one to perpetuate their memory. This was different at a later period, when the Byzantine Christians had settled in Palestine; indeed, it was different in the Mohammedan time, when the genealogical spirit of the Arabian immigrants seized upon the old names which had been current in the southern country, and restored them to currency again. The district between Safed, northwest of the Sea of Galilee, Kedesh and Hunin on the east, and the Litany on the west, is even yet a complete *terra incognita;* for, with the exception of Robinson's route from Safed, by way of Bint Jebeil and Tibnin, no thorough exploration has ever been made in modern times: for Pococke and Stephen Schulz are so incorrect in their names, as to be of little service; and unfortunately the discoveries in this field of my honoured friend Dr E. G. Schultz have not yet been placed at the public service. Relating though they do to the condition of Galilee, they would unquestionably have thrown much light on the ancient topography of the country. In the lack of adequate authorities, we must content ourselves with the slight materials at our command.

The chartographical delineation of the territories held by those Israelite tribes who settled in Galilee must be imperfect in the best biblical atlases, among which Kiepert's[1] occupies the first place. This arises partly from our lack of knowledge about Galilee as it is, and partly from the meagreness of the data in the book of Joshua.

[1] Dr Kiepert, *Bibel Atlass*, Pl. iii.

In this book (xix. 10-16), the territory of Zebulon,[1] comprising twelve cities, is described according to its boundaries; but these remain undetermined, because the position of most of the cities is unknown to us. For example, the first-named places, Sarid and Maralah, of which it is said that they "reached to Dabbasheth, and to the river that is before Jokneam," are unknown; and all we have to guide us is the probability that the river mentioned is the Kishon, since in Josh. xii. 22 Jokneam is said to be on Carmel. In this case Maralah must be supposed to be in that neighbourhood; but, like Dabbasheth, its precise locality must remain unknown. We know that Sarid was somewhere in the west, because we read that the line ran eastward to Chisloth-tabor. Mount Tabor plays as a boundary-mark an important *rôle* in the tribes of Zebulon at the north-west, Naphtali at the north-east, and Issachar at the south; and it is probable that Chisloth-tabor was a place upon the north-west base of the mountain. This is made more certain, that the Daberath, the next point in the border line, is identical with the modern Deburieh, and that Japhia is to be seen in the present Jaffa, a half-hour's distance north-west of Nazareth. Gath-hepher, or Gittah-hepher, the birth-place of the prophet Jonah (2 Kings xiv. 25), appears to be the village of el-Mejed, two hours north-east of Nazareth; and Remmon is supposed to be Rummaneh, two and a half hours north of Nazareth. The location of the five places last mentioned in the account of the boundaries of Zebulon is entirely unknown. It is impossible even to conjecture the northern boundary of Zebulon; and all we know definitely of its outline is, that it began in the western part of the plain of Esdraelon, and ran north-eastward to Tabor. From that point its course extends indefinitely away among the mountains of Galilee. The most important place mentioned in the Bible in connection with Galilee is Nazareth. Its name, however, does not occur in the whole Old Testament; but we know from its relation to Mount Tabor, that it must have been included within the territory of Zebulon.

[1] Keil, *Comment. zu Josua*, pp. 337-342.

The domain of Issachar[1] (Josh. xix. 17-24), lying south of the preceding, embraced the remaining portion of the plain of Esdraelon, and with its sixteen cities extended to the Jordan. It is impossible, however, to designate accurately its boundaries. They ran, we are told, past Jezreel, the present Zer'in, Chesulloth, and Shunem. It appears probable that the fine fruitful valley of Wadi Beisan, lying farther to the east, was embraced in the territory of Issachar, although Beth-shean was given to Manasseh (Josh. xvii. 11). Chesulloth seems to have been the Chosalus of Jerome, the Xaloth of Josephus, and the Iksal of the present day. Shunem is probably Salam, two hours north of Zer'in. West of that was Hapharaim, the Aphrain of Jerome. Shihon is unquestionably the place of the same name which Jerome found on the southern side of Tabor. Almost all the other places named in Issachar are unknown, although the Engannim mentioned in ver. 21 have been conjectured by Robinson and Wilson[2] to be the Jenin of the present day.

The third tribe, that of Asher[3] (Josh. xix. 24-32), occupied the country on the west slope of the Galilæan mountainland, from the Carmel range northward to Tyre and Sidon. But although the territories adjacent to both of these cities were apportioned to Asher by lot, yet the tribe never came into possession of them; in fact, the whole line of plain along the coast resisted their attacks: at least the cities, although nominally given to the Israelites, were always considered as Sidonian. The position of most of the places in Asher remains unknown to us; and the few whose location has been determined are far away from the coast. The Beten of ver. 25 is perhaps to be identified with the Bathne of the *Onomasticon*, which was eight Roman miles east of the city Ptolemais. The three cities named in ver. 26 are known in their topography; yet of the third, Misheal, it is said it reaches to Carmel and to Shihor-libnath. It is uncertain whether this name applies to one of the coast streams

---

[1] Keil, *Comment. zu Josua*, pp. 342-344.
[2] Wilson, *Lands*, etc., ii. p. 84.
[3] Keil, *Comment.* p. 344.

in the north, or in the south near Carmel. The boundary, then, is traced to Zebulon; but whether a tribe or city be meant, is undetermined. All the names of places that follow are unknown to us, till we come to Cana, a large village near Tyre; then follow Tyre and Sidon; and the account closes with the unknown names of Ramah and Hosah, inland cities, near which the boundary-line turns southward to Achzib, the present ez-Jib. The twenty-two villages mentioned without being named had no relation to the boundary-line, and only make the fact more certain, that Asher occupied the mountainous coast-line of Galilee.

Naphtali[1] (Josh. xix. 32–40). This district, comprising nineteen cities and villages, occupies almost the whole of the mountainous region of Lake Gennesaret and the waters of Merom. At the south it borders on Issachar, at the west on Zebulon, farther north on Asher, and at the east on "Judah upon Jordan." By the last expression not Judah proper was meant, but the country of Havoth-jair, which was reckoned as belonging to Judah, because Jair, the possessor of its sixty cities, was connected with that tribe (Num. xxxii. 41). Since it is impossible to determine the northern limits[2] either of Issachar or Zebulon, the southern boundary of Naphtali cannot be designated. From Heleph, whose location is unknown, and from the oak of Zaanannim, which according to Judg. iv. 11 was near Kedesh, the line ran by a number of unknown places to Lakum, also unknown, and terminated at the Jordan. This shows that the northern boundary of Naphtali ran in a general north-east direction to the sources of the sacred river. Kedesh, the city of refuge, lay in the heart of the country. Aznoth-tabor forms one point in the eastern boundary; and the cities of Hammath, Rakkath, Chinnereth, and Migdal-el, indicate unquestionably the shore of Lake Gennesaret, for Hammath cannot be looked for on the east side of the Orontes, as some have thought. Rakkath has been preserved by the rabbis as the name of a place near Tiberias, while Migdal-el is unquestionably the later Magdala. If the territory of Naphtali really

---

[1] Keil, *Comment.* p. 851.   [2] Von Raumer, *Pal.* p. 408.

extended as far southward as it seems to have done, from the fact that places near Tiberias were embraced in it, yet there is no doubt it never extended beyond the Jordan.

Although there are many breaks in the topographical descriptions of the book of Joshua, yet it is not to the want of fulness and exactness there, as we have so often learned, while investigating the geography of southern Palestine, but to our own imperfect knowledge regarding the Galilee of the present day, that we must chiefly attribute our inability to follow the boundary-lines designated by Joshua.

Later in the history of the Jewish nation, only special localities of the northern Galilæan province are brought into distinct sight; as, for instance, in the passage which tells us of the arrangements made during the reign of Solomon for the supply of the king's table from the produce of the fertile plains of the north country (1 Kings iv. 12). Among the twelve men chosen to supervise this department, we read that Baana the son of Ahilud had charge of Taanach, Megiddo, Beth-shean, and Jokneam; Ahimaaz had the jurisdiction of Galilee; Baanah the son of Hushai had control of Asher; and Ahinadab of Mahanaim (not the place of the same name on the east side of the Dead Sea, but, as Reland[1] conjectures, an unknown place of the same designation). Since Josephus speaks of a sixth official of the same character, as having charge of Itabyrium in Lower Galilee, and a seventh as controlling the plain of Esdraelon, neither one of whom is mentioned in the first book of Kings, we may infer that the province of Galilee played a very important part in the duty of providing for the wants of a luxurious court. At the time of Pekah king of Israel, which had separated itself both religiously and politically from Judah, the Assyrian king Tiglath-pileser was very severe in his treatment of the northern province; for we are told in 2 Kings xv. 29, that "he took Ijon, and Abel-beth-maachah, and Janoah [perhaps Januah, north-east of Acre], and Kedesh, and Hazor,[2] and Gilead, and Galilee, and carried them captive to Assyria." In the book of Judith we are told of Nebuchadnezzar, that

[1] Reland, *Pal.* p. 182.   [2] Von Raumer, *Pal.* p. 186, Note 214a.

he sent messengers to all the inhabitants of Damascus, Carmel, and Kedar, also to those of Galilee and the great plain of Esdraelon, and to all in Samaria, and as far south as Jerusalem. The venerable Tobias, who in the time of Shalmaneser was sent as a captive to Nineveh, was one of the few godly men of the tribe of Naphtali, and belonged to a city of Upper Galilee; and in the account of him we have a very instructive picture of his home and of his times. The aspect of Galilee appears in a very different light in the later epoch, when it was held up as the prize for which the Greek Seleucidæ of Syria and the Greek Ptolemies of Egypt were contending. In the long strife, whose object was to destroy the last vestiges of the Jewish name and fame, to plunder the temple and to pillage the whole land, the heroic family of the Maccabees[1] arose to defend the oppressed northern country, and to repel with a succession of glorious victories the attacks of the oppressors. When in great danger of destruction, the Galilæans sent with all haste to call the Maccabees to their assistance. The messengers came, were received, in token of their affliction rent their clothes, and told their story. They said (1 Macc. v. 14-23) that the heathen from all the country round, from Ptolemais, Tyre, and Sidon, had overrun Galilee, and filled the land with people who purposed nothing else than the extinguishment of the name of Israel. The glorious defence offered by Judas Maccabæus and his brothers is known to all readers of history. The enemy was driven from the land, and the nation was preserved. One thing is to be noticed—namely, that in the account of the whole defence, we see no traces of the scorn with which the Jews subsequently looked upon the Galilæans. The Maccabees themselves were of the tribe of Judah, but in the time of national distress they did not hesitate to call the Galilæans their brethren.

At a later period, Josephus, who was for a long time governor of Galilee, and was acquainted with the whole province to a degree of familiarity which no one before or after him has equalled, and who in his official capacity

[1] S. Salvador, *Ges. der Römerherrschaft in Judäa*, Pt. ii. p. 40.

was entrusted with the duty of defending the country against the Romans, has given us a sketch of the province as it existed in his days. There were, he says, two Galilees, an upper and a lower, both of which were bounded by Phœnicia and Syria. On the west the country is terminated by Ptolemais and by the Carmel range, the latter of which once belonged to Galilee, but in the time of Josephus was under the control of the Tyrians. Gaba, the "city of horsemen," was on the south,—so named because Herod had given the city to his cavalry. Samaria and the district of Scythopolis extended along the southern boundary as far as the Jordan; and on the east were Hippene, Gadara, and Gaulonitis, where the territories of Herod Agrippa lay. On the north the frontier touched the possessions of Tyre.

Lower Galilee, Josephus tells us, ran from the Sea of Tiberias to Zebulon, which was near Ptolemais. In breadth it extended from Xaloth, in "the great plain,"[1] to Bersabe; while in length it reached from Thella, a place near the Jordan, to Meroth. Unfortunately, the location of all these places is unknown to us, and we are able to form only a conjectural notion of the boundaries of the province. Only Gaba, the "city of horsemen," is known to us with any degree of certainty: Reland has established its identity with the present Haifa,[2] south of Acco.

The cities which Josephus caused to be walled, in order to defend them from the Romans, are known to us only very slightly, as no thorough investigations have yet been set on foot to enable us to determine their localities. He gives their names in the following order:[3] Jotapata, Bersabe, Selamin, Kaparreccho, Jafa, Sigo, Itabyrion (Tabor), and the great caves on Lake Gennesaret. In Upper Galilee he mentions the rocks of the Achabaroi, Seph, Jamnith, and Mero.

In his *Life*, where he repeats the list, he writes the names differently: Jamnith becomes Jamnia, Mero becomes

---

[1] Reland, *Pal.* pp. 367, 1062, 624, 612, 1034, 560, 896; v. Raumer, *Pal.* pp. 108, 125.

[2] Reland, *Pal.* pp. 769, 770; v. Raumer, *Pal.* pp. 139, 140.

[3] Joseph. *de Bello*, ii. 20, ed. Haverc. fol. 208.

Amerida, Achabaroi becomes Charabe. These Reland sets in their true form—Jamnith, Meroth, and Acharabe.

Josephus informs us in his further description of Galilee, that although the province was so extensive, and surrounded by hostile nations, yet the inhabitants made a valiant defence of their country, being a courageous, resolute, and warlike people. Their land is very fertile, rich in trees and forests, everywhere under cultivation, densely populated, and full of towns: in one passage in his *Life* he states that there were two hundred and four of these, each of them having at least 15,000 inhabitants, including those of the adjacent villages. Whether this estimate of between 3,000,000 and 4,000,000 of souls as the population be not exaggerated, I will not venture to decide; but of this there is no doubt, that the number of inhabitants was very large, and I do not hold Josephus' statement as improbable, that in Galilee he could call together an army of a hundred thousand men.[1]

The chief interest of the New Testament centres at Nazareth,[2] a village of Galilee, where Jesus lived, and silently cherished the thought of His great work till the time came for Him to make His revelation known to the world. The lives of many of the apostles also bring the readers of the gospel into close connection with Galilee. The repeated journeyings and wanderings of Jesus, too, have made the names of Cana, Capernaum, Cæsarea, and many others, for ever memorable. The first Christians were stigmatized by their enemies, both Jews and Gentiles, as Galilæans, and continued to bear that name down to the time of Marcus Antoninus and Julian the apostate, the terrible persecutor of the followers of Jesus. Isaiah's word (ix. 2), "The people that walked in darkness have seen a great light," had a specially fitting application to the inhabitants of the land which is called in Matthew (iv. 15, 16) Galilee of the Gentiles; and Galilæa Gentium remained for centuries, and even as late as to the time of Jerome, the special designation of the northern portion of the province, extending from lake Gennesaret to the Phœnician frontier; the origin of which term is to be traced

[1] Von Raumer, *Pal.* Ap. ix. p. 430. [2] *Ibid.* p. 104 et sq.

to the immigration of people from beyond the borders. The number of these foreigners and aliens gave the country the name Γαλιλαία ἀλλοφύλων in the time of the Maccabees. Reland makes this discrimination[1] between Upper and Lower Galilee, that no sycamore trees are found in the former, although they thrive in the latter. The name Sycaminos, applied to one of the cities on the lowlands of the coast, seems to be a sign of this physical difference.[2]

We now commence our wanderings through the parts of Galilee which have not yet come under our notice, examining the country from the Jordan basin westward to the Mediterranean coast, and first of all considering the great plain of Esdraelon, lying at the south.

DISCURSION I.

THE SOUTHERN PORTION OF GALILEE—THE GREAT PLAIN OF ESDRAELON, OR THE JEZREEL OF THE JEWS—THE BROOK OR RIVER KISHON.

The great plain which extends from Zer'in at the southeast to Acco (Ptolemais) at the north-west, the present Acre on the gulf of the same name, and from Tabor at the northeast to the promontory of Carmel at the south-west, and whose direct breadth is from the recess in which Jenin lies to the hills around Nazareth, fully deserves the name μέγα πεδίον which Josephus applied to it; for it is really the greatest, and at the same time the most fruitful, plain of Palestine, owing in large measure to the thorough distribution in every part of the tributaries of the Kishon.

Jezreel is the name (with some slight differences in the spelling[3]) given in the Hebrew writings to the great plain and the city which overlooked it, the former of which was the territory of the tribe of Issachar, and the latter the royal residence of the kings of Israel. The city of Jezreel lay at the eastern entrance to the plain, upon one of the low

[1] Reland, *Pal.* pp. 183, 184.
[2] *Ibid.* p. 806.
[3] Winer, *Bib. Realw.* i. pp. 580, 581.

western spurs of the Gilboa mountains; and its site is marked even at the present day by the ruins of walls, sepulchres, and a watch-tower. The name now given to the place by the Arabs is Zerin; among the crusaders it was known as Parvum Gerinum. The Greeks gave it the name of Esdrelom; and in the middle ages the designation Stradela came into vogue, either from the number of highways which cross it, or as a contraction of the Greek form. In the time of Jacob and Joseph, as well as in that of Gideon (Judg. vi. 33), the plain of Jezreel was a great thoroughfare for the caravans of the Midianite and Amalekite merchants on their way to Egypt; and in the time of Nebuchadnezzar, the Assyrian army, under the leadership of Holofernes, passed by Bethulia and Dothan (Dothain), traversing the great plain on their way to Jerusalem. And from the time of Josephus down to the present day, all the caravans and all the armies which have passed between Damascus and Jerusalem have been compelled to cross Esdraelon, and to pass Jenin and Zerin. But regarding this I have elsewhere fully spoken, and need not dwell longer upon the "open gate" which connects the great plain on the east with the Jordan valley; making a passage so easy and accessible, that the height of the water-shed is but about four hundred and fifty feet above the level of the sea.

There are three principal valleys which meet not far from the village of Zerin, and by their confluence form the broad plain. I have already spoken of each, using the results of Robinson's researches, who examined and described them all. He calls them the three great arms into which the plain divides on the east. The northern one[1] comes from Tabor: it is about two and a half miles in breadth; it is distinguished by the decided character of the hills which hem it in. It begins at the northern base of Tabor, runs northward as far as Hattin, and then diverges, a part forming the Wadi Bireh and running down to the Jordan; the other turning westward, passing the villages of Murussus, Endor, and el-Fuleh, and forming the chief channel of the Kishon.

[1] Robinson, *Bib. Research.* ii. p. 330.

The middle arm[1] begins at the western extremity of the Beisan valley, near the Gilboa range, and close by the village of Zerin and the Ain Jalud (Goliath's spring), and runs for some distance between two parallel ridges, the continuation of Little Hermon in the north and the mountains of Samaria at the south, taking a direction slightly north-west. It is called Wadi Jalud.[2]

The third and southernmost arm of the Kishon comes[3] into the plain at Jenin from the northern mountains of Samaria, where a narrow pass allows the waters of the southern slope of the Gilboa range to increase the bulk of those which flow from the Samarian peaks, producing a stream of considerable bulk. The wadi or valley through which this stream runs is about three-quarters of an hour broad, and has a general north-westerly direction. The mountains which overhang it on the east of Jenin are, according to Robinson, more rocky and bare than those which accompany the other two arms: they run along close to the southern border of the plain, and really form a part of the Carmel range, though they do not take that name. Yet one can see the physical unity of the chain which borders the Kishon basin on the south, from the source of the river to its mouth. West of Jenin the range is less elevated and less bold[4] in its character than on the east or on the north, in the neighbourhood of Nazareth.

The lofty ridges which culminate at Jenin may be traced southward through the mountain-land of Samaria as far as to Ebal. As I have already spoken at sufficient length of the arm which runs up to the north of Tabor, and of that which runs east to Wadi Beisan, it only remains that I should devote some attention to the position of Jenin, and the southern opening into the hill country.

Djenin or Jenin has been considered to be the En-gannim of Josh. xix. 21, which was in the territory of Issachar; but the reasons for believing it to be so do not seem to me to be decisive. The place is called Ginæa in Josephus. It lies at

---

[1] Robinson, *Bib. Research.* ii. p. 319.
[2] Euseb. de Salle, *Peregrinations ou Voyages en Orient, etc.*, T. i. p. 344.
[3] Robinson, *Bib. Research.* ii. p. 316.   [4] *Ibid.* ii. p. 315.

the mouth of a wadi[1] which, hemmed in by gentle eminences, may be traced far into the plain. The place is a flourishing one even now: it is veiled from sight by thick orchards of fruit trees, among which a few palms are to be seen: the hedges of the gardens are cactus. A fine spring behind the village is so skilfully cared for, that its waters form a pleasant fountain in the place, and are distributed through a number of rills, which carry fertility everywhere. Wilson thinks that the name En-gannim (Ain-gannim, spring of the garden) is derived from this source. The village contains at the present time about 2000 souls.[2]

From Jenin most travellers of the present day[3] are accustomed to go northward to Nazareth, whither three different routes lead. The most western is that by way of Ta'anuk and el-Leijun, taken by Wolcott and Barth. The most direct and comfortable is the one which leads across the middle of the plain, over which Robinson and Eli Smith sent their luggage; while they took the third or most easterly route, leading along the base of the Gilboa range and by way of Zerin. The last named was also the one taken by von Schubert and Wilson. This one has already been made the subject of our careful inquiry. All travellers agree in testifying to the extraordinary fertility of the country through which it passes. Von Schubert[4] tells us that on entering the plain his eye could not be satisfied with gazing: it was the season of spring, the air was most bland and grateful; the blue mountains around Tabor, Gilboa, and Carmel, rose in stateliness, and the words of the eighty-ninth Psalm seemed to be fulfilled, "Tabor and Hermon shall rejoice in Thy name." The plain he called a "field of grain which no man's hand sows, and which no man's hand reaps." The different varieties of grain all seemed to him to perpetuate their own kind, and that in so lavish a way that the mules

---

[1] Robinson, *Bib. Research.* ii. p. 313; Wilson, *Lands*, etc., ii. p. 84; v. Schubert, *Reise*, iii. p. 163.

[2] Buckingham, *Trav. in Pal.* ii. p. 383.

[3] Robinson, *Bib. Research.* ii. p. 318.

[4] Von Schubert, *Reise*, iii. p. 164.

who walk through the grain seem to be half-hid from sight. Notwithstanding what Schubert says about the corn sowing its own seed, Robinson discovered that a portion of the plain was regularly tilled. The former traveller saw the herds of oxen and flocks of sheep and goats ranging at large, and treading down more grass than they consumed. The wild boars of Tabor sometimes come down and wallow in the rich bosom of the plain, and the leopard not unfrequently finds here his easy prey. Between variegated flowers, many of them lilies of varieties which he had never before seen, Schubert saw the water of many almost stagnant brooks slowly making their way to the Kishon, but so leisurely in their course, as to transform the whole adjacent district into swampy ground. The river itself could hardly be called by that name, so very torpid is its flow, and in many places its waters collected themselves in basins or muddy pools, leaving a thick deposit of slime when dry. On the farther side of a soft ford the way was very steep up the mountain side: on the elevation thus attained there is an excellent spring. The way then is a short and pleasant one to the Convent of Nazareth.

Robinson, who traversed[1] the same road, but at an advanced stage of the summer, found the bed of the Kishon, which in Deborah's time was such a wild and dashing stream (Judg. v. 21), and powerful enough to wash the bodies of Sisera and all the slain to the sea, a dry channel. Still in the neighbourhood he discovered patches of green grass, of growing cotton, and of grain, which seemed to his eye like the figures of a variegated carpet. No trace of villages was to be seen in the plain, but on the heights which border it there may be descried the little hamlets of Leijun, Um el Fahm, Ta'anuk, Sileh, and others.

On the northern road from Zerin through Solam (the ancient Shunem, the home of the Shunammite widow who befriended Elisha (2 Kings iv. 8–37), the corn-fields are more productive and the soil more fertile. From that point the way leads over the western end of el-Duhy or Little

[1] Robinson, *Bib. Research.* ii. p. 318.

Hermon, and crosses the third and most northern arm of the great plain. It then passes the ruins of el-Mezraah, the little villages of Nain and Iksal, and then reaches the mountain wall which bounds the plain on the north. This is traversed by a path which leads direct to Nazareth, but the muleteers avoid it, and take an easier but more circuitous one. The view from the height thus gained gives a new impression of the beauty of the plain, and affords a noble prospect of the cone of Tabor. It is true, very few cultivated plains can be seen thence, but the dense grass discloses the great fertility which everywhere prevails.

Wilson, who passed through Esdraelon in the spring, confirms[1] the account given by his predecessors of the exuberant fertility of the soil. He could not see a single tree over the whole extent of the Merj beni Amir, or meadows of the sons of Amir, as the Arabs call it; but all over the plain he could see barley, wheat, beans, pease, flax, and cotton growing; and where these were not found, a rank thick grass was universal. A large part of the soil he found to be that black earth which in India yields the finest crops of cotton. Wilson attributes the great fertility of Esdraelon to the basaltic rock which forms the basis, and which, wherever found, gives when it crumbles a fine soil. He tells us that the Kishon bears at present the name el-Mukatta, apparently the Megiddo of the past, somewhat changed.[2] Freitag, however, does not admit that the modern name is a corruption of the older word, but asserts that it means a ford.

O. v. Richter[3] and Buckingham[4] pursued another route, running from Sanur northward to Nazareth, but their account is very meagre. The last-named traveller estimated the breadth of the plain from south to north to be about five hours, and its length to be about eight hours. He remarks that the plain, although so called, is no perfect level, but that it is only so in contrast with the very mountainous country

---

[1] Wilson, *Lands*, ii. pp. 85, 302.
[2] Buckingham, *Trav. in Pal.* ii. p. 384.
[3] O. v. Richter, *Wallf.* p. 57.
[4] Buckingham, *Trav.* ii. p. 468.

which surrounds it, its own surface being undulating,—a fact which makes its fertility greater than it would be were it an unbroken plain.

Dr Barth[1] took a more westerly route still, which led him from Jenin past Sileh and Leijun. The last-named place did not seem to him a true village: he could not find that more than a half-dozen people were living there. Their occupation was to superintend the mills, which were set in motion by a small tributary of the Kishon flowing in from the Carmel range. On the banks of this stream Barth saw the ruins of an ancient city, but such relics were especially numerous at the foot of a rock called the Ras el Ain, where a fine spring emerges from the earth. This he thought to be the waters of Megiddo, and the ruins to be the ancient city of that name. The walls were too far destroyed to afford many interesting observations, yet they seemed to him to be the relics of mere dwelling-houses of no remarkable pretensions. The grass around them was too rank to enable one to form a very correct judgment without making excavations. The most remarkable feature was the rock-hewn chambers excavated in both sides of Ras el Ain. Of the good Khan Leijun which Maundrell[2] praised in 1697 as one of the best on the road from Egypt to Damascus, there remain only the ruins in the form of a square.

Wolcott visited the same ruins subsequently, and held the place to be the site of the ancient Legio. He as well as von Wildenbruch[3] declare the place to be the ruins of a city of some importance, and the hill on which it stood bears marks of being once terraced with some skill. He observed two or three limestone pillars very much worn by the hand of time; also a building with polished granite shafts still standing, between which there were limestone ones. The finest of these structures stood near the south-west corner of the ruins, near the brook, and was characterized by two marble pillars with

[1] Dr Barth, MS. communication.
[2] Maundrell, *Journ.* March 22, p. 57.
[3] Von Wildenbruch, *Reiseroute in Syrien*, in *Berliner Monatsb. der geog. Ges.* Pt. i. p. 253.

Corinthian capitals. Wolcott saw also an arched gateway, which, like the houses, was decorated with pillars of various patterns. Passing over the stream by a small bridge, he found the remains of the khan, built evidently in the Saracenic style, and surrounded by six or eight arches. On one side is a tower forty feet high, ascended by a staircase inside. From it he took a number of measurements[1] with his compass. But I will not cite them here, nor enter into a consideration of the data which have been collected in order to determine the position[2] of the ancient Megiddo, and whether Leijun, *i.e.* Legio, occupies its site. That question, regarding which the learned investigations of Robinson and von Raumer[3] have led to exactly opposite conclusions, derives its chief difficulty from the varying statements regarding distances to be found in the itineraries of various widely-separated epochs. These statements it is almost impossible to apply to a place as little fixed as a military position generally is. From the time of the old Syrian colony of Hadad-rimmon, when, according to Zech. xii. 11, there was great mourning in the plain of Megiddon, down to the time when the bishops of Maximianopolis entered their signatures at Nicæa and Jerusalem, there might have been great changes in the location of a place whose mission was to guard a mountain pass.[4] That the old name might easily be changed to that of the Legions of the Roman army that guarded it, is easily understood from the modern origin of the name given to the entire plain, Merj beni Amer, unquestionably recording the fact that a tribe of Arabs known as the Beni Amer[5] once held possession of Esdraelon.[6]

Before leaving this fruitful but neglected region, let me devote a few words to the river Kishon, which traverses the

---

[1] Wolcott, in *Bib. Sacra*, 1843, p. 78.

[2] Reland, *Pal. sub Legio, Maximianopolis, Megiddo*, pp. 873, 891, 893.

[3] Robinson, *Bib. Research*. ii. 329 et sq.; *Bib. Sacra*, ii. p. 220; von Raumer, *Pal.* p. 402 et sq.

[4] Maundrell, *Journey, etc.*, p. 57.

[5] Dr Barth, MS. *Reise*.

[6] Wilson, *Lands, etc.*, ii. p. 86, Note; Winer, *Bib. Realw.* i. p. 452.

plain and terminates in the Bay of Acre or Akka. This is the stream regarding which it is written, after Barak and Deborah had gained their victory over Sisera, "The river of Kishon swept them away, that ancient river, the river Kishon. O my soul, thou hast trodden down strength." Although it is now no insignificant stream, yet it needs heavy rains to make it really considerable in magnitude: it is very unequal in size, and seems to be only temporary in its character. At any rate, when Robinson passed its head waters in midsummer, he found the channels all dry, and they had been so for a whole year. On the other hand, in the winter the waters are often exceedingly abundant, particularly in the northern and southern chief tributaries; so that in 1799, at the time of the French invasion, many of the vanquished Turks perished in the floods which swept down from Deburieh, and which inundated the plain. It was a scene like that described in Judg. v. regarding the fate of Sisera's hosts. The streams which run in from the Carmel seem to bring important accessions of water; and hence the little lakes and the swamp land which travellers frequently notice on the south side. Von Wildenbruch,[1] who went from the Khan Leijun and the ruins there across the bed of the Kishon, probably in the summer, speaks of it being dry even there. The *Onomasticon* calls it a "winter stream" ($\chi \epsilon i \mu \alpha \rho \rho o \varsigma$): its three head tributaries—those of Jenin, Gilboa, and Tabor, of which the last is the longest—may indeed sometimes supply no water, but it is very rare that the lower course of the river is entirely dry. Schubert found it forty feet wide in its middle course, and Monro crossed it lower down and near the mouth, where it was thirty yards across. Whether the waters of the fine spring of Leijun (Megiddo) reach the Kishon channel the whole year through, is an unsettled question.

[1] Von Wildenbruch, as above, p. 233.

DISCUSSION II.

THE MOUNTAIN RANGE AND PROMONTORY OF CARMEL.

Carmel is first mentioned in Josh. xix. 26 as a part of the southern boundary of the territory of Asher; but, at the same time, in ver. 11 it is spoken of in connection with the frontier of Zebulun,[1] and seems to have a relation to the domain of three tribes, equally important with that sustained by Tabor at the north-east. Its name, which in the Hebrew signifies " a fertile field," is used by Isaiah, like that of Lebanon and Sharon, to typify the future glory of Israel, when the waste places should put on beauty (xxxv. 2): "It shall blossom abundantly, and rejoice even with joy and singing: the glory of Lebanon shall be given unto it, the excellency of Carmel and Sharon." The sublimity, and at the same time the beauty, of Carmel furnishes Solomon with an illustration of the beauty of the head of his bride (Song of Sol. vii. 5): "Thine head upon thee is like Carmel." The prophet Jeremiah introduces the same name in his promises to the Jews when in captivity, to return to Jerusalem if they should hold fast to Jehovah. Isaiah refers to it to picture the fallen estate of Israel (xxxiii. 9): "Bashan and Carmel shake off their fruits;" and Amos, in announcing the judgments of God upon His people (i. 2), says, "The Lord will roar from Zion, and the top of Carmel shall wither."

But it was not for beauty and fertility alone that Carmel was celebrated: it had a reputation for sanctity; it was considered a holy mountain, an altar of Jehovah, as well as a place where Baal might be worshipped. The latter notion was especially prevalent during the reign of Ahab, who had married Jezebel, a Sidonian princess, and had introduced the Phœnician worship upon Carmel, building an altar to Baal there, and consecrating a grove to his honour. The whole account is familiar to the reader of Elijah's history, and may be found in 1 Kings xvi. 32, 33, xviii. 19-46, in which the relation of the Israelite worship and that of the Phœnicians

[1] Keil, *Comment. zu Josua*, p. 345.

is brought into the most marked contrast, and the prominence of Carmel as the scene is brought into powerful relief. The whole is most vividly brought before us, and affords a fine view of the times in which it occurred. It is plain that Carmel was regarded by both classes of worshippers as a place of special sanctity; and it appears that at a very early period the Phœnicians had come thither to worship Baal, just as Abraham went up into Moriah,—just as David sought the high threshing-floor of Araunah as the place for his altar,— just as Moses once, and Elijah at a much later time, went up into Sinai and Horeb (1 Kings xix. 8). At the time of Deborah, Tabor was a sacred mountain (Judg. iv. 6, 12); when Samuel lived, the heights of Mizpeh were sought as a place of worship and confession. And Jehovah himself dwelt upon Carmel, as we are told in the impressive words of Micah vii. 14.

Although this attributing of holiness to the sacred mountains gradually disappears in the Hebrew history, and only manifests itself occasionally in its poetry,—as, for instance, in Ps. lxviii. 16, "This is the hill which God desireth to dwell in; yea, the Lord will dwell in it for ever," and in the allusions to Hermon, Lebanon, Horeb, etc.,—yet the fine wood-crowned range of Carmel was a place of lasting sanctity among even the Gentile nations. Scylax of Karyanda, in his *Periplus*, calls Carmel a shrine of Zeus even in Greek eyes. Tacitus says that it received the same veneration which was paid to God; just as Mount Casius, according to Movers,[1] had neither temple nor statue, yet had an altar standing upon it, and received great reverence. Tacitus informs us also, that on Carmel was an oracle of wonderful wisdom, whose priests once assured Vespasian that he should become the master of the world; and Suetonius confirms the account of the great authority possessed by the oracle of Carmel. Jamblicus, in his *Life of Pythagoras*, tells us that the latter spent some time on Carmel in contemplation, because that mountain was considered especially sacred. Movers, who has specially investigated the Phœnician history, informs us that that nation did

[1] Movers, *Phönizier*, Pt. i. p. 670.

not believe that any particular Numen dwelt on Carmel, but rather that the Divine Nature (what Scylax meant by Zeus) which manifests itself in the phenomena of the outward world made Carmel his favoured seat. This is confirmed by the story of Elijah and the priests of Baal, and their appeals to their respective deities.

In modern times, the name of Carmel, applied physically to the range which bounds the great plain of Jezreel on the south, extending from Jenin north-westward to Haifa,[1] has generally been used to designate the lofty promontory which runs eastward into the Mediterranean close by the mouth of the Kishon. With even that limited part of the whole range, however, we are not familiarly acquainted; and what knowledge we have is owing to the fact that two important passes exist there,—that of Megiddo and Leijun, and that of Haifa, close by the sea. The time is not yet come when it is possible to make any thorough exploration of the Carmel range. The perils of passing through it are too great, for now, as at Strabo's time, the whole ridge is the favourite resort of robbers: its limestone caverns,[2] which are many hundreds in number, are even now inaccessible to civilised travellers, owing to the refuge which they afford to savage Beduins. The words of Amos ix. 3, "And though they hide themselves in the top of Carmel, I will search and take them out thence," are perfectly intelligible even to-day; and the multiplicity of those hiding-places has exercised an influence on the history of the whole land up to the present time.

About the eastern Carmel pass at Megiddo, we have no fuller accounts than those given by Barth[3] and Russegger. The former intended to pass directly across the ridge Leijun to Cæsarea, but lost his way, and has given us consequently a brief and unsatisfactory narrative. His ascent was made in a few minutes, and, crossing the chain, he pursued a circuitous path to the sea-coast, crossing a region

---

[1] Robinson, *Bib. Research.* ii. p. 362 et seq.
[2] O. v. Richter, *Wallf. i.a.l.* p. 65.
[3] Barth, MS. *Reise.*

not remarkable for beauty, but well watered, and abounding in fertile land. The only two places of any importance which he passed were Chobaese and Sendiyane. He discovered, however, the ruins of a fine old castle on the way; to which he could find no name attached.

The westerly pass, or that which leads from Haifa around the base [1] of the promontory of Carmel, has been much oftener taken; and many travellers have given us accounts of it, as well as of their visits to the convent which stands on the promontory, and which is reached from this road; but no one has ever extended his inquiries into the character of the inland district. Most travellers leave Acre [2] at the north, and follow the path which leads along the sea-side, from which there is a very fine view of Carmel, which, seen hence, justifies the words of Jeremiah: "As Tabor is among the mountains, and as Carmel by the sea." Seen, too, from the sea, it is a majestic object, and always awakens great interest.

Russegger,[3] who explored the whole country with the careful and practised eye of a geologist, tells us that this coastland over which the western route runs is largely arable, and different in character from the Jura limestone of the mountains of Samaria and Galilee. The soil is mostly of an alluvial character, and it is only upon reaching the base of the hills that stones begin to appear. There are, it is true, singular masses of a rock-shaped appearance, which in more superstitious times were supposed to be melons, gourds, and the like, turned to stone by the curse of God; but they are now found to consist of animal petrifactions, bound together by a kind of limestone cement. On leaving this alluvial and mixed formation, containing some flint and crystals, with its other component parts, and ascending the mountains, the traveller is at once introduced to the jurassic limestone, and to the fine dolomite which compose almost exclusively the mountains of Palestine. On the north side of the Carmel

---

[1] *Nautical Mag.* 1841, p. 1.
[2] D'Arvieux, ii. p. 9 ; v. Prokesch, *Reise*, p. 24.
[3] Russegger, *Reise*, iii. p. 257.

ridge, however, there is a very large basalt dyke which has broken through the limestone, and which may be traced north-eastward to Tabor and the Sea of Galilee.

In 1820, Dr Scholtz[1] ascended Carmel from the south side. He discovered many caverns, which during the Byzantine time had been the dwelling-place of hermits. The largest one of these grottos was called "Elijah's School," and was held in high veneration by both Jews and Mohammedans: it was eighteen feet long by ten broad, and had several divisions, in the rear of the largest of which there were banners and lamps, the place being a resort for Moslem pilgrims. This cave, together with another which was called the grotto of Elisha, seemed to Scholtz to have a palæographical interest, on account of the Greek inscriptions found on the walls. These are not valuable on account of their contents, as they are only requests on the part of the writers to be remembered; but the form of the letters, which are angular, and not rounded, as in modern times, shows that their antiquity is great, and that they must date from the first epochs of Christian pilgrimages.

Scholtz visited the ruins of the once celebrated convent from which the order of the Carmelites take their name. A hundred years before it was a new and elegant edifice, but in 1820, when he was there, hardly one stone lay upon another. At the time of the French invasion it had been used as an hospital, and many wounded soldiers had been brought thither from Acre. After the withdrawal of the French, Jezzar Pasha's troops plundered the convent and unroofed the church.

When Turner was there five years before (1815), he found the convent in a much better state of preservation than Scholtz describes it in 1820; yet even then Jezzar Pasha had taken away many of the marble stones to build his mosque in Acre. Turner rode in two hours from Haifa up to the ruins, which were not more than two hundred feet above the level of the sea. The way thither was beautified with the rarest flowers. In the convent he found standing

[1] Dr Scholtz, *Reise*, p. 151.

a French inscription, stating that it served as an hospital in 1799; and he learned that an Egyptian Mameluke had begun the destruction of the buildings fifty years before his visit, so that all the blame could not be laid upon the Turks and French.

Among the most instructive accounts of visits to Carmel is that of von Schubert,[1] who was there in 1837. With his accustomed appreciation of natural beauty, he describes the ascent from Haifa, the number of caverns which he saw in the mountain, and the charming prospect which he enjoyed on the way. Above, as he ascended, was the new and elegant convent, open to all comers,—a monument of the zeal and piety of a single man (John Baptista), the fruit of many years' anxious toil. He had been driven from Haifa, and had fled to Rome, when in 1819 he received the order to return to Carmel, and to employ his knowledge of architecture in rebuilding the convent, that it might be a blessing to pilgrims to the Holy Land, and that it might be there to travellers what the Convent of St Bernard is to those who have occasion to cross the Alps in inclement seasons.

Arriving in Syria, however, to carry out his mission, he found himself thwarted by the opposition of Abdallah Pasha, who pretended to suppose that, in case of any more trouble with Christian nations, the new convent might become the means of doing great harm, and of communicating intelligence to the enemy. So far from allowing Baptista to proceed, Abdallah gave orders that the old ruins should be utterly destroyed; and the good man was compelled to witness the complete destruction of the shattered remains. The whole country was then given over to the wild beasts and to the savage Arabs, and for some time no traveller ventured to visit Carmel. Still Baptista remained unshaken in his purpose to re-erect the convent, and the manner in which he fulfilled that purpose will long remain a proof of remarkable devotion united to a wise discretion. In 1826, while in Constantinople, he received, through the assistance of the French embassy, a firman which permitted him to

[1] Von Schubert, *Reise*, iii. p. 209 et sq.

re-erect the convent. He hastened back to Carmel; and on the ruins of the perished one he drew up a plan of a new, spacious, and elegant structure, the expense of erecting which would be not less than 350,000 francs. He had no means of raising this great sum, and began immediately to devise plans for its being done. Between Carmel and Nazareth he discovered, either on the Kishon or on one of its tributaries, two unused mill privileges which he believed to be valuable, and which he thought, if he could get possession of them for a small sum, might be turned to very profitable account. But they were owned by a Druse, who, out of his hatred to Christians, refused to sell them for any price. Nothing daunted, Baptista sought the friendly intercession of a Turk whom he had formerly known, and borrowed of him nine thousand francs, payable without interest, and to be refunded out of the profits of the mills if they could be hired. This was accomplished; the Druse received as his share a third of the profits, another third went to repay the Turk, the third remained to Baptista. Such was the success of the good man's plan, that in twelve years the 9000 francs loan was paid, and a good beginning was made. He then ventured to lay the foundation of the convent. His next step was to spend a half-year in making a tour through Asia Minor, Egypt, the Archipelago, and to Constantinople, which brought him in 20,000 francs. This sum was soon consumed, however, and then at the age of sixty the indefatigable man began to make the tour of Europe. The building had then so far progressed, that many distinguished men—among them Lamartine, etc.—had been entertained at the Hospitium, without any expense to themselves. The introduction of steamship navigation had made it necessary even then to found the institution, in order to provide for the wants of the increasing number of travellers. The venerable man visited Vienna, London, Paris, and Berlin: he presented his suit at the courts of princes, and at length had the satisfaction of seeing his work crowned with ample success.

Schubert praises very highly the orderly arrangement of the institution, and compares it with the Convent of Banz on the Main. The elegance of the apartments, the size of the halls, the imposing beauty of the view from the windows —embracing the Lebanon and the whole northern coast, Acre seeming to be close at hand, and the fertile valley of the Kishon lying still nearer, while the blue Mediterranean extends far away to the west,—the innumerable variety of plants which grow all around, the clear and delicious waters, the neatness and orderly management of the house,—all made a very deep impression on the mind of von Schubert. These all prove an alluring feast to strangers, and the convent does not lack for guests.

The flora of Carmel is remarkable for its profuseness, because it embraces all the products of such a valley as Esdraelon, and all those of a mountain ridge. The number of insects is so great, too, that a year might be spent there in collecting them, and yet the field not be exhausted. The new convent lies, according to the measurements of von Schubert, five hundred and eighty feet above the level of the sea, and stands on a bluff projecting from the north-west side of the Carmel promontory, which at its highest point is about 1200 feet in elevation. I have already alluded to the extensive and charming view which is gained from it.

Near the convent is an attractive edifice erected by Ibrahim Pasha, and given by him to the Carmelite monks, to serve as an hospital for the sick, and as a place of accommodation for pilgrims. In the immediate neighbourhood of the buildings are gardens, trees, and vineyards. From Carmel to Nazareth is a journey of little less than ten hours.

DISCURSION III.

THE BAY OF ACRE AND THE PORTS OF HAIFA (ΠΕΡΗΑ) AND AKO (AKKO, ST JEAN D'ACRE), OR PTOLEMAIS.

Among the better known of the coast towns which lie near the opening of the plain of Esdraelon at the sea, are Haifa and Akka; and among the more important localities among the mountains of southern Galilee are Nazareth,

Sefurieh, and Kefr Kenna, and the places in their neighbourhood. With these places it is important to gain a certain preparatory acquaintance, before we enter the labyrinthine maze of Northern Galilee, and seek to penetrate the few regions which have been opened there to the knowledge of the world.

1. *Haifa, Khaifa (Cayphas), Hepha, Kepha of the Hebrews, Sycaminos.*

Of this place, the ancient Hepha or Kepha, Reland[1] has no further account than that, situated at the foot of Carmel, it was afterwards called by the names Caiphas, Sycaminos, Porphyreon, and Gabe.[2] At the time of Edrisi[3] (the twelfth century) it must have been a large and flourishing place, and the harbour appears to have been superior to what it is at the present time, since large ships are said to have lain safely at anchor there. According to the accounts of Jerome and Eusebius,[4] the place stands on or near the site of the ancient Sycaminos; but William of Tyre disputes this, and states it was the location of the ancient Porphyrion, in which he is unquestionably wrong, that place having been situated north of Sidon. Wilson inclines to the opinion that Haifa occupies the site of the old Mutatio Calamon. This would locate Sycaminos where the ruins now stand, which are seen north-west of the present town. These ruins are not at all remarkable in their appearance: they have never been carefully explored; and the sand which has blown up from the sea-shore has, in fact, nearly hidden them from observation.

The present town is supposed to contain a population of about 3000 souls, mostly Turks from Barbary: a tenth part of the whole, however, are Catholics, a few are Greeks, and about ten families are Jews. The country immediately around Haifa

---

[1] Von Raumer, *Pal.* p. 139.
[2] Comp. v. Schubert, *Reise*, iii. p. 208; Irby and Mangles, p. 193; Bové, p. 386.
[3] Edrisi, in Jaubert, i. p. 348.
[4] Euseb. de Salle, *Pérégrinations, etc.*, i. p. 396.

is not especially attractive: a few bold bluffs rise in the south, and a few olive and fruit trees are to be seen. The glory of the place, however, as it was in Edrisi's time, is all past. The Bay of Acre is one of the most dangerous places on the whole coast for a ship to anchor in a time of storm. The bottom is of fine sand, and the effect of powerful west winds may be seen in the great dunes which line the shore. Yet the south side of the Bay of Acre is safer than the north side; and this fact gives Haifa a certain degree of importance, and consuls of most of the leading European governments are stationed there. Fishermen may be seen there plying their trade without any boats, merely wading out into the water like the Indian fishermen, and then casting their nets. They are said to be successful, however, notwithstanding their rude method. The place is surrounded on the land side with walls and towers: there are, however, but two gates.

2. *Ako, Acre, Ptolemais.*

Ako—Akko of the Greeks and Romans, Ptolemais, Accon of the crusaders, St Jean d'Acre of the Knights of St John, Tholemais of William of Tyre—is a place of humble pretensions in its commencement, and of low estate at the present day, but has been a city of such importance and splendour, and has exercised such an influence on the whole of Christendom, that the destruction of it produced terror all over Europe; for, with its fall in 1291, the power of the Christian nations of the West lost its last hold upon the East.

The first mention of the place seems to be in Judg. i. 31: " Neither did Asher drive out the inhabitants of Accho, nor the inhabitants of Zidon," etc. In Josh. xix. 24–31, however, where the statement is made of the territorial limits of the tribe of Asher, Ako or Accho is not alluded to. In the time of Shalmaneser, Ako stood in a certain dependent relation to Tyre. The name of the place has been very variously explained,[1] and the etymologies put upon it differ widely. It is safe to say that the question is not yet[2] satisfactorily settled.

[1] Hitzig, *Die Philistäer*, pp. 138-142.
[2] Steph. Byz. *Ethnicorum*, ed. Meineke, p. 59.

Scylax in his *Periplus* calls the place Ἄκη πόλις; Strabo, however, speaks of it under the name of Ptolemais, remarking that previously it had been designated as Aka. Pliny fully corroborates the account of Strabo. Steph. Byz. makes the additional statement that it was a Phœnician city; and others, in speaking of the name of the place, have asserted that the name Ake or Aka was originally applied only to the citadel, but was afterwards transferred to the entire place. The name Ptolemais it probably owes to an extension of the ancient Ako by the first of the Ptolemies, who was for a considerable time the master of Cœle-Syria. Diodorus, Nepos, and others, agree with Strabo in the assertion that the harbour of Ptolemais was formerly of great service to the Persian armies at the times of the expeditions against Egypt.

Josephus gives the location of the place with great exactness, as on the sea border of Galilee, standing in a great plain surrounded by mountains, one hundred and twenty stadia north of Carmel, and a hundred stadia south of Scala Tyrorum. Two stadia from the city was the river Belus, on whose banks stood a statue of Memnon. The sands which compose these banks Josephus speaks of as admirably adapted to the manufacture of glass. During the long-protracted wars between the Syrians and the Egyptians, Ptolemais played a very important part, and was successively[1] the object and the prize of both. Josephus speaks of the inhabitants of the place as not well inclined towards the Galilæans.

The account of Pliny, that the city was held by a Roman colony at the time of the Emperor Claudius, is confirmed by the coins of that ruler's reign, as well as those of Trajan and Hadrian, among which are some which bear the impression of a rock standing near the waves of the sea, houses standing close by, three corn-ears at the side as a sign of fertility, and at the bottom the river god Belus with outstretched hands.

Ptolemais is mentioned in the history of the apostles (Acts xxi. 7). Later it became the seat of a Christian

[1] Comp. Strabon. traduct. fr. Paris, T. v. p. 224, Note.

bishopric. The harbour—which, at the time when the fame of the city was at its height, extended even into the city,[1] huge excavations having been made for that purpose, ensuring safety to the ships which visited the place—made the port the most desirable one on the coast for Christian pilgrims to land at. Edrisi[2] was able to say with perfect correctness, that the city comprised a harbour within itself. This, however, could not be permanent, as the wash of sand at length removed every trace of it. Edrisi wrote at the time when the city was in its prime, having a crowded population, and surrounded with villages and with tilled land. Abulfeda,[3] the other eminent Moslem geographer, saw it when in ruins, after its entire destruction by his fellow-religionists.

Even as early as during the rule of the Egyptian sultans, Acco, generally known as Akka or Accon, had again become the most important port of Syria, as it had been long before under the sway of the Ptolemies, and when it bore a name derived from their own. When the Christians first gained possession of it, they obtained an immense store of gold, jewels, and all kinds of precious goods. During the existence of the Christian kingdom of Jerusalem, Acre was next to the capital in power, importance, and splendour, for it could offer far greater advantages than the nearer but poorer haven of Joppa; and when Jerusalem had passed, as it repeatedly did, into the power of the Moslems, Acre was the capital and royal residence. In its harbour were gathered the fleets of the Pisanese, Genoese, and Venetians, laden with crusaders; along its streets and quays were immense buildings for the storage of merchandise, as well as for the accommodation of the thronging crowds of pilgrims. The plan of the city given by Marin Sanuto,[4] with the walls in some cases of three thicknesses, and with its massive towers, shows the former strength of the place. In the year 1148, Acre

---

[1] Wilken, *Gesch. der Kreuzzüge*, Pt. ii. p. 194.
[2] Edrisi, in Jaubert, i. p. 348.
[3] Abulfedæ *Tab. Syr.* ed. Koehler, p. 82, Note 26.
[4] Marin Sanuto Torsellini *Liber Secretorum Fidelium Crucis, etc.*, Tab. v. Comp. Pococke, Plan viii. in his *Travels*, Pt. ii. p. 76.

was the city where the convocations[1] of the king and barons were held, and was the central point of commerce between the East and the West. When the city fell without a blow into the hands of Saladin in 1187, the Moslems gained possession of booty whose value was inestimable. As the single key[2] to Syria, the Christian leaders felt themselves compelled to make a gigantic effort to recover Acre, and for two years the plain around was the scene of the most heroic endeavours on both sides. On the 12th of July 1191, Acre, with its uncounted stores of gold, silver, merchandise, and ammunition, fell into the hands of the Christians.

The hill Turon, east of the city, was the chief camping-ground of the crusaders: the hills Ajadiah and Mahumeria were also good places for encampment, while the great plain from which they rise was always the field of battle,—a large triangle, whose western side was formed by the sea, and whose longest side, that on the north-east, was formed by the mountains of Galilee. At the northern apex of the triangle stood the Accursed Tower, on the sea side the Castle of the Templars, at the south-east corner the Patriarchs' Tower, and in the midst of the city the Citadel, the royal residence, and the Hall of the Knights of St John.[3] The entrance of the harbour, which was protected by a short breakwater, was fortified by two castellated towers: the whole was very much strengthened during Saladin's short possession. Ptolemais suffered exceedingly from an earthquake[4] in 1202, which affected all Syria from Egypt to Damascus, Antaradus being the only city that was spared. Acco recovered itself meanwhile; and after the downfall of Jerusalem in 1229, it became the only safe capital of the kingdom. St Louis and Philip Augustus of France, and Richard the Lion-hearted[5] of England, expended great pains upon the strengthening of the place, and increasing its splendour. The kings of Cyprus,

[1] Wilken, *Gesch. der Kreuz.* iii. pp. 236, 292.
[2] *Ibid.* iv. p. 254.
[3] Sebast. Pauli, *Codice diplomatico*, fol. i. p. 436, ii. p. 486.
[4] Wilken, *Gesch.* vi. pp. 6, 515.
[5] *Ibid.* vii. pp. 37, 285, 357.

the Templars, the princes of Antioch, many counts and barons of Joppa, Tyre, Cæsarea, and other places, built their palaces here. The highest tribunal was transferred from Jerusalem to Acco; Venetians, Pisanese, and Genoese built sumptuous shops in the heart of the city, and on streets which often bore familiar European names. They were broad and spacious, overhung with silks and mottled stuffs, to ward off the rays of the sun; every corner was guarded by a tower with an iron gate and a strong chain; even the harbour could be closed in the latter way. All the merchandise of the Orient and the Occident was displayed for sale in the storehouses; all languages were heard in the streets. Luxury of every kind abounded: tournaments, encounters with the lance, parades, and festivities of all sorts, belonged to the order of the day in Acco; and the only palace that could bear comparison with it was the luxurious and industrious Colonia on the Rhine (Cologne), which Petrarch praised so highly. The city was full of churches and towers, the harbour was full of ships and masts. The largest houses were built of stone, were provided with glass windows, were adorned with pictures and coats-of-arms, while the flat roofs were covered with the most beautiful flower gardens. The palaces of the leading men were built in great splendour. But there was a dark side to the picture; for notwithstanding the wealth, the luxury, and the power of the city, there was the most bitter enmity between the Genoese and the Venetians,[1] and constant encounters took place both within and without the walls. The reputation of the inhabitants of Acco was not the best; they were accused of siding with the Saracens against the Christians. This, as well as the unhealthy climate, which struck down brave European warriors almost instantly, and frequently carried them off, prevented many stout hearts from entering the armies of the Crusades, and kept them back from the Holy Land. At last, after all the severe losses which had been experienced in Palestine, this last asylum of the brave Knights Templar Hospitallers, which had three times withstood the attacks of the fanatical Saracen troops, after a

[1] Wilken, *Gesch.* vii. pp. 37, 383.

most heroic defence, passed out of the hands of the Christians. In May[1] 1291 the city was taken, and all its defenders put to the sword, only a few escaping by ship. The city was fired at the four corners, and was burned to the ground. This step completed the expulsion of the Christian kings from Palestine.

Of the old buildings of Acco, Prokesch,[2] who visited the place in 1829, found many a trace. The modern appearance of the city, however, had been very much improved during the residence there of the tyrannical Jezzar Pasha, who died in 1824. He had built a large and expensive but tasteless mosque there, and had plundered the ruins of Cæsarea, Askelon, and Carmel, to obtain pillars and ornaments to decorate his capital. The new part of the town he found to be not over five hundred paces in extent, and washed on two sides by the sea;[3] on the land side there was a double wall. The fortress of the place he found to be one of the best in the Levant. Almost all the buildings within the city were surrounded by high walls. At the east Prokesch found many traces of the ancient Ptolemais: these extended for a half-hour's distance along the shore, as far as the river Belus. Among those dating from the Crusades, Prokesch discovered an ancient Cathedral of St Andrew, the Convent of the Hospitallers, the palace of the Grand Master, and the remains of a large nunnery. The water-gate led to the little narrow harbour, which is now wholly unprotected: ships of war have to lie at anchor outside the roads, which are very dangerous on account of the prevalence of the west winds. The number of the inhabitants he estimated at ten thousand, two thousand being Christians.

Wilson,[4] who visited Acco in 1843, regarded the city not as one of the strongest, but as one of the most regularly built, in the land. Its greatly improved character in respect to the strength of its fortifications, it owed to the engineers employed

[1] Wilken, *Gesch.* vii. pp. 731, 770, 774.
[2] Von Prokesch, *Reise*, p. 145.
[3] Irby and Mangles, *Trav.* p. 194.
[4] Wilson, *Lands of the Bible*, ii. p. 233.

by Mohammed Ali; yet it could not withstand the attack of the combined English and Austrian fleets, and the Egyptians were obliged to withdraw to Damascus.

Among the most noteworthy edifices of the city is unquestionably the costly but tasteless mosque built by Jezzar Pasha. Of the harbour which used to be within the city there remains not a trace. Wilson estimated the population at from eight to ten thousand souls, most of whom were Turks. These exercised a very oppressive authority over both the Christians and Jews. The number of the latter was only about one hundred and fifty souls.

The broad plain lying south and east of Acco has, when considered more closely, a waving surface and a row of large dunes, the result of the powerful western storms. The dunes lying most inland seem formerly to have been wooded.

We close these remarks with a sensible observation made by the Duke of Ragusa,[1] which was the result of his journey through the plain of Esdraelon. It serves to throw much light upon other places of similar character in the East. The extraordinary fertility of the plain of Esdraelon, he remarks, is a gift of nature which can profit no man, for it is entirely destitute of human life. Not a twentieth part of its admirable soil is under cultivation: its tall grasses wither, without supplying any herds with nourishment; they only add new fertility to the plain every year. This is the result of men's mismanagement continued for centuries. Population withdraws from the places most liberally endowed with the gifts of nature, where man would with the least labour obtain the largest results, simply from the fact that those places are generally so open to attack. No kind of country is so easily conquered, none with so much difficulty defended, as a fertile plain. It is different with mountain-land, guarded by crags, ravines, and valleys; and men always choose these locations first, because, though less productive, the results of their labour are secure. This, too, is one reason why the villages in the East are often removed from springs and brooks of sweet

[1] *Voyage du Marechal Duc de Raguse en Syrie, Palestine, etc.*, iii. p. 22.

water, though so necessary to the inhabitants : the most sterile localities are chosen in preference to the most favoured ones, in order to attain security for the products of labour. This must always be the case where there is a lack of a well-administered government. Instinct, therefore, has always taught people in barbarous countries to seek mountain homes; and all through the Orient the most fruitful plains are of little more service to man than though they were sandy deserts. It is so with the fertile west bank of the Jordan, the fine soil around Lake Tiberias, the fruitful vale of Baalbec in Cœle-Syria, the plain of Antioch, the most productive of all, and that of Esdraelon, while the rough and wild mountain-land of Samaria is crowded with population.

Should it ever be proposed to colonize these fertile plains with European settlers, it would be necessary, so long as the country remains in its present unsettled and misgoverned state, to furnish every such colony with at least five hundred armed men, whose only business and care should be to protect the agriculturists in their labour.

DISCURSION IV.

NAZARETH AND ITS NEIGHBOURHOOD.

Nazareth is at present the most hallowed place of all Galilee,—a name which before the birth of the Saviour is nowhere mentioned, but which since that event has been carried to the ends of the earth as no other has, and is intimately associated with every thought of that eternal life and salvation which Jesus revealed. For here He passed His childhood and youth, secluded from that darkened world into which He was soon to pour new light, long foretold indeed by the prophets, but whose radiance was to lighten the hearts of but few until His resurrection should take place, and the light of the gospel should illumine the world.

The place where such a youth as Jesus' was spent will always be consecrated ground to the believer; and the preservation of its old charms must always awaken pious thoughts and quickened feelings in the pilgrims who have thronged, and still throng, to visit it. It is but natural that foolish legends and superstitions should gather round such a place; and thus have arisen the idle fancies of the middle ages, which in the minds of the ignorant have taken the place of positive facts, and being poured by the monks[1] into the ears of travellers, have at last received a certain measure of currency. It is different with the nature of the country around, which here as elsewhere has remained unchanged, and whose surpassing loveliness must have exercised an influence on the opening of the young spirit who was trained there, although the whole secret of His spiritual development must remain one of the divine mysteries. Yet it is impossible to prevent the mind sweeping from Nazareth over land and sea, and imagining that even the very features which go to make up the landscape there may have had an influence on the destinies of the world.[2]

The present en-Nasirah of the Arabs, the Nazareth of the Christians, is a place of small importance, with a population at the highest of three thousand souls, among whom there are only seven hundred and eighty men who pay taxes. These are divided ecclesiastically in the following manner: Greeks, 260; Greek Catholics, 130; Roman Catholics, 120; Maronites, 100; Mohammedans, 170.

The city lies on the western side of a long and narrow basin-like valley,[3] running from N.N.E. to S.S.W. Its houses stand in the lower part of the western slope, which is steep, and rises high above them. This hill is covered with aromatic herbs and flowers: at the very top stands a wely, called Neby Ismail. This lies, according to Robinson, four or five hundred feet above the valley, which itself is not far from a thousand feet above the level of the sea: the measurements vary. The mountains which lie north and north-west of Nazareth[4] are

---

[1] Burckhardt, *Trav.* Gesenius' ed. p. 583.

[2] Robinson, *Bib. Research.* ii. p. 333 et sq.

[3] *The Christian in Palestine*, Plates 9, 11, 12; Roberts, *The Holy Land*, Book xix. Plate 54.

[4] Russegger, *Reise*, iii. p. 130.

from 1200 to 1300 feet high. The loftiest lie north-west; those less elevated more to the north: they sink towards the east and south-east, till they rise suddenly again in Tabor. Towards the south-east the valley of Nazareth becomes narrower, and ends in a winding path leading to the plain of Esdraelon. There are also roads leading east to Tabor and Tiberias, south-east to Jenin, south-west by way of Yafa and the plain to Carmel, north-east to Kafr Kenna, and north-west to Sefurieh and northern Galilee Both of the latter run east of the Wely Neby Ismail, whence a magnificent panoramic view[1] may be taken, which serves to supply the deficiencies in the records of personal travel in the Galilæan hills.

Here Robinson surrendered himself on Sunday morning, June 17, 1838, to the enchanting prospect before him, embracing the beautiful cone of Tabor, Little Hermon, and Gilboa in the east; the mountains of Samaria at the west; the whole plain of Esdraelon, the battle-field of ancient and modern times, at his feet. Beyond the plain he could see the long wooded Carmel ridge, reaching to the new convent, and to Haifa, washed with the sea, which the rays of the morning sun lighted up with great splendour. The city of Acca was hid behind the hills. Toward the north there stretched away another of the beautiful plains that adorn this part of Palestine, el-Buttauf, which runs east and west, and sends its waters into the Kishon. On the northern limit lies the large village of Sefurieh (Diocæsarea), near to the foot of a solitary peak, on which stand the ruins of a castle. Beyond the plain of el-Buttauf there are long ridges running east and west, and advancing in height till the mountain of Safed (the city set upon a hill, Matt. v. 14) is reached. Farther eastward lies an ocean of larger and smaller peaks, beyond which the higher ones in Hauran are discernible; and north-east the majestic Hermon, with its cap of snow, is in full view. South-west, but far nearer, the noble promontory of Carmel projects into the silver mirror of the Mediterranean. South-eastward, one standing on the

---

[1] Wilson, *Lands, etc.*, ii. pp. 93-99; v. Raumer, *Pal.* pp. 119-122.

heights in the rear of Nazareth can see the nature of the country which connects Carmel with the mountains of Samaria; that it consists of a large number of low wooded hills, separating the Esdraelon plain from the fertile valleys at the south of Samaria. The same rich supply of woods and low bushes gives the Carmel range an attractive appearance, remarkably in contrast with the naked hills of Judæa. The beauty and grandeur of the view from the Wely Neby Ismail, together with the almost infinite number of recollections connected with localities in view, make this prospect one of the most sublime and most deeply interesting that the world affords.

The city of Nazareth consists of stone houses with flat roofs, among which the citadel-like Franciscan Convent, inhabited mainly by Spanish monks of birth, and surrounded by its detached but dependent buildings, rises as the most prominent structure in the place. The church of the Annunciation is very small; but, according to some travellers,[1] it is decorated with pictures of great beauty, and with finely-wrought marble. It is reputed to stand on the site of the house of Mary. The house itself is said to have been transported by angels to Loretto. The church and convent were begun in 1620, out of materials which remained from the ruins of former structures of a similar character: in 1730 both[2] were enlarged, and received that castellated form which characterized them at a later day. In 1837 the buildings were destroyed by the great earthquake;[3] but they have been completely re-erected since, and the so-called *Casa nuova*, used for the reception of pilgrims, is considered one of the finest buildings for its purpose in the East. In Burckhardt's time (1812) the convent had an income of £900, a part of which came from Jerusalem, while a part was derived from the lands around Nazareth, and from the rent of houses. At the time of his visit ten Franciscan monks were living in the convent.

---

[1] O. v. Richter, *Wallf.* p. 63; W. Turner, *Journal*, vol. ii. p. 132.

[2] Burckhardt, Gesenius' ed. p. 583.

[3] Russegger, *Reise*, iii. p. 130; Thomson, in *Miss. Herald*, 1837, p. 440.

According to Burckhardt, the monks of Nazareth live under a less strict discipline than is usual in institutions of the same kind, and do not wholly abstain from the manners and customs of the world. They have, too, enjoyed a milder treatment at the hands of the Turkish rulers than Christians in any other part of Palestine have received. To this the personal influence of the superior of the convent at the time of Burckhardt's visit, Father Catafago, may have contributed, —a man who had hired of the pasha the ground of two villages, paying £3000 rent therefor, but who was managing his own interests so skilfully, that he had already enriched his whole family, had become the signal protector of all Christian travellers, and had earned the gratitude of a large circle of intelligent men, who have thankfully recorded their obligations to him.[1]

The little church of the Maronites stands in the southwestern part of the city, beneath a rocky mountain wall forty or fifty feet high. Many similar precipices of the same character may be seen west of the town: one of them may have been the place to which allusion is made in Luke iv. 28, 29, although the one which the monks assign as the place of that occurrence lies two miles to the south-east. The legend, however, does not go further back than to the time of the Crusades: the older writers make no allusion to it. The claim of the Greeks that their church is the true Church of the Annunciation, the story about the fine spring which is shown to travellers being that of Mary, and the legends connected with the praying stations around the town, are all the outgrowth of the last few centuries.

That must have been an unimportant place indeed about which the question could be asked, "Can any good thing come out of Nazareth?" and which gave the first nickname applied to the Christians—the sect of the Nazarenes (Acts xxiv. 5). And even yet the title is retained among the Arabs, who designate the people of the whole Christian world as en-Nusara.

The name of Nazareth does not occur in the Old Testa-

[1] Von Prokesch, *Reise*, p. 130.

ment nor in Josephus. Nor is it met after the time of the Saviour, till Eusebius, writing in the fourth century, speaks of the place as a village lying fifteen Roman miles from Legio (el-Lejun), and in the neighbourhood of Tabor. From a document written by Epiphanius[1] in the same century, stating that up to the time of Constantine only Jews had lived in Nazareth, it seems probable that in his time Christians were resident there. Still no bishopric was erected there during the Byzantine supremacy, although the place was much visited by pilgrims. The manner in which Antoninus Martyr, who visited Nazareth about 600, speaks of it, shows the reverence in which the place was held at that time, for he compares it to a paradise on earth. He praises not only all the gifts of nature—the fruit, wine, oil, honey, grain—as of remarkable excellence; but he extols the beauty of the women of Nazareth as far beyond the beauty of other women,—an advantage which he ascribes to the personal favour of Mary to her sex ("in civitate vere illa tanta est gratia mulierum Hebræarum, ut inter cæteras pulchriores inveniantur, et hoc a Sancta Maria sibi concessum dicunt"[2]).

After the crusaders had taken possession of Jerusalem, Tancred was invested with the governorship of Galilee from Tiberias to Haifa. Under his direction, Nazareth, which had been entirely destroyed by the Saracens, was rebuilt, and the province was ruled with a degree of kindness and justice which has not been found in all his successors.[3]

With the new ecclesiastical arrangements which followed, the bishopric, whose seat had before been at Scythopolis, was removed to Nazareth. The exact time of this transfer is unknown; but in the year 1112 Nazareth had a controversy with the Benedictine Convent at Tabor regarding their respective rights of jurisdiction. An appeal was made to Jerusalem, and the decision made, that the consecration of the abbot and the monks should take place on Tabor, and be done by the patriarch, but that the bishop of Nazareth

[1] Reland, *Pal.* p. 905.
[2] *B. Antonini Martyris Itinerar.* ed. Juliomagi Andium, p. 4.
[3] Wilken, *Gesch. der Kreuz.* ii. pp. 37, 365.

should exercise all other episcopal functions over the convent.[1]

With the fall of all Palestine, after the battle of Hattin in 1187, into the hands of the Saracens, Nazareth relapsed into decay,[2] but was continually visited by pilgrims; and in 1620, through the intervention of the philanthropic Druse, the celebrated Fakhr ed Din, the Franciscans obtained permission to rebuild the convent anew. Since that time the place has renewed its old charms, and has become a favourite resting-place for Christians who seek holy places.

Within our own times the American missionaries labouring at Beirut as a centre have succeeded, by their diffusion of the Bible, their schools, and preaching, in awaking an active interest in the gospel, and a desire to receive it. A Greek Christian of Nazareth,[3] who had visited the mission schools at Beirut with great pleasure, took an active part in introducing them among his own townspeople. At the time of Robinson's visit he had succeeded in establishing a school of fifty boys, and a new room was already needed to accommodate twenty more. He had even ventured to send his daughter to be instructed: she was the first of her race in Nazareth who learned to read, but others were not slow to follow her example. Of course there was no lack of difficulties and hindrances, and he would gladly have seen the mission undertake the charge of the schools, but funds were lacking for this purpose.

The observation of Burckhardt[4] is an interesting one concerning the inhabitants of Nazareth, compared with that of Antoninus already quoted. The latter praises highly the beauty of the Nazarene women, and it may be that his laudation is not unfounded. The inhabitants of this place differ, says Burckhardt, in physiognomy and colour from all their Syrian neighbours: in the contour of their face they resemble the Egyptians. In western Palestine the people

---

[1] Comp. Sebast. Pauli, *Cod. dipl.* i. p. 179.
[2] Wilken, *Gesch. d. Kreuz.* iii. p. 230, vii. p. 461.
[3] Robinson, *Bib. Research.* ii. pp. 334, 338.
[4] Burckhardt, *Trav.* p. 341; comp. Russegger, *Reise*, iii. p. 131.

have far more similarity to the natives of Egypt than the inhabitants of northern Syria. Eastern Palestine shows just the reverse: the inhabitants of Jerusalem, Hebron, and Nablus have the genuine Syrian form and contour of face, although their speech differs from that of the north. It is apparent at once that the physical character of the eastern and western mountains of the country has had an influence on the people, and the whole history of the country confirms it. Very interesting would it be, says Burckhardt,[1] the practised ethnographer, to collect representations of the various classes of Syrians, and compare them, the Aleppines, the Turkomans, the natives of the Lebanon, the Damascenes, the coast people from Beirut down to Acre and Joppa, the Beduins, the inhabitants of the hills of Judah. They all have, he says, a national physiognomy; and yet, although all living within the same country, and that not a large one, they have minor ethnographical differences, like those which distinguish French, English, and Italians. It would be well to secure these varieties now by the aid of the photograph, for the tendency of time is to cause them to disappear.

Places in the immediate neighbourhood of Nazareth.

In the direct vicinity of Nazareth, on the side towards the sea, belong Yafa and Jebata on the south-west, Semuniah on the west, and Sefurieh[2] on the north-west, together with Shefa Amer, Abilin, Cabul, and other places. Yafa, a half-hour's distance south-west of Nazareth, has been considered since the time of Marin Sanudo to have been the home of Zebedee and his two sons John and James. It is a village of about thirty houses and some palm trees, and may perhaps indicate the site of Japhia, mentioned in Josh. xix. 12 as one of the terminal cities of Zebulun: it is also probably identical with the Japha alluded to in Josephus as a place which he fortified, and which was at the time of Titus' attack upon it a large and populous town.

[1] Burckhardt's *Trav.* p. 341; comp. Russegger, *Reise*, iii. p. 131.
[2] Robinson, *Bib. Research.* ii. p. 344 et sq.; Keil, *Comment. zu Josua*, p. 339.

Jebata, lying a little farther towards the south-west, is probably the Gabatha of Eusebius and Jerome, which lay near the boundaries of Diocæsarea (Sephoris), near the plain of Legio, *i.e.* Esdraelon. The place is not mentioned in the Scriptures. Nor do we find in the sacred record the name of Semunieh (the Simonias of Josephus, *Vita* 24), a little Mohammedan village west of Yafa,[1] upon one of a row of hills north of Esdraelon, where the Romans sought to fall unawares upon Josephus in the night and take him prisoner.

Sefurieh, the Sephoris of Josephus and the Tsippori[2] of the rabbis, is not named in the Bible. It is at present a little village situated at the foot of a castle-crowned eminence, one and a half hour's distance north-west of Nazareth, and at the southern limit of the plain el-Buttauf. It has retained its old name, although during the earlier centuries of the Christian era it was known as Diocæsarea. Plundered by Herod the Great, burned by Varus, rebuilt and fortified by Herod Antipas, it was called by Josephus[3] the most important place in Galilee, and for a time had precedence even over Tiberias. Several synagogues and one of the provincial sanhedrim were established there by Gabinius. In 339, as the result of repeated insurrections of the Jewish inhabitants, the place was levelled to the ground. It was afterwards rebuilt by the Christians; and Antoninus Martyr tells us of a basilica which was erected on the spot where the popular superstition asserted that the Virgin received the salutation of the angel, of which Robinson remarks that it probably grew out of the old legend that in this place lived the parents of Mary. For convenience sake, the site was afterwards transferred to Nazareth, and the place of the salutation is now pointed out there. Sefurieh was subsequently noted for its fine spring, which often became the place of encampment for Christian armies.[4] It was last used for that pur-

---

[1] See Eli Smith's narrative in Robinson's *Bib. Researches.*
[2] Reland, *Pal.* pp. 999-1003; von Raumer, *Pal.* p. 123.
[3] Joseph. *Vita*, 41.
[4] Wilken, *Gesch. der Kreuz.* iii. pp. 208, 231, 273, 274, 292; Sebastian Pauli, *Codice diplomatico*, p. 439.

pose at the time of the retreat after the battle of Hattin; but on the approach of the victorious Saladin, the Christians were compelled to surrender, and the place was sacked. Since that time its importance has diminished: it has become a mere village, visited by pilgrims merely on account of the legends connected with it, but in itself not specially noteworthy. It remained entirely unaffected by the great earthquake[1] of 1837, which was so severely felt at Nazareth, Safed, and throughout Galilee. The northern road from Nazareth to Akka passes by Sefurieh: the one more commonly taken, however, runs through Abilin,[2] near the village of Kabul. This road has been seldom traversed. Buckingham[3] and Barth have described it, however.

The situation of the villages on the east of Nazareth—Iksal (Chesulloth), Deburieh (Deberoth), Lubieh, Hattin, Khan el Tujjar, and Kefr Sabt—has already been depicted, and need not be referred to again here. They lie generally on the great Damascus road leading by Tabor to Tiberias or Capernaum and the basin of the Jordan.

Between the road running eastward and those leading westward, there lies, directly north of Nazareth, and beyond the fine plain el-Buttauf, the interior province of northern Galilee,—a true *terra incognita*, but well worthy of the attention of travellers. Death has taken away the patient, conscientious, and thorough explorer of this region, Dr E. G. Schultz, before he could publish the results of his investigations, and it is yet uncertain whether his manuscripts have been left in a state to be used. From some of his personal communications to me, I shall be able to gather some valuable facts, but they will not make good his loss. We have, it is true, the names of many places in northern Galilee, of which several have been identified with ancient localities; but it is names alone that we have: with the character of the population we have at present no acquaintance. It is to be hoped that this *hiatus* will soon be filled, and that

---

[1] Thomson, in *Oriental Herald*, 1837, vol. xxxiii. p. 440.
[2] Eli Smith, in *Bib. Researches*.
[3] Buckingham, *Trav.* i. pp. 135-142.

the north of Palestine will be as well known to us as the south.

DISCUSSION V.

THE INTERIOR OF GALILEE—THE UPPER AND THE LOWER PROVINCES, THE HIGHLANDS AND THE LOWLANDS.

On Robinson's second ascent from Nazareth to the Wely Neby Ismail, Abu Nasir, his guide, who was thoroughly acquainted with the whole neighbourhood, pointed out, at a considerable distance northward, the village of Kefr Menda;[1] and east of this, on the extreme northern edge of the plain el-Buttauf, a village which was called by the inhabitants of the district Kana el Jelil. A little farther eastward was the hamlet Rummaneh, perhaps the Remmon of Josh. xix. 13, one of the frontier stations of Zebulun. It could not, however, be seen by Robinson. From the high point where he stood he took the bearings of Sefurieh, Kefr Menda, Kana, and Safed. The last-mentioned place he visited, the other three he did not. But he did go to the Kana or Cana, lying an hour and a half eastward of Nazareth, on the way to Hattin, the Kefr Kenna, more strictly written, where the monkish legend asserts that Jesus transformed the water into wine, and which, for a very long time, has been held by devout pilgrims to have been the scene of the first miracle of the Saviour. The village lies upon the watershed of the Galilæan mountain chain, where the waters part, which find their way on the one side into the Jordan, and on the other pass through the plain of el-Buttauf and enter the Mediterranean.

Even up to the present time, the village of Kefr Kana is visited by crowds of pilgrims, who enter the house pointed out by the monks as that of Bartholomew, and look at the fragments of the jars which held the wine (readily replaced from the convenient pottery in the village). Burckhardt[2] even saw no reason to doubt the authenticity of the tradition

---

[1] Robinson, *Bib. Research.* ii. p. 340; Keil, *Comment. zu Josua*, p. 339.
[2] Burckhardt, *Trav.* p. 336; Reland, *Pal.* pp. 680, 681.

which makes the place the scene of the Saviour's miracle. Yet Pococke[1] did not fail to observe that the Greek legend assigned another site to Cana. Robinson, surprised by the remark of Abu Nasir, that in the distance northward lay the village of Kana el Jelil, *i.e.* Cana in Galilee, was led to the conviction that there, and not in the Kefr Kenna at the east, was the scene of the transformation of water into wine. The name uniformly given to the latter village is both written and spoken differently from the other; and what is still more to the purpose, the uniform name applied to the newly discovered Cana was "Cana of Galilee," a term exactly corresponding to that always used in the New Testament. Robinson conjectured that the monks had arbitrarily changed the locality, in order to suit the convenience of pilgrims who might wish to take the great road leading to Tiberias and Tabor, and whom it would put to inconvenience to visit the northern and authentic Cana. On examination, his conjecture was confirmed by the want of any authority for the modern view older than the sixteenth century. Quaresmius[2] speaks of two Canas in Galilee, Kana el Jelil and Sepher Kana. He describes their location, and makes his decision in favour of the latter, on the ground that it lay nearer Nazareth; but he does not wholly venture to discard the other. Since his day, this has been the view inculcated by the monks, and generally received by travellers. Robinson's investigations showed him conclusively, that the more northerly Cana was formerly held to be that in which the wedding was celebrated, and of which John says that the mother of Jesus was there, whither her Son followed her. This view is supported by the authority of Breydenbach, Marin Sanudo, Brocard, Saewulf, Willibald, and the *Onomasticon*. There is no ground of any value, according to Robinson, for identifying Kefr Kenna with the Cana of the marriage.[3]

That Kana el Jelil, or Cana of Galilee, is therefore, without much doubt, the interesting place of which John says

---

[1] Pococke, *Travels*, Pt. ii. p. 77.
[2] Quaresmius, *Elucidatio Terræ Sanctæ*, ii. p. 85.
[3] Comp. Sebast. Pauli, *Codice diplomatico*, No. clvi. p. 200.

(ii. 11), " This beginning of miracles did Jesus in Cana of Galilee, and manifested forth His glory; and His disciples believed on Him." It was to this place also that Jesus came on His return from Judæa, when the nobleman came and besought Him to heal his son, and to whom Jesus said, " Except ye see signs and wonders, ye will not believe" (John iv. 48). In this Cana, too, Nathanael was born (John i. 47, xxi. 2.)[1] Another and third Cana, which lay in the territory of Asher, and near Sidon (Josh. xix. 28), must not be confounded with this; and at the present day there is to be found a village south-east of Tyre, bearing the same name. The New Testament Cana is nowhere mentioned in the Old Testament.[2]

If now we leave these places which have been thoroughly explored and described by travellers, and go northward into the interior of Galilee, we shall have to guard every step with great care, since we have but few guides to direct our steps, and the confusion of the maps tends rather to perplex than enlighten. We have the account of Josephus, drawn up at the time when he was governor of Galilee, and obliged to traverse the whole country, in order to defend it from the Romans. We have also some brief notices dating from the time of the Crusades, when the country was distributed by the King of Jerusalem, and assigned to counts and barons, who dotted the country with fortresses and castles. Still, not

[1] Keil, *Comment. zu Josua*, p. 347 ; Robinson, *Bib. Research.* ii. 455.
[2] A writer in the *Athenæum* (Feb. 10, 1866) has some excellent remarks on the site of Cana, in which he adopts the older view, that Kefr Kana was the scene of the first miracle. He states, that " on the spot there is no contest. The natives have not heard of the controversy. The Arabs have an immemorial tradition in favour of a particular site, as that on which the great Nazarene Prophet turned water into wine. The Greeks have more than a tradition: they have memorial stones, the ruins of a church and convent, going back to a remote antiquity. Arabs and Greeks agree that the miracle took place at Kefr Kana, village of Cana, standing on a low hill close by the Roman road from Sephoris to Tiberias and Capernaum. Everything is in favour of that site; local tradition, material evidence, and literary testimony. Kefr Kana stands between Nazareth and the Sea of Galilee ; and every reader of Josephus and St John must see that that Cana lay on the road between Nazareth

much is known of these; for, with the expulsion of the Christians from Palestine, the Frankish names were also driven out, to be supplanted by corruptions from the ancient designations, or by new names given by the Mohammedan conquerors. It is difficult, therefore, in Galilee, to identify what is ancient, what is modern, and what belongs to the middle ages; for no Eli Smith and Edward Robinson have traversed this region in all directions, studied the popular habits and traditions, and made the world acquainted with the result of their strenuous efforts. Robinson was only acquainted with the east side of Galilee, as far into the interior as the west tributaries of the Jordan extend, and to the road customarily taken, running from Safed north-west by way of el-Jish (Giscala), Bint Jebail, Tibnin (Turonum), and Kanah, as far as Tyre. On the west side, all travellers, from Richard Pococke (1737) down, have taken the coast road through the plain reaching from Acre to Tyre; and no one has ventured to plunge into the interior mountain-land of Galilee. There are only two men, as far as I am aware, who constitute an exception to this: Stephen Schultz, the Halle missionary among the Jews, who traversed the country in the middle of the last century; and E. G. Schultz, Prussian consul at Jerusalem, who thoroughly explored Galilee a century later. The first was a thorough orientalist, and wrote with little attention to matters of topography and anti-

and the lake. Christ comes to Cana on His way from Nazareth to Capernaum; the centurion comes to it on his way from Capernaum to Nazareth; Josephus hurries from Cana to Tiberias by a secret night march;—evidence that it stood on the Roman road, with no walled city between it and the lake. The Syrian Christians never lost the knowledge of this sacred place: early in date they built a shrine in honour of the marriage feast, which shrine St Willibald visited and described in 722. There can be no question of the locality which he indicates; for he says in express words, that he went to Cana on his way from Nazareth to Mount Tabor. Four centuries later (1102) Saewulf described the same village and shrine. From generation to generation the Church of the Marriage Feast remained in evidence: it was mentioned by Quaresmius in 1629; and its foundations may still be seen by those who seek them. It would seem, then, that the evidence in favour of Kefr Kana being the real site of Cana of Galilee is of its kind perfect."—ED.

quity; the second a keen investigator into everything which could elucidate history. The lamented death of the latter, occurring as it did before his material was ready for publication, has cut us off from the use of materials collected during many journeys between 1845 and 1848, in the course of which he thoroughly explored northern Galilee.

1. *Upper and Lower Galilee, according to the narrative of Josephus: the plains of Zebulun, Batthauf, and Asochis, discriminated from the great plain of Esdraelon.*

An important result of the investigations of Schultz,[1] is the better understanding of the division which Josephus makes of the whole country into Upper and Lower Galilee, the latter of which was the scene of all the events which he describes in his history of his defence of the land against the Romans. Lower Galilee he designated as "the great plain," —an expression which has been usually supposed to refer to Esdraelon, around which, therefore, antiquarians have sought for traces of the places which he mentions. As lately as 1842, Wolcott[2] looked for the site of the celebrated fortress of Jotapata, which Josephus so bravely defended until he was overpowered and taken prisoner, on the northern edge of Esdraelon, near the present Yafa, a short distance southwest of Nazareth. Here von Raumer and Reland located it also, while D'Anville supposed it farther north than Safed. The plain spoken of as pre-eminently Μέγα πεδίον in all general descriptions of Palestine, is unquestionably[3] Esdraelon or Jezreel; but that is an uninhabited district, and was so in Josephus' time. It was a neutral district; it had no intimate relations with the war, or with the seat of war. The province of the governor of Galilee lay, in a strategic sense, wholly outside of it. Esdraelon had no part to play in those tumultuous times, which stirred all Galilee to its centre, and

---

[1] Dr E. G. Schultz, *Mittheilungen, über eine Reise durch Samarien und Galiläa*, in *Zeitsch. der deutsch. Morgenl. Ges.* iii. pp. 46-62.

[2] Wolcott, in *Bib. Sacra*, 1843, p. 79; v. Raumer, *Pal.* p. 115, Note 32.

[3] See *Zeitschrift*, as above, p. 59.

whose most crowded centres of population were in the fruitful district in the southern part of the province, and which then and now was dotted with villages. This part of the province, compared with the elevated watershed between the Mediterranean and the Jordan valley, is not, strictly speaking, a plain; and yet Josephus may have had in mind, when calling it by that name, the contrast between the low and slightly rolling Buttauf, and the tracts which run off towards the east and the high and inaccessible hills of the north-west. This district of Lower Galilee was very fruitful, uncommonly well tilled, and densely populated; while the mountain-land of the north-west served as a strong barrier against the attacks of enemies coming from the sea. Among the fortresses which crowned the line of hills on the west of what Josephus called the great plain of Lower Galilee,—meaning, as it now seems, the comparatively level region in the southern part of the province,—he mentioned the strongholds of Jafa, Sepphoris, Chabolo, Jotapata, Sogane, Selamin, Achbara, Seph, and Mero (reckoning from south-west towards the north-east), which are with the highest probability identified with the present Yapha, Xaloth (Iksal), Tarichæa, Tiberias. The "great plain" spoken of by Josephus, at its south-western portion, approached Sepphoris, and was hence sometimes known by that name. And since it lay in the territory of Zebulun,[1] whose eastern limit was Tabor, it was also called the plain of Zebulun. Since the time of the Crusades, its fertility has procured for it the name el-Butthauf or Buttauf.[2]

On Robinson's map there is depicted, north of Nazareth and Sefurieh, a valley coming down between Kefr Menda and Cana el Jelil. This valley divides itself at the north into two smaller ones. Schultz tells us that, at the point of bifurcation, a steep ragged cliff rises, called Jebel Jefat, on which stand the ruins of a place which seems, with the highest probability, to have been the ancient Jotapata, in which Josephus took his final refuge, and where he made his brave stand against the Romans under Vespasian. As he has woven so

---

[1] J. v. Hammer, in *Wien. Jahrb.* 1836, vol. lxxiv. p. 57.
[2] Sebast. Pauli, *Codice dipl.* i. p. 162.

many particulars regarding the situation of the city into the exact and circumstantial history of its siege, the identity of the present ruins of Jeffat with Josephus' Jotapata may be seen not only in the surviving though contracted name, but in the physical character of the place, and its distance from other known points. Schultz, after leaving Nazareth, came, after passing a few small villages, to Rummaneh. From this place the great plain el-Buttauf extended to the eastward. In this plain, and lying near its southern border, he discovered two places known as el-Ozair and Beni, or, more correctly, el-Buaineh. Kefr Menda lies near the north border of the plain, and east of it is Cana el Jelil. Schultz's course took him north-eastwardly over this plain, and then northwardly over the ridge, upon whose farther side he discovered Arabeh. Thence he went to Sakhnin, an hour's distance westward, or the twenty stadia which Josephus described Sogane as being from Araba. Both of these villages are in a tolerably flourishing state.

INDEX.

Aaron's tomb on Mount Hor, i. 448.
Abadiyeh, ii. 289, 296.
Abarim, ridge of, iii. 2.
Abel, ii. 213.
Abhira, the supposed Ophir, i. 111-120.
Abiela, ii. 300.
Abil, ii. 303.
Abil el Kamh, ii. 213.
Abu Dis (Bahurim), iii. 5, iv. 213.
— Duweir, iii. 327.
— Fares, ii. 338.
— Kusheibeh, i. 422.
— Obeidah, ii. 293.
— Selime, harbour of, i. 337.
— Shusheh, ii. 268.
— Suweirah, i. 178, 365, 371.
Abulfeda, i. 14.
Aceldama, iv. 165.
Acre (Ako, Ptolemais), its history and present state, iv. 361-368.
Adorain, iii. 222.
Aduan Arabs, iii. 65, 78.
Adummim, iii. 10.
Æla (Elath), i. 23, 24, 32, 35.
Ælanitic Gulf, i. 71, 77.
Afieh, ii. 318.
Afuleh, ii. 319.
Ai (Chai, Gai), iv. 222.
Aijalon, iv. 221.
Aijun Musa (Wells of Moses), i. 364, 365.
Aila, i. 421.
Ain Abus, iv. 300.
— Akabe, ii. 282.
— Bedija, iii. 74.
— Belat, ii. 209.
— Dekar, ii. 285.
— Duk, iii. 18, 35.
— el Akhdar, i. 371.
— el Asal, iv. 303.
— el Barbiereh, ii. 209.

Ain el Berideh, ii. 262.
— el Feshkhah, iii. 61, 130.
— el Hazuri, ii. 200.
— el Masiah, ii. 209.
— el Meiyiteh, ii. 325.
— el Mundanwarah, ii. 268.
— el Weibeh. See Kadese.
— er Radghah, ii. 338.
— es Serab, ii. 209.
— es Sultan, iii. 16, 18, 33, 34.
— et Tabighah (supposed Chorazin), ii. 277.
— et Thahab, ii. 209.
— et Tin, ii. 266, 271.
— Eyub (Spring of Job), ii. 272.
— Ferchan, ii. 191.
— Fit, ii. 196.
— Ghuweir, iii. 81, 115.
— Hajla, iii. 18, 47, 49.
— Haramiyeh, iv. 295.
— Jalud, ii. 325, 326.
— Jedi (Jeddi, Engedi), iii. 61, 101, 110-113.
— Kades, i. 433, 443.
— Karim, iv. 215.
— Kaun, ii. 338.
— Keir, ii. 285.
— Maliha, i. 427.
— Sara, iii. 120.
— Sgek, iii. 78.
— Shakhab, ii. 300.
— Shems (Bethshemesh), iii. 237, 239, 241.
— Silwain, the fountain of, iv. 151; the village of, 171.
— Sinai, iv. 294.
— Sitti Mariam (spring of Virgin Mary), iv. 152-158.
— Terabeh, iii. 81, 114, 115, 134, 146.
— Yebrud, iv. 294.
Ajalon (Jalon), iii. 243.

Ajja, iv. 329.
Ajlan (ancient Eglon), iii. 247, 248.
Akaba, described by Ruppell, i. 50, 58; the island of Faroun, and its ruins, 72, 73; the Castle on mainland, by Robinson, 74.
Akab Jabar, iii. 8.
Akab Jahor, iii. 16.
Akir (Ekron), iii. 221, 242.
Akrabah, ii. 345.
Alawin (Aluein) Arabs, i. 410.
Alba Specula Castle, iii. 223.
Alexandrium, ii. 343.
Aleygat Arabs, i. 389.
Almug trees. See Sandalwood.
Altir, iii. 107.
Altitudes in Palestine, drawn up by Van der Velde, ii. 373-390.
Alyka—Chapel of the Burning Bush at Convent of St Catherine, i. 235.
Amalekites, their descent, country, and history, ii. 141-144.
Ameer Arabs, ii. 293.
Ameime (City of Cisterns), ruins of, i. 424.
Ammonites, their descent, country, history, ii. 157.
Amorites, their country, ii. 124; their contests with Israel, 126.
Amwas (Emmaus, Nicopolis), iii. 222, iv. 236.
Anab, iii. 107.
Anabah, iv. 235.
Anderson, Dr, note by, on Dead Sea, iii. 171.
Anathoth, iii. 10, iv. 217.
Anizee Arabs, ii. 293.
Antipatris (Kefr Saba), iv. 244, 250.
Antonia of Josephus, iv. 108.
Antoninus Martyr, i. 9, 10.
Antus, convent on Om Shomar, i. 193.
Avites, their country, ii. 133.
Aphik, ii. 283.
Apostles' Spring, iii. 7.
Apples of Sodom, iii. 19, 20.
Arab tribes of Sinai Peninsula, and their characteristics, i. 377-413.
Arabs in the Ghor between Sea of Galilee and Dead Sea, iii. 55-57.
Arab bards, iii. 78.
Arabbunah, ii. 330.
Arabia Petræa. See Sinai Peninsula.
Arabian writers on Sinai Peninsula, i. 12.
Arad, i. 34.
Aramœa (Aram), ii. 105.
Araneh, ii. 330.

Arar (Aroer), springs of, i. 430.
Arbela, caves of, ii. 266.
Arboth Moab (Araba, Shittim), iii. 2.
Archelais, ii. 345.
Ard el Hammah, ii. 295, 310.
Ard el Mejel, ii. 267.
Ard et Tor, or Peninsula of Tor, i. 372.
Areopolis (Rabbath Moab), i. 25, 33.
Argob, ii. 283, 284.
Arimathæa, iv. 215.
Arindela, i. 27, 33, 53.
Arnon river, iii. 74, 75.
Aroer (Ararah), i. 24, 35.
Arsuf, iv. 267.
Ashdod (Esdud, Azotus), iii. 221; its present state, 223; its history, 225-228.
Ashkelon, iii. 213; excavations in ruins by Lady Hesther Stanhope, 214-216; birth-place of Herod, 216; its history, 217; its idol Dagon, 219.
Ashkenazim Jews, ii. 262, iv. 209.
Ashtaroth, ii. 125.
Asluj (Kasluj), i. 431.
Atara, iv. 229, 294.
Ataroth, iii. 73.
Athlit, iv. 281; its ruins, 282-287.
Attah, ii. 291.
Attarus, iii. 73.
Attir (ancient Jattir), iii. 284.
Authorities on Palestine, list of, ii. 22-78.
Ayun spring, ii. 209.
Azmet, ii. 350.
Azmut, iv. 301.

Babel Mandeb, i. 57.
Balsam or balm tree, iii. 22.
Balua, iv. 245.
Banias, spring of, a source of the Jordan, ii. 161, 193; castle of, now Subeibeh, 199.
Barada river, ii. 17.
Barghaz, ii. 189.
Barth, Dr, excursions between Jordan and Nablus, ii. 347.
Barygaza in Abhira, supposed port of Ophir, i. 112, 116, 118.
Bashan, ii. 125.
Bathing-place of pilgrims in the Jordan, iii. 40-44.
Bathn-nachl, i. 42.
Baths, warm, of Hammam Musa, i. 156; of Hammam Faroun, 339; of Gadara, ii. 305.
Beduin Arabs, i. 240, 337, 383, 397, 400; in the Ghor, iii. 54-57.

INDEX. 387

Beir Zeit, iv. 245.
Beisan (Bethshean, Scythopolis), ii. 291, 324, 331-336, 415.
Beersheba, i. 28, iii. 288.
Beit Ainun (Bethanoth), iii. 327.
— Hagar (House of Hagar), i. 432.
— Hanina, iii. 234, iv. 239.
— Ilfah (Bethulia), ii. 330.
— Jibrin (Eleutheropolis), iii. 249, 252-256.
— Nettif, iii. 238, 239.
— Nusib (Nussib), iii. 237, 239, 256.
— Nuba, iv. 235.
— Sahur, iii. 84.
— Taamar, iii. 81.
— Tamar, iii. 5.
— Ummar, iii. 326, 329.
Beit-Ur (Bethhoron), iv. 241.
Beitima, ii. 169.
Beit el Janne, ii. 169.
Beitin (Bethel), iii. 36, iv. 26, 225, 226.
Bekka, valley of, ii. 164.
Belad Beshara, ii. 189, 213.
Belameh (Belmah), ii. 331.
Belad Shukif, ii. 189.
Bell Mountains (Jebel Nakus), i. 161.
Belled en Nassara, i. 153, 154.
Beni el Sham Arabs, i. 403.
Beni Hameide Arabs, iii. 66.
Ben Hinnom valley, iii. 81.
Beni Naim village, iii. 101, 102.
Beni Sakker Arabs, ii. 289, 293.
Beni Salem Arabs, ii. 350.
Beni Wassel Arabs, i. 392.
Berein, supposed Eboda, i. 54.
Berket el Khulil, iii. 142.
Bet Dejan, ii. 351.
Bethany, iii. 5, iv. 24, 214.
Betharamphtha Julias, ii. 257.
Bethel. See Beitin.
Bethesda pool, iv. 144, 156.
Bethhogla, iii. 47, 48.
Bethlehem, iii. 135, 339-350; Church of the Nativity at, 340; its situation and climate, 340, 341; its inhabitants, 342-345; Church of St Mary at, 345-349.
Bethsaida: two cities of this name, ii. 233, 234. See Khan Minyeh, 269.
Bethshemesh. See Ain Shems.
Bethulia (Beit Ilfah), ii. 330.
Bet Zur, iii. 329, 330.
Biblical authorities on Palestine, ii. 27-29.
Bint Jebeil, ii. 164.
Bireh, ii. 311.
Bir, wells of, i. 154.

Bir el Malekh, i. 430.
Bir es Ozeiz, iv. 242.
Bir es Seba (Beersheba), iii. 288.
Bir es Zaferaneh, iii. 101.
Bir et Themed, i. 427.
Birket el Haj (Pilgrim's Pool), iv. 30.
Birket el Khalil, iii. 118.
Birket er Ram, ii. 172, 178, 179.
Birket Faroun, bay of, i. 337.
Birket Hammam Sitti Marjam, iv. 30.
Birlahairoi, i. 432.
Birsama (Bethshemesh), i. 32.
Birsebhub, iii. 108.
Birshonnar, Well of, i. 196.
Bishoprics, early, of Sinai Peninsula, i. 7.
Bostra, ii. 198.
Botany of Mount St Catherine, i. 201; of Jericho, iii. 19-26; of Hebron, iii. 296; of Jerusalem, iv. 184.
Botthin, ii. 331.
Boundaries of tribes of Israel, iii. 184-190.
Bozrah, i. 26, ii. 137.
Budj, ii. 300.
Bukah, ii. 296.
Burckhardt, author's opinion of him, i. 52; his journey across Sinai Peninsula, 51-55; his ascent of Om Shomar, 192; of Mount St Catherine, 194-200; his knowledge of Arab tribes, 382.
Burka, iv. 327.
Burkin, iv. 329.
Burj el Faria, ii. 345.
Burj el Humma, iii. 84.

Cæsarea Palestinæ, port of Jerusalem formed by Herod, its church history and its ruins, iv. 243, 269-277.
Cæsarea Philippi. See Banias, ii. 193.
Callirrhoe, baths of, iii. 67-69.
Cana of Galilee, its site; three towns of that name, iv. 378-380.
Canaan, origin of name, ii. 106; its inhabitants related to the Phœnicians, 106-112; southern and eastern boundaries of, 112-115; primitive population of, 115-129; tribes living outside of, 130-159.
Capernaum. See Tell Hum.
Capitolias, ii. 281, 300.
Caphthorim, the, iii. 262-268.
Caravan route from Aleppo to Medina, ii. 12.

INDEX.

Carmel, Mount, iv. 352-359; its geology, 355; its convent, 355-359.
Casium, i. 40.
Castle of Doves, ii. 266, 268.
Catherine, St, Convent of, i. 3, 6, 178; described by travellers, 231-246.
— Mount of, 194-202.
Caves of Adullam (Chereitun), iii. 96.
Chapel of Moses on Sinai, i. 210.
Characmobra (Kerek), i. 25, 419.
Cherbit Szammera, ii. 282.
Cherith, brook of, iii. 8.
Chinnereth, city of, ii. 257; Sea of. See Galilee, Sea of.
Chirbet Fassail (Phasaelis), ii. 346.
Chisloth Tabor, ii. 312.
Chorazin, ii. 277.
Churbel, ii. 268.
Churbet Sammer, ii. 351.
Cities considered sacred by the Jews, ii. 260.
Climate of Sinai Peninsula, i. 247, 248; at Sea of Galilee, ii. 240, 252; of Jericho, iii. 28; at Dead Sea, 140; of Bethlehem, 341; of Jerusalem, iv. 182, 183.
Colchians, origin of the, iii. 260.
Colzum, i. 47, 365.
Conies, the, iii. 79.
Convents in Sinai Peninsula, i. 227, 231, 239, 314, 449; on Mount Karantal, iii. 39; in plain of Jericho, 43-45; of Mar Saba, 86-91; of Carmel, iv. 354-359; of Nazareth, iv. 371.
Copper mines of Wadi Nasb, i. 348.
Coral reefs in Red Sea, i. 162-166.
Cosmas Indicopleustes the first traveller in footsteps of Israel, i. 7.
Costigan, his attempt to navigate the Dead Sea, iii. 125.
— Point, iii. 139.
Crusades, 'Palestine during the, ii. 39-43.
Crusaders in Sinai Peninsula, i. 6, 415-418.

Dabira (Deberath), ii. 314.
Daer Senin, iii. 213.
Dagon, idol of, iii. 219.
Dahab, supposed Eziongeber, i. 62-64.
Dahlak, i. 98.
Dalmanutha, ii. 263.
Damascus, roads to, ii. 167-176.
Dan, i. 28, ii. 205-207.

Dandora (Tantura, the ancient Dor), its mussels from whence came the purple dye, iv. 278-281.
Dareya, ii. 170.
Darfureck, plain of, i. 49.
David's Well, iii. 340, 341; Grave of, iv. 56.
Dead Sea, shores of, iii. 58-62; Seetzen's journey along shores of, 64-79; water of, 60, 112; attempts to navigate the sea, 124-130; official report of Lynch's expedition, 130-150; sulphureous smell of, 136; soundings and temperature of, 147, 148; depression of surface, 150, 151, 168, 169; salt of, 161; vapour clouds of, 159, 161; general results from our knowledge of, 150-173.
Debbe, pass of, i. 49.
Debbet en Nasb, i. 330.
Debbet er Ramleh, plain of, i. 343.
Debir, iii. 256 (Kirjath Sepher).
Deir Diwan, iv. 222.
Deir Dosi Convent, iii. 81.
Deir Dubban, the supposed Gath, iii. 249-252.
Deir el Aades, ii. 286.
Deir el Hatab, ii. 301, 350.
Deir Ibu Obeid, iii. 84.
Delhemiyeh, ii. 296.
Derakit, ii. 187.
Derb Serieh (Path of Moses), i. 189, 190.
Desert et Tih Beni Israel, i. 360.
Dhafory, i. 337.
Dhoheriyeh (ancient Beth Zacharia), iii. 193, 288, 289.
Diban, iii. 74.
Diodorus Siculus, i. 19, iii. 152.
Dionysius of Alexandria, i. 5.
Dizahab, i. 63.
Docrayan, ii. 283.
Dothan, ii. 331.
Dshurf el Gerar, i. 430.
Dukah, ii. 232.
Dura (ancient Adoraim), iii. 258.

Earthquake of 1837, its effects in north of Palestine, ii. 248.
Ebal, iv. 302.
Eboda (Ebuda, Abdah), i. 38, 373.
Ed-Dahy, ii. 318.
Ed-Daumeh, village, iii. 290.
Ed-Deir, convent on Mount Hor, i. 449.
Ed-Dirweh (Bethzur), iii. 328.
Ed-Dhoheriyeh, iii. 193.

INDEX.

Edomites (Idumæans), i. 26, 429; their descent, country, and history, ii. 135-141.
Edrei, ii. 125.
Edrisi, i. 14.
Ehrenberg on vegetable life of the Red Sea, i. 163; on animal life of Dead Sea, iii. 169.
Ehteim Arabs described by Seetzen, iii. 37-39, 85.
Egyptian ruins in Wadi Nasb, i. 353.
Ekron, iii. 213.
Elah valley, iii. 240.
El-Aal, ii. 284.
El-Ahedar, i. 302.
El-Ahsa river, iii. 123.
El-Ain, valley of, i. 70, 375.
El-Akaba, spring of, iv. 227.
El-Alya, iv. 225.
El-Arbain, valley and convent of, i. 173, 174, 184, 227, 246.
El-Arish, i. 40.
El-Aradj, ii. 232.
El-Ahtha, plain, i. 366.
El-Aujeh, ii. 346.
El-Aziriyeh (Bethany), iii. 5, iv. 214.
El-Batiheh, ii. 231.
El-Bekaah (Bohah), where are the ruins of Baalbec, ii. 185.
El-Bireh (Beeroth), iii. 229, iv. 227.
El-Birka, i. 199.
El-Botthin (Batanæa, Bashan), ii. 281.
El-Bueb, or the Gate, i. 181, 301.
El-Buk'ah, ii. 289, 299.
El-Bukeiah, iii. 86.
El-Burj Azzil, iv. 294.
El-Buttauf, iv. 370.
El-Daba, i. 421, 422.
El-Derb Serbal. See Wadi Alciat.
El-Dhelel, i. 199.
El-Djoze, i. 199.
El-Ge'ah, i. 299.
El-Gennain, i. 302.
El-Ghuwein (ancient Anim), iii. 284.
El-Ghuweir, or Little Ghor, ii. 267.
El-Ghujar, ii. 212.
El-Hesmih, ancient Hashmonah, or Azmon, i. 75.
El-Hessue, i. 329, 330, ii. 286.
El-Hössn, ii. 221, 283.
El-Hudhera (Hazeroth), i. 371.
El-Huleh (Lake Merom), ii. 209, 210; its level, 226.
El-Humr, i. 184.
El-Jib (Gibeon), iii. 229.
— plain of, iv. 241.
El-Jish, iv. 381.
El-Kaa, plain of, i. 157, 200.
El-Kebur, ii. 285.

El-Kerma, ii. 292.
El-Khalil, iv. 64.
El-Khiyam, ii. 212.
El-Khude, iii. 332.
El-Kordhye, plain of, i. 366.
El-Korriat, iii. 72.
El-Kubab, iv. 235.
El-Kura, iii. 73.
El-Kustul, iii. 234, 235.
El-Lejjun (Legio or Megiddo), iv. 268.
El-Mellahah, spring of, a head-water of the Jordan, ii. 209.
El-Mersed, iii. 112.
El-Mesadiyih, ii. 232.
El-Mezraah, peninsula, iii. 123; village of, iv. 348.
El-Milh (Molada, Malatha), iii. 283.
El-Muchna, ii. 352.
El-Mukrah, i. 375.
El-Mureikhy, pass of, i. 375.
El-Nakhl, i. 371, 372.
El-Noweyba, i. 69.
El-Odjme, i. 199.
El-Oja, ii. 337.
El-Rabua, ii. 294.
El-Rakineh, pass of, i. 375.
El-Shder, iii. 74.
El-Szanamein, ii. 300.
Eli Smith's itinerary from Jaffa to Jerusalem, iv. 234.
Elim. See Wady Gharundel.
El-Tor or Tur, which see.
El-Ujah (Aujeh), ii. 348.
Elusa (el-Kulasah), i. 34, 373.
Emmaus, iv. 215.
Endor, ii. 316, 319.
Engedi, wilderness of, iii. 109.
— city of (Hazazon Tamar), 111-113.
En-Taamirah, iii. 81.
Ephraim (Ephron), iii. 10.
Episcopates in Arabia Petræa, i. 28, 33.
Er-Raha, plain of, supposed by Robinson to be the place where Israel encamped when the Law was given, i. 178, 180, 182, 226.
Er-Ram (Ramah), iv. 216, 217, 230.
Er-Rameh, iv. 328.
Er-Ramleh, i. 371.
Er-Ruhaibeh, i. 373.
Esdud, iii. 213, 224.
Esdraelon (Jezreel), plain of, ii. 314, 315, 317, 322, iv. 333, 343-350.
Eshcol, iii. 258, 298.
Eshtemoh, iii. 285.
Etam (Etham), iii. 93, 337.
Et-Tell, ii. 230, 232, iv. 295.
Et-Teym, iii. 73.

Et-Tih, range of, i. 42, 199.
Eusebius, ii. 31.
Ezbuba, iv. 330.
Eziongeber, i. 64, 91, 92.

Faran (Pharan), i. 17, 304.
Fassail, ii. 337.
Fath Allah, ii. 293.
Feik, ii. 283, 284.
Fellahs or fellahin Arabs, i. 383, 412.
Fendekumieh, iv. 328.
Fineh, iv. 235.
First churches in Palestine, ii. 33.
Frank expeditions into Arabia, i. 416-418.
— mountain, iii. 95.
Fukuah, ii. 330.
Fuleh, ii. 319, 320.

Gadara (Om Keis), ii. 299-303.
Galilee, Sea of (Cinnereth, Tiberias, Gennesareth), its names, ii. 235; its level, extent, depth, 237; its climate, 240, 252; geology of district, 241-245; hot salt springs, 246-248; fish of lake, 250; storms on, 251, 252; west and north-west shores of, 253; south and south-east side of, 278.
— a division of Holy Land, physical character of, iii. 198-200; limits of province, and boundaries of tribes of Israel in it, iv. 332-340.
Galilee of the Romans, 341-343; the country of, from Jordan basin to Mediterranean coast, 343-382; our present knowledge of its interior, 380-384.
Garden of Sinai Convent, i. 427.
Gath, iii. 213, 222, 250.
Gaulonitis, ii. 196, 284.
Gaza, i. 40; its history, iii. 205-211; its port Majumas, 212.
Gebim, iv. 218.
Gennesareth, plain of, ii. 267, 410.
— Sea of. See Galilee, Sea of.
Geography of Palestine, ii. 1-21; early Gentile authorities on, 23, 26; Jewish authorities on, 27-30.
Geographical positions of localities in, according to Van de Velde, iii. 359-372.
Geology of Sinai, i. 265.
— of upper route from Suez to Sinai, 346-352.
— of Galilee, ii. 241-246.
— of district between Jerusalem and Jericho and of south of Palestine, iii. 12-14.

Geology of Dead Sea coasts, iii. 76, 77.
— of Mount Carmel, iv. 355.
Geological character of Palestine, iii. 196.
Gerar, i. 30, 374, 430.
Gerizim, Mount, the ruins on it, and the sacrifices offered there by the Samaritans, iv. 302-309.
Gethsemane, its olive trees, iv. 169.
Ghawarineh Arabs, ii. 232.
Ghor, ii. 281, 289, iii. 1.
Ghor, Lower, ii. 298.
Ghor el Belka, iii. 64.
Ghor es Safieh, iii. 76.
Gibeah of Saul (Tell el Full), iv. 217, 219, 231.
Gibeon (Djeb), ii. 124, iii. 230.
Gibeonites, a remnant of the Amorites, ii. 125.
Giblites of Gebal, ii. 215.
Gihon, valley and pools, iv. 69-77,164.
Gilboa (Jelbon, Jelbun, Jebel Fukuah), ii. 328, 329.
Gilgal, iii. 40, 45, 46.
Gilgoul, iv. 268.
Gipsies in Magdala, ii. 264.
Girdan, plain of, i. 364.
Girgashites, their country, ii. 127.
Gold of Ophir, i. 80, 81, 127-134.
Gomorrah, iii. 138.
Gomsude, i. 164.

Hadji en Rukkab (Knights' Rock), i. 364.
Hadj Musa, i. 230.
Hadra (Hazeroth), i. 67.
Haj stations between Suez and Akaba, i. 43-45.
Hajaja Arabs, iii. 65.
Haifa (Khaifa, Kepha), iv. 360.
Haiwat Arabs, i. 49, 405.
Hammameh, iii. 221.
Hammam Faroun mountains, i. 338; hot springs, 339.
Hammam Musa hot springs, i. 156.
Hammet er Rih, ii. 304.
Hammam es Shefat baths, iv. 87.
Hammet es Sheikh hot springs, ii. 304.
Haram of Hebron (cave of Machpelah), iii. 291, 305-316.
Harbours on west coast of Sinai, i. 159.
Harde, i. 302.
Hasbeya, ii. 161, 165; account of, 186-190.
Hasmeh, iv. 218.
Hattin, ii. 310.
Hauran, ii. 221, 300, 303.

INDEX. 391

Hawara, iv. 301.
Haydar, plain of, i. 66.
Hazuri (Hazor), ii. 200, 214; opinion of Ritter as to its site, 221-225.
Hebron, ii. 122, iii. 193; its present state and history, 290-323; supposed cave of Machpelah, 291, 305-310; visitors to, 311-316; its grapes, 297; Abraham's oak, 298; legends of old sites there, 296, 302; the Luar, 303; present population and their occupations, 317-323.
— vale of, iii. 256.
Hebrews, land of, why and when so called, ii. 105.
Heights, absolute, of localities on west side of Jordan, ii. 355; relative, of do. above surrounding districts, ii. 356.
— in Samaria, iv. 292.
Helena, the Empress, in Palestine, ii. 33.
Helu-ford on Jordan, iii. 44, 49, 52.
Hercir, river, ii. 300.
Hereibe, ii. 190.
Hererat, ruins of convent, i. 303.
Hermon. See Jebel es Sheikh.
Herodium, iii. 95, 96.
Herod, his birth-place, iii. 216.
Heshbon, river of, iii. 49.
Hibel el Hawa, ii. 190.
Hinnom valley, iv. 164.
Hippicus, tower of, iv. 66.
Hippos, ii. 281, 283.
Hiram of Tyre, i. 137.
Hittites, the, ii. 121-123.
Hivites, ii. 123, 124.
Hor, mount, i. 447; view from, 448, 451; ed-Deir, convent on, 449.
Horeb (Chorif), Mount, ascent of, i. 204, 207, 226, 328.
Horites, their country, ii. 133.
Hormah (Zephath), i. 431.
Hot springs, i. 156, 339, ii. 245, 304, iii. 66, 77.
Howara, well of, supposed Marah, i. 367, 368.
Howeytat Arabs, i. 52, 408.
Huj, iii. 246.
Hulhul, iii. 326.
Hunin, castle of, ii. 214.
Husasah, iii. 114.

Ibl (Hibl, Abel), ii. 212.
Idna (Jedna), iii. 256.
Idumæans, i. 2.
Ijon (Merj Ayun), ii. 213.
Iksal, iv. 348.

India, i. 122, 142.
Irbid (Irbil), supposed Arbel or Beth Arbel, ii. 266.
Issachri, i. 12.
Itinerarium Antonini, i. 27, 39.
Ivory, i. 122, 142.

Jabbok ford (Kalaat Serka), ii. 228.
Jabneh (Jebna), iii. 222, 242, 244.
Jacob's bridge, ii. 174, 228.
— well, iv. 301, 317-319.
Ja'dch or hyssop, i. 190.
Jaffa, iii. 245, iv. 243.
Jalije Arabs, i. 202, 242, 384.
Jalud, ii. 344, 349.
Jamea Elabidh, or Church of the Forty Martyrs, iv. 263, 264.
Jattir, iii. 107.
Jeba, iii. 238, iv. 219, 231.
Jebata, iv. 376.
Jebein, village of, ii. 284, 285.
Jebel, village of, i. 153, 154.
Jebel Ajloun, ii. 294.
— Araif, i. 374, 375.
— Arbel, ii. 185.
— Attarus, iii. 73.
— Attika, i. 369.
— Belka, iii. 31.
— Beyane, i. 54.
— Chalil, iii. 194.
— Debbe, i. 303.
— ed Deir, i. 185, 191.
— el Dahy (Lesser Hermon), ii. 309, 316, 318, 319.
— el Fureidis, iii. 94.
— el Ghubsheh, i. 184.
— el Khirm, i. 432.
— el Kods, iii. 194.
— esh Sharkie, ii. 165.
— es Sheikh (Hermon), ii. 161, 163, 167, 181, 184.
— Ebestemi, i. 181.
— Fera, i. 179.
— Fureia, i. 180, 266, 269.
— Gilboa, ii. 309.
— Guddus, ii. 348.
— Hallal, i. 432.
— Hardhe, i. 302.
— Hammam, i. 340.
— Hauran, ii. 300.
— Heish, ii. 161, 166, 284.
— Hemam, i. 200.
— Homr, i. 346.
— How (el-Haui), i. 182.
— Hunin, ii. 214.
— Jura (Jeidur), ii. 190.
— Kula, i. 52.
— Kulcib, ii. 221.
— Menega, i. 302.

Jebel Merura Jubba, ii. 198.
— Mokatteb or Himam, Valley of Inscriptions, i. 160; described by Burckhardt and others, 331-336.
— Musa, group of, described, i. 177-188.
— Musa, paths traversing, 188-191.
— Nakus, or Bell Mountain, i. 161.
— Nasb, i. 347.
— Oef, i. 269.
— Rakab, i. 55.
— Richa, iii. 176.
— Safed, ii. 166.
— Sanin, ii. 165.
— Sebaijeh or Meraga, i. 180.
— Serabit, i. 330.
— Serbal, i. 176.
— Shera, i. 423.
— Shereyk, i. 182.
— Tih, i. 347, 372, 375.
— Tor (Tabor), ii. 311-318.
— Usait, i. 340.
— Wutah, i. 343.
— Yelek, i. 432.
Jebusites, the, ii. 128.
Jedur (Iturea), ii. 286, 302.
Jeddur, iii. 326.
Jedye, ii. 286.
Jefna, iv. 293, 294.
Jehalin Arabs, i. 407.
Jehoshaphat, valley of, iii. 81, 82, iv. 168, 176.
Jelamch, ii. 330.
Jelbon, ii. 330.
Jenin (Ginœa), ii. 329, iv. 328.
Jericho, its early history, iii. 1, 33; roads to, from Jerusalem, 4-10; geology of district, 12-15; castle of, and aqueducts, 17, 29-32; botany of, 19-27; climate, 28; changes in, 32, 33; present inhabitants of, 35.
Jerusalem, its ancient names, iv. 3.
  Its topography, earliest authorities on, 1-8; latest authorities on, 8-18; plans and maps of city, 11-14; site of city, 18-21; view of, from Mount of Olives, 21-28.
  Circuit of present walls, 29-98, and objects there seen, as Jews' wailing-place, 50, 51; Mount Zion, 54; grave of David, 54-58; Jews' quarter, 58; Armenian convent, 62; castle of David, 62, 64-67; Hippicus tower, 66; Gihon valley and pools, 69-77; Hammam es Shefat baths, 87.
  Gates of city. Stephen's gate, 31;

Jerusalem—
  Golden gate, 32; Dung gate, 49; Zion gate, 53; Jaffa gate, 63; Damascus gate, 79, 83-85; Herod's gate, 97.
  Interior of city, 99-121; its streets, 100-104; Antonia of Josephus, temple site, and discussion thereon, 106-112; Omar Mosque, 112-121.
  Christian quarter. Site of Holy Sepulchre and Golgotha, discussion thereon, 122-142.
  Water supply of city, 87-96 and 142-158; Bethesda, 144, 156; well of Rogel, 145-148; fountain of Siloah (Siloam), 148-151; Ain Silwain, 151; Ain Sitti Mariam, 152-158. Objects in city worthy of further exploration, 159.
  Ancient necropolis around Jerusalem, 161-181.
  Climate and soil, 182, 183; botany, 184-187; animals, 187, 188.
  Its inhabitants and sects, 192-212; Mohammedans, 192; Greek Church, 194; Georgians, 196; Armenians, 198; Syrians, 200; Copts and Abyssinians, 201; Roman Catholic, 204; English Church, 206; American Mission, 207; Jews in city, 209-212.
Jesor (Jazur), supposed Azor, iii. 245.
Jezreel, spring of, ii. 321.
— plain of. See Esdraelon.
— city of. See Zerin.
Jezzin, ii. 189.
Jibea, iv. 294.
Jibna, iv. 245.
Jiljilia, iv. 295.
Jiljulieh, iv. 249.
Jilaad es Szalt, ii. 336.
Jimzu (Gimzo), iv. 241.
Job's fountain, iv. 235.
Jolan (Gaulonitis), ii. 196, 284, 300.
Joppa, the port of Jerusalem, its past history and present condition, iv. 253-259.
Jordan, river, ii. 14, 20.
— its sources: (1.) the Nahr Hasbany river, 161, 186, 203; (2.) the Banias spring, 193; (3.) Tell el Kadi spring, 201; (4.) other head-waters of, 209.
— boat exploration of, by Molyneux, 288-294; by Lynch, 294-299; tributaries of, 300.
— bathing-place of pilgrims in, iii. 40-44.

INDEX.

Jordan—its inundations and fords, 50-53.
— its junction with Dead Sea, 54, 154.
Josephus, ii. 29, 154, iv. 106, 246, 340.
Joseph's tomb, iv. 319.
Joshua, his conquests in Palestine, iii. 325.
Jsar el Medjamea, ii. 280, 290.
Judæa, physical aspect of, iii. 194-196.
— hill cities of, iii. 324-331.
Jumah, ii. 298.
Jurish, ii. 344.
Jurmuk (Jarmuth), iii. 239.

Kaabineh Arabs, iii. 108.
Kabelan, ii. 350.
Kadese (Kadesh), site of, discussed, i. 425-433.
Kadesh, wilderness of, i. 432.
Kadmonites, the, ii. 147.
Kafr Berdoweil, ii. 284.
— Hajla, iii. 18, 47.
— Kallin, iv. 301.
Kakon, iii. 31.
Kakun, iv. 268, 269.
Kalaat el Dem (Adummim), iii. 10.
— el Hossn, ii. 281.
— Ibn Maan, ii. 265.
Kalla et Tor, i. 153, 154.
Kanaby, ii. 185.
Kannir, iv. 268.
Kanneytra, ii. 167, 172, 280.
Kanoytor, rock inscriptions, i. 177.
Karn Surtabeh, ii. 343, iii. 53.
Karijut (Koreæ), ii. 342, 345, 349.
Karyat el Kurd, iii. 6.
Karyet el Enab, iv. 238.
Karyat el Chan Hudrur, iii. 6.
Katar Hadije, iii. 47.
Katieh, i. 40.
Kaukabah, ii. 189.
Kedesh Naphtali, ii. 217.
Kedron river, iii. 81.
— valley of, iv. 25.
Kefarat, ii. 301.
Kefr Addan, iv. 329.
— Hareb, ii. 283.
— Istunah, ii. 342, 350.
— Kenna, ii. 310.
— Kud, iv. 329.
— el Kuk, ii. 184.
— Kulin, iv. 301.
— Menda, iv. 378, 384.
— Saba, iv. 249-252.
— Sabt, ii. 310.
Kenites, history of, ii. 144-146.
Kenizzites, account of, ii. 146.

Kerak, village of, ii. 279.
Kerek (Petra Deserti), capital of Edom, i. 25, 418, iii. 119-123, 145.
Kerek, river, iii. 119.
Khalassa (ancient Chesil), i. 431.
Khan el Akabeh, ii. 280, 282.
— Denur, ii. 286.
— Ezzeiat, ii. 286.
— Hashbeya, ii. 189.
— Hathur, iii. 10.
— el Hatrum, iii. 9.
— el Houl, iii. 9.
— Hudhrur, iii. 7.
— el Lubban, iv. 296.
— Legoun, iv. 269.
— Minyeh (Bethsaida), ii. 269-271.
— es Sahil, iii. 10.
— Tudjar, ii. 310, 312.
Khasneh, or rock treasury of Petra, i. 438-440.
Khulaseh (Elusa), i. 427.
Khureitun, labyrinth of, iii. 96-98.
Khuweilifeh, well of, iii. 288.
Kilkel, iii. 340.
Kirbet el Gerar, i. 431.
Kirjathaim, iii. 73.
Kirjath-jearim (Kuriet el Enab), iii. 229, 233, 242, iv. 233.
Kirjath Sepher, iii. 257.
Kishon stream, ii. 311.
— its tributaries, iv. 343-350.
Kolzum, gulf of, i. 13, 64.
Koros, plain of, i. 48.
Kosem, village of, ii. 286.
Kosseir, i. 57.
Kubatiyeh, iv. 329.
Kubeibeh, iii. 248.
Kubelan, iv. 360.
Kudeirah, iv. 222.
Kudna, iii. 252.
Kulensawe, iv. 268.
Kulonieh, iii. 229, iv. 239.
Kumieh, ii. 318.
Kurahy, iii. 77.
Kuriat el Kurd, iii. 6.
Kuriat el Chan Hudrur, iii. 6, 9.
Kurmul, its ruins, iii. 105.
Kurun Hattin, ii. 266, 310.
Kuza, iv. 300.

Laborde, his journey from Aila to Petra, i. 421.
Lanneau's itinerary from Jerusalem to Jaffa, iv. 234.
Latitude and longitude of places in Palestine according to Van de Velde, ii. 359-372.
Lebanon, ii. 17-19.

Latron, iv. 236, 238.
Ledja, vale of, i. 94, 189, 227, 230.
Legends of Arabs in connection with Sinai, i. 210, 211, 227, 228.
Legio, iv. 349.
Leimun Lut (Lot's Lemon), iii. 21.
Library of Sinai Convent, i. 237.
Lifta, village of, iv. 239.
Litany, river of, ii. 165.
Lot, use of, in early division of land, iii. 181, 182.
Lubban, i. 299.
Lud (Lydda), iv. 240.
Lynch, his boat voyage on lower Jordan, ii. 294-299; voyage on Dead Sea, iii. 130-150.

Maarath, iii. 329.
Maaz Arabs, i. 408.
Machpelah, cave of, iii. 305-316.
Macrizi, i. 17, 18.
Madeba, iii. 73.
Madmenah, iv. 218.
Magdala, ii. 263, 264.
Malays, i. 119.
Manadra, valley of, ii. 261.
Manna of Sinai desert discussed, i. 271-292.
Maon (Main), i. 424, iii. 105, 286.
Marah, well of, i. 306.
Mar Saba, convent of, iii. 31, 86-91.
Masada (Sebbeh), iii. 116, 117.
Maundrell, Henry, travels of, ii. 50.
Maximianopolis, ii. 324.
Mazarah, ii. 196.
Medan, ii. 174.
Medj Ayun, ii. 189.
Megiddo, ii. 317, iv. 330.
Meithalon, ii. 345.
Mejel, village of, ii. 168, 263.
Mejdel, iii. 221.
Mejdel Yaba, iv. 248.
Mellahah, ii. 187.
Menadhere Arabs, ii. 300.
Menetisheh, iii. 94.
Menoida, i. 32.
Merassrass, ii. 308.
Merj Ibn Amer, ii. 317.
Merj Ayun, ii. 207, 208.
Merj Ibn Omeir, iv. 235, 242.
Merom, lake of (el-Huleh), ii. 209, 210.
Meshrae, iii. 132.
Mesraa es Safieh, iii. 77.
Mezeine Arabs, i. 390, 397.
Mezar, ii. 330.
Mezereib, ii. 300.
Mines of Wadi Nasb, i. 348-350.
Minna Dahab, i. 64.
Minyeh Khan, ii. 266, 269-271.

Mirzah, ii. 316.
Mizpeh. See Neby Samwil.
Mkaur (Machaerus), iii. 65, 70.
Mkes (Omkeis), ii. 281.
Moab, plains of, ii. 152.
Moabites, their history, ii. 148-156.
Mohala, mountains of, i. 179.
Mohilahi Hajar (Hagar's well), i. 432.
Mojet Nimri, iii. 76.
Mokad Seidna Musa (Moses' resting-place), i. 261.
Moladah, i. 36.
Molyneux, boat exploration of Jordan, ii. 288-294; of Dead Sea, iii. 128; Point Molyneux, 139.
Momur, brook of, ii. 345.
Monks of St Catherine Convent, i. 244.
Moore and Beke's attempt to navigate Dead Sea, iii. 126.
Morkha, well of, i. 337.
Mosque el Aksa, iv. 41.
Mosque of Omar, iv. 31.
Mountain cities of Judah, iii. 324.
Mountain groups of Sinai Peninsula, i. 177; their heights, 180.
Mountains of Ephraim and Judah, iii. 229.
Mountains of Palestine, ii. 16, 17.
Mount Nebo, iii. 2.
— Okra (Cassius), iii. 194.
— of Olives, iv. 21.
— Quarantania, iii. 18, 36, 37.
— St Catherine, its height, i. 180; described by Seetzen and Burckhardt, 195-200.
— Tabor, ii. 311-318.
Mreir, ii. 349.
Muchalid, iv. 269.
Muchna, plain of, ii. 350.
Mukhna, plain of, iv. 300.
Mukhmas (Michmash), iv. 221.
Muldaan, ii. 348.
Murussus, ii. 318.

Nabathæans, i. 2, 11, 19, 423.
Nablus (ancient Shechem), iv. 300, 302; Samaritans at, 308-320.
Nahr Abu Yabura, iv. 269.
— Arsuf, iv. 266.
— Aujeh, iv. 266, 300.
— el Belka, iv. 278.
— es Serayib, ii. 198.
— es Serka, iv. 278.
— Husban, iii. 49, 64.
— Kuriyum, iv. 302.
— Lejjun, iv. 330.
— Rokad, ii. 300.

INDEX. 395

Nahr Rubin, iii. 221, 228.
Nakb Badera, i. 336.
— Egani, i. 179.
— el Hawy (Pass of the Winds), i. 169, 170, 181.
— el Raha, i. 177.
Nakhl, visited by Ruppell, i. 48.
Names, old, in Sinai Peninsula retained, i. 67, 68.
Nain, ii. 316, 319, iv. 348.
Nar Hashbany, river, ii. 161, 186, 203, 204.
Nazareth, iv. 368-375; Church and Convent of the Annunciation, 371; history of town and its inhabitants, 368; villages in neighbourhood, 377.
Nebek, thorn plant, iii. 21, 135.
Nebi Musa, iii. 5, 8, 85.
Neby Samwil (Mizpeh), iii. 229, 231, iv. 25.
— Ismail, iv. 369.
Nejemeh, ii. 348.
Nekb mountains, i. 215.
Netopha, iii. 239.
Niebuhr, commended as a geographer, i. 176; his observations on Sinai range, 177, ii. 52.
Nob, iv. 218.
Notitia Dignitatum, i. 27, 28.
Nowairi, i. 15.
Nowa (Neve), ii. 284, 285.
Nowaran, ii. 173.
Nuris, ii. 330.
Nuweibi, i. 69.

Oijmeh, i. 179.
Om el Mezabel, ii. 286.
Omkeis (Gadara) hot springs, ii. 247, 301-308.
Om et Taybe, ii. 308.
Omran Araba, i. 407.
Omar Mosque, iv. 112-121.
Om Shomar, Mount, ascended by Burckhardt, i. 191-193.
Onhol, ii. 286.
Ophel, iv. 146.
Ophir, opinions as to its locality, i. 78; historical data of the route to, 81-89; identity of name with places in Arabia, Africa, and India, 91-116; commodities from, shown to be Indian, 116-134; views of such as consider Ophir to have been in Yemen or Safala, 134-151.
Ophra, iii. 7.
Orontes, ii. 11, 14.
Oreb, iii. 121.

Osher plant (apples of Sodom), iii. 19, 135.

Palestine, general views of land and people, ii. 2-9; physical character of Syria and Palestine contrasted, 10-20.
— geography of, from Greek, Roman, Jewish, and early Christian sources, 23-39.
— during the period of the Crusades, 39-43.
— visits to, from fourteenth to eighteenth centuries, 43-53.
— oriental writers on, 54-60.
— travellers of nineteenth century to, with Ritter's opinion of their works, 60-74.
— fragmentary contributions to a knowledge of, 74-78.
— maps of, 78-86.
— supplemental list of most recent works on, by the editor, 86-103.
— Tobler's list of works on, ii. 391-409.
— ancient boundaries of and divisions among the twelve tribes, iii. 175-190; later division into Judæa, Samaria, Galilee, physical basis of this, 191-200.
Paneas, ii. 206.
Parah, iv. 219.
Paran, wilderness of, i. 63, 69, 428, 432.
Pella, ii. 281.
Pentapolis, i. 30.
Perizzites, ii. 121.
Petra, the Nabathæan capital, Sultan Bibor's visit to, i. 16; Romans at, 20, 21; Laborde's visit to, 422; entrance arch, 436; Khasneh, 438-440; tombs, 440-446; temple, 443.
Peutinger Tables, i. 23.
Pharan, village of, i. 256.
Pharanites, i. 22.
Phasaelis, ii. 346.
Phiala, lake of, ii. 177.
Phik, ii. 281.
Philistines, iii. 259-268; proper, their history, 268-281.
Phœnicians, ii. 108.
Pietro Della Valle, travels in Sinai Peninsula, i. 172, 173, ii. 49.
Pilgrimages to Palestine, ii. 35-39.
Pillar of Salt at Usdum, by the Dead Sea, iii. 139.
Pools of Solomon (el-Burok), iii. 93, 333.

INDEX.

Punon, i. 36.

Rabbath, Ammon and Moab, i. 25.
Radjoin el Abhor, ii. 284.
Rafia, iii. 205.
Rakkath (Hammath), ii. 257.
Ram, ii. 285.
Ram Allah, spring of, iv. 227, 228.
Ramleh, ii. 213, iv. 234; described, 260.
Ramah, iv. 217, 230.
Ramathaim (Saba), iii. 229.
Rana, iii. 252.
Raphia (Rapha), i. 40.
Ras el Ain, ii. 348, iv. 248, 266, 303, 349.
Ras el Balka, promontory on Dead Sea, iii. 129.
Ras el Feshchah, iii. 31, 60.
Ras el Ghuweir, iii. 80.
Ras el Kerah, iii. 129.
Ras el Tafila, iii. 129.
Ras es Susâfeh, i. 208.
Rasheya, ii. 181, 184.
Rasheyat el Fuchar, ii. 191.
Ras Hish, iii. 138.
Ras Mohammed, i. 59.
Rawak valley, i. 48.
Red Sea, charts and surveys of, i. 56-60.
— coral reefs and islands in, 162-166.
Remthieh, iv. 268.
Rephaim or giants, ii. 131.
— plain of, iv. 27.
Rephidim, its locality discussed, i. 323-328.
Rhinakorura, i. 372.
Riha, ancient Jericho, which see.
Rithem, or juniper plant, i. 345.
Rithma, i. 427.
River of Egypt, i. 372.
Roads or paths of Sinai, i. 188.
— from Lake Tiberias to Damascus, ii. 284.
— from Jerusalem to Jericho, iii. 4.
— Jericho to Bethel, iii. 36.
— from Egypt and from Sinai Peninsula into Judæa, iii. 201.
— from Jerusalem, iv. 25.
— from Jerusalem to Mediterranean, 232.
— from Bethel to Nablus, 293.
— from Nablus to Sebaste, 320.
— in south of Galilee, 343.
— to the coast at Acre, 359.
— Nazareth and its neighbourhood, 368.
Robinson as a traveller, Ritter's opinion of, i. 5, ii. 70-74, iv. 9, 16.

Rock inscriptions in Kannoytor, i. 177.
— in Wady Sehab, 270.
— in Wady Mokkateb, 331.
— in Wady Nasb, 349, 350.
Rogel, well of, iv. 145-148.
Romans, the, in Palestine, ii. 25-27.
Routes: early routes from Ælа to Jerusalem, i. 24; from Gaza to Pelusium, 39-41; from Suez to Akaba, 41-44; cross routes of travellers, 45; from Gulf of Akaba to Sinai Convent, 60, 71; from Suez to Sinai, 338, 339; northern routes in Sinai Peninsula, 361; from Akaba to Petra and Hebron, 374, 421; from Sinai to Palestine, ii. 1. See Roads.
Rubtat el Jamus, iii. 141.
Rumon (Rimmon), ii. 348.
Rummon, iv. 219.
Ruppell in Arabia Petræa, Ritter's opinion of him, i. 46, 50.
Russegger on Sinai mountains, i. 180, 247.
— levels from Red Sea to St Catherine, i. 202.
— geology of Wadi Sheikh, 265.
— geology of upper route to Sinai, 346-349.
— his account of Red Sea, 368.
— geology of district around the Sea of Galilee, ii. 241-245.
— geology of south of Palestine, iii. 12-14.
— his account of asphaltum of Dead Sea, iii. 156.
Ruins in Sinai Peninsula, list of, given by Seetzen, i. 419.
— on road from Banias to Damascus, ii. 168.
Ruhaibeh, i. 374, 431.
Rummaneh, iv. 378.

Saada, i. 68.
Saba, Convent of, iii. 340.
Safed, ii. 164.
— city and castle of, 219-221.
Sahel Hattin, ii. 265.
Sahhnin, iv. 384.
Sair, iii. 101.
Salem (Salim), ii. 350, iv. 330.
Salt of Dead Sea, iii. 161.
Samaria (Sebastieh), city of, iv. 321-326.
Sebaijeh, plain of, 218-224.
Samaria, district of, physical character of, iii. 197.

INDEX. 897

Samaritans, iv. 287-291 ; at Nablus, 313-320.
Samireh, ii. 384.
Sandalwood, or almug, i. 124.
Sanoah, iii. 239.
Sanur, iv. 328.
Sara, iii. 68.
Saracens in Sinai Peninsula, i. 265.
Sarbat el Chadem, Egyptian ruins in Wady Nasb, i. 352-359.
Sarbout el Jemel, i. 199, 342.
Sartabah, horn of, ii. 344.
Sasa, ii. 169, 171, 175.
Sath el Akaba, i. 45.
Sawich, iv. 300.
Schimper on botany of Mount St Catherine, i. 201.
Scopus, iv. 216.
Sebbeh (Masada), iii. 110, 114-117, 137.
Sebunta (Heshbon), i. 26.
Seetzen, discoverer of Petra, i. 419.
— Ritter's opinion of him, ii. 61, 62.
— on coasts of Dead Sea, iii. 64-79.
Sefurieh, iv. 376.
Seguia, ii. 293.
Sehab plateau, i. 266.
Seilun (ancient Shiloh), iv. 295-299.
Seir, i. 429, ii. 135, 136.
Semneim, ii. 286.
Semua (Esmua, Eshtêmoh), iii. 107, 283, 285.
Senjol, ii. 267.
Sepata (Zepath Hormah), i. 431.
Sephardim Jews, ii. 261, iv. 210.
Serbal, Mount of, its ascent, 292-300.
— a mount of heathen worship, and its identity with Sinai and Horeb, 313-328.
Serka Maein, iii. 66.
Sersaf, vale of, i. 191.
Seybarany river, ii. 170.
Shadlic, i. 153, 154.
Shafat, iv. 226, 231.
Shamor, ii. 349.
Sharon, plain of, iv. 265.
Sheba, queen of, and her gold, i. 82, 129, 145.
— water of, ii. 178.
Shech Muntar, iii. 205.
Shechem, iv. 308.
Sheik Saleh's (Szaleh's) tomb, i. 262; Arab pilgrimages to, 262-265.
Sheik Othman el Hazur, ii. 200.
Sheme, plain of, i. 47.
Shemskein, ii. 300.
Shera, range of, i. 52.
Sheriat el Menadra, ii. 280, 299, 301.
Sherm Sheik and Sherm el Moyah, arms of the Red Sea, i. 59.

Shur, i. 369, 432, ii. 104.
Shuweikeh (Socoh), iii. 108, 237, 239, 241, 286.
Sich el Udhar, ruins of, described, i. 312.
Sickha el Hejas, i. 59.
Sidumad, ii. 298.
Sikka Tekruri, i. 324, 329.
Siloam or Siloah fountain, iv. 148-157.
Sileh, iv. 349.
Simoon, the, i. 248.
Sin, wilderness of, according to Robinson, i. 322, 337, 428.
Sinai Peninsula (Arabia Petræa), its boundaries and original inhabitants, i. 1-4; its early Christian history, 4-12; its early Moslem history, 12-18; Greek and Roman accounts of, 18-21; its topography, 21-51.
Sinai, Mount, its height, i. 180; called also Jebel Musa and Jebel et Tur, 184.
— pilgrims' path to summit, and view from, 209-216.
Sinjil, iv. 296, 297.
Sittim or Shittim, plain of acacias, iii. 1.
Sobah (Rama, Ramathaim), iii. 234.
Sofala in Africa supposed by some to have been Ophir, i. 147.
Solam (Shunem), ii. 320, iv. 347.
Solomon, pools of, iii. 332-337.
Sophor in Arabia supposed to be Ophir, i. 96.
Soristan, ii. 17.
St Anna valley, iii. 256.
Strabo, iii. 152.
Subieh, ii. 310.
Succoth (Sukkot), ii. 338, 340, 341.
Suez, i. 159.
Sufiah, iii. 107.
Sukkariyeh, iii. 246.
Summakh, village of, ii. 289.
Sundela, ii. 323.
Sur Babil, iii. 84.
Sura (Zohar), iii. 239.
Surafend, iv. 260, 281.
Surveys of eastern and southern coasts of Sinai Peninsula, i. 56-60.
Symonds' exploration of Jordan and of Dead Sea, iii. 127.
Syria, its extent, ii. 7; physical character of, 11-20; origin of name, 104.
Szammagh, ii. 282.
Szamra, ii. 348.
Szanamein, ii. 300.

Szemmak, ii. 279.
Szermadin, ii. 246.
S'zuema, brook of, iii. 64.

Taamira Arabs, iii. 81, 100.
Taanuk (Taanoch), iv. 329.
Tabor (Jebel Tor), Mount of, view from, ii. 311-318.
Tadmor, i. 138.
Taiyibeh (Ephron, Ophra), ii. 348, iii. 36, iv. 224.
Tali, ii. 352.
Tamarisk or tarfa tree that yields manna, i. 260.
Tantur, horn worn as head ornament, ii. 184, 188.
Tarichæa, ii. 278, 279.
Tarshish, discussion about, i. 82-89; ships of, 83, 84, 89.
Tartessus in Spain, i. 84, 85, 89.
Tauros, iii. 31.
Tekoa, wilderness of, iii. 99.
Tekua (Tekoah), iii. 98-100.
Tell Dilly, ii. 300.
— el Faras, ii. 285.
— el Hasy (supposed Ziklag), iii. 246, 247.
— el Horn, ii. 286.
— el Kadi, spring of, ii. 201.
— el Khanryr, ii. 173.
— es Safieh, iii. 221, 222.
— Hum (Capernaum), ii. 272-277, 283.
— Jabye and Jemera, ii. 285.
— Khaibar (Hepher), ii. 345.
— Moerad, ii. 286.
— Shakhab, ii. 286.
— Tawaneh, iii. 107.
— Zechy, ii. 285.
Teman, i. 37, ii. 137.
Terabin Arabs, i. 405.
Terkumieh, iii. 256.
Terrace culture of the Hebrews, ii. 21.
Thamara (Kurnub), i. 35.
Thrax, iii. 31.
Tiberias (Tabaria), city of, its history, ii. 256-259; its present condition, 259-262.
— Sea of. See Galilee, Sea of.
Tibneh (Timnath), iii. 239, 241, iv. 246; the home and burial-place of Joshua, 247.
Tih, range of, i. 69, 343; plateau, 370-376.
Tiyaha Arabs, i. 404.
Tobler's list of works on Palestine, ii. 391-409.
Tophet, iii. 82.
Tophila (Tophel), i. 26, 63.

Towara Arabs, protectors of Sinai Convent, i. 243, 388, 393, 397.
Travellers in Sinai Peninsula most frequently quoted by Ritter, i. 12.
— who have ascended Mount St Catherine, 189.
— to Sinai Convent, 240.
— across Tih desert, 361.
— from Dead Sea to Gulf of Aka, 420.
— to Palestine, most recent, ii. 60-77.
— to Jericho, iii. 3, 4.
Tribes, primitive, of Canaan, ii. 121-128.
— outside of Canaan, 131-159.
Tubas, ii. 341.
Tumrah, ii. 318.
Turan, ii. 310.
Tur (el-Tor), history of, i. 152-160.
Tur el Hammer, iii. 64.
Turmus, ii. 352.
Turmus Aja, ii. 342.
Turmus Aya, iv. 296, 297.

Ulad Said Arabs, i. 262.
Ulad Soleiman Arabs, i. 391.
Um el Amad, iii. 104.
Um el Orszas, iii. 74.
Um et Taiyibeh, ii. 318.
Um Lakis (Lachish), iii. 247.
Urtas (Etam), iii. 93.
Usdum mountains, iii. 123, 138.

Valley of the Convent, i. 225.
— of Jehoshaphat, iii. 81, 82.
Valleys of Palestine, ii. 17.
Vapour clouds of Dead Sea, iii. 159, 161.
Via Dolorosa, iv. 103.
Volney's travels, ii. 53.
Von Schubert's travels, ii. 69.
Van der Velde, latitude and longitude of localities as drawn up by him, ii. 359-372; altitudes of localities as given by him, 372-385.

Wadi Aallan, ii. 300.
— Abu Obaideh, ii. 339.
— Abu Sadra, ii. 339.
— Ahmed, iii. 340.
— Akhdar, i. 345.
— Ain Tuleib, iv. 245.
— Aleiat, i. 295, 296, 304, 311.
— Ali, iv. 236.
— Amora, i. 366.
— Araba (Lower Ghor), i. 49, 51, 53, 375, 421, ii. 301.
— Ararar (ancient Aroer), iii. 283.
— Ashdod, iii. 221, 222.
— Attuerwik (Tuerwik), i. 364.
— Aujeh, iv. 248.

INDEX.

Wadi Azariyye, iii. 5.
— Badera, i. 330, 336.
— Barak, i. 345.
— Beishan, ii. 309, 321.
— Beit Hanina, iii. 229.
— Belat, iv. 245, 294.
— Beni Salim, iii. 328.
— Berah, i. 345.
— Ber el Kulah, iii. 86.
— Bkia, iii. 339.
— Bozeirah, i. 68.
— Chomille, i. 345, 342.
— Dabus el Abed, iii. 9.
— Debbe, i. 303.
— Diab, ii. 339.
— el Abyad, iii. 339.
— el Agaba, i. 372.
— el Ahsa, iii. 77.
— el Ain, iv. 224.
— el Amud, ii. 266.
— el Arab, ii. 337.
— el Arish (Sihor or river of Egypt), i. 372, 375, iii. 203.
— el Atiyeh, i. 427.
— el Aujeh, ii. 339.
— el Beka, i. 331.
— el Beyanah, i. 374.
— el Beydhan, ii. 337.
— el Bireh, ii. 308, 311.
— el Chambeh, iii. 340.
— el Chan, iii. 10.
— el Delbeh, iii. 290.
— el Dharfory, i. 337.
— el Fariah, iii. 338, 339.
— el Fasail, ii. 339.
— el Fejas, ii. 309, 310.
— el Ghor, i. 184, iii. 109.
— el Gurabeh, iii. 86.
— el Hamd, iii. 7.
— el Hemar, ii. 338.
— el Humam (Hammam), ii. 265.
— el Jarafeh, i. 375, 426.
— el Jeib, i. 425.
— el Jib, iv. 294.
— el Khulil, iii. 287, 288.
— el Kid, ii. 197.
— el Koszeir, ii. 337.
— el Leban, iii. 83.
— el Makhfurijeh, ii. 345.
— el Malih, iii. 337, 338, 341.
— el Mutyah, iv. 224.
— el Nachal, i. 155, 302.
— el Seklab, ii. 337.
— el Sheik, i. 258; described by Burckhardt, 260; geology of, by Russegger, 265; topography of, by Lepsius, 269.
— el Wezy, iii. 83.
— el Wuttaijah, i. 269.

Wadi en Nar (Fire Vale), iii. 60, 83.
— en Nawaimeh, ii. 340.
— Ensous, i. 329.
— er Rahib (Monk's Vale) iii. 60, 81.
— es Safieh (Clear Ravine), iii. 139.
— esh Shaar, iv. 300.
— esh Shaib, iii. 53.
— es Sumpt, iii. 237, 240, 241.
— es Syk, i. 436.
— et Taamira, iii. 84, 340.
— et Taybe, ii. 337.
— et Teim, ii. 165, 183, 191.
— et Tih, i. 364.
— Farah, iv. 219, 230.
— Franshi, iii. 119.
— Fassail, ii. 346, iii. 8.
— Fatun, ii. 338.
— Feiran, described by Niebuhr and others, i. 255-258; characteristics and ruins of, by Burckhardt and others, 301-311.
— Ferra, ii. 351.
— Fyadh, ii. 339.
— Genne, i. 180, 345.
— Gharundel (supposed Elim), described by Burckhardt, i. 53, 367; by Niebuhr, 364.
— Ghoyer, i. 426.
— Ghurbeh, i. 169.
— Hadji, i. 47.
— Hamy Sakker, ii. 284, 285, 300.
— Hanina, iv. 229.
— Hebran, i. 168.
— Heshbon, iii. 2.
— Hodh, iii. 10.
— Hommer, i. 342.
— Humeir, iii. 140.
— Ismael, iii. 234.
— Ithm (Getune), i. 75, 424.
— Jabis, ii. 331.
— Jalud, iv. 345.
— Jamel, ii. 339.
— Jeremiyeh, i. 269.
— Kaddum, iii. 83.
— Kedum, iii. 7.
— Kelt, iii. 8, 9, 17, iv. 221.
— Kerak, iii. 76.
— Khan Hachurah, iii. 7.
— Kobeyshe, iii. 119.
— Kubarah, iii. 118.
— Kuneitrah, iii. 86.
— Kuweilifeh, iii. 268.
— Kyd, i. 61, 62.
— Lahyane, i. 54.
— Lubban, iv. 296.
— Machara (Magara, Keneh), Valley of Caves, i. 330, 335.

INDEX.

Wadi Mejedda, ii. 337.
— Melaha, i. 254.
— Moakkar, ii. 285, 300.
— Mohsen, i. 261.
— Mojeb, iii. 136, 146.
— Mokkateb (Valley of Inscriptions), described by travellers, i. 329-337.
— Morra, i. 371.
— Muhariwat, iii. 140.
— Musa, where Petra is, i. 434.
— Musurr, iii. 237.
— Muttiyah, ii. 352.
— Nabk, i. 61.
— Nahr Musrara, iv. 268.
— Nakhl described by Ruppell, i. 47.
— Nasb, i. 344; its mines and rock inscriptions described by Ruppell and Lepsius, 348-352.
— Nawaimeh, ii. 348.
— Nedjil and Nisrim, i. 329.
— Oesche, ii. 308, 316, 318.
— Osh, i. 260; its height, 352.
— Owass, i. 192.
— Rabadiyah, ii. 266, 268.
— Rahab, i. 68.
— Rakmah, i. 430.
— Ram, iv. 230.
— Retemat, i. 427.
— Romman, i. 331.
— Rudwah, i. 202.
— Rustuk, iv. 329.
— Rymm (Rimm), i. 270, 294, 301.
— Sabia, i. 423.
— Sabra, i. 423.
— Sal (Sayal), i. 66.
— Salakha, i. 70.
— Samghy, i. 68.
— Santa, iii. 18.
— Saasaf, iii. 119.
— Schubert, i. 169.
— Seba, iii. 287.
— Sebaijeh, i. 185, 186; the camping-ground of Israel when the law was given, 215-224.
— Seder (Sudr), i. 366.
— Sehab, i. 270.
— Selaf, i. 169, 177, 181, 202.
— Semek (Szemmak), ii. 284.
— Seyal, iii. 116, 118, 137.
— Seyde, ii. 285.
— Sheriah (brook Besor), iii. 248.
— Shellal (Valley of Cataracts), i. 336.
— Shoeib (Vale of Jethro), i. 184.
— Shubash, ii. 338.
— Shubeikeh, i. 338, 342.
— Sidr, iii. 7, 10, 17, 112.
— Simsin, iii. 222.
— Sinein, iii. 116.

Wadi Sinjil, iv. 295.
— Sittere, i. 338.
— Sudeir, iii. 135, 141.
— Suleim, iv. 216; Wadi Suleiman, 232, 241.
— Sumt, iii. 240.
— Surar, iii. 221, 228, 237.
— Sur Bahil, iii. 83.
— Suweinit, iv. 221.
— Szadeke, i. 421.
— Szemmak, ii. 230.
— Taamirah, iii. 135, 333.
— Tarfa, i. 270.
— Taijibe (Taibe), i. 274, 338, 341.
— Thal, i. 338, 341.
— Tullah (Tula), i. 184, 202.
— Tyh, i. 45.
— Um Rathama, i. 302.
— Urtas, iii. 93, 332.
— Usait, i. 338, 341.
— Wara, i. 61.
— Wardan, i. 364-366.
— Werd, iv. 27.
— Wetir (Outir), i. 70.
— Wutah, i. 343.
— Yetma, iv. 300.
— Zerka, iii. 133.
War Ezzaky, ii. 286.
Watershed line between Jerusalem and Tabor, ii. 352-354.
Waters of Dosh, ii. 348.
Well of Howara (supposed Marah), i. 366-368.
— of Moses at St Catherine Convent, ii. 234, 235.
— of Jacob, iv. 301, 317.

Yafa, iv. 375.
Yalo (Aijalon), iii. 243, iv. 235.
Yamon, iv. 329.
Yebna, ii. 213, iii. 222.
Yemen, i. 93.
Yermak (Hieromax), river of, ii. 297-299.
Yitma, iii. 300.
Yutta (Juttah), iii. 107, 286.

Zaliane Arabs, iii. 10.
Zebier range, i. 199.
Zeiteh, iii. 328.
Zemaraim, Mount of, ii. 349.
Zered brook, ii. 149, iii. 78.
Zerin, ancient Jezreel, ii. 322-325, 327, iv. 344.
Ziph, hill of, iii. 103.
Ziph, city of, iii. 103, 104.
Zoar, i. 27, 30, iii. 76, 144.
Zodocatha, i. 32.
Zuweirah, pass of, iii. 123.

INDEX OF TEXTS.

Reference	Vol.	Page
Gen. i. 7-13,	vol. iii.	page 48
ii. 11, 12,	i.	97, 121, 133
ix. 20,	iii.	297
x. 4,	i.	84, 89
x. 6, 15-19,	ii.	106
x. 7, 26-30,	ii.	141
x. 14,	i.	316
x. 14,	iii.	260
x. 15,	ii.	110, 119
x. 18,	ii.	112
x. 19,	i.	30, 39
x. 19,	ii.	112
x. 19,	iii.	262, 268
x. 20,	iii.	208
x. 21,	ii.	105
x. 21,	iii.	262
x. 28,	i.	97
x. 29,	i.	94, 96, 103
x. 30,	i.	96
xi. 16,	ii.	105
xi. 31,	ii.	106
xii. 6,	ii.	106
xii. 6,	iv.	308
xii. 8,	iv.	26, 223
xiii. 9,	iv.	227
xiii. 18,	iii.	293
xiv. 2, 8,	i.	27
xiv. 3-6,	ii.	131
xiv. 5,	ii.	149
xiv. 6,	i.	432
xiv. 6, 7,	i.	426
xiv. 6, 7,	ii.	134, 135
xiv. 7,	ii.	125
xiv. 7, 13,	ii.	121
xiv. 13, 24,	ii.	120
xiv. 13, 24,	iii.	298
xiv. 15, 18,	ii.	120, 180
xiv. 20,	iii.	183
xv. 18,	ii.	105
xv. 19,	ii.	146, 147
xv. 19, 21,	ii.	144
xv. 21,	ii.	127
xvi. 7,	i.	426, 431
xvi. 7,	ii.	142
Gen. xvi. 12,	vol. i.	page 383, 402
xvii. 8, 9, 23-27,	iii.	300
xviii. 10,	iii.	8
xviii.,	iii.	299
xix. 20,	i.	31
xix. 28,	iii.	8, 85
xx. 1,	i.	30
xx. 1,	iii.	277
xx. 2,	i.	317
xx. 21,	iii.	275
xxi. 14-21,	i.	426
xxi. 20, 21,	i.	432
xxi. 22,	iii.	278
xxi. 28-30,	i.	28
xxi. 32,	i.	317
xxii. 17,	i.	399
xxii. 20-23,	ii.	104
xxiii. 2,	ii.	120
xxiii. 5-7,	ii.	119
xxiii. 7,	ii.	122
xxiii. 17-19,	iii.	298
xxiii. 19,	ii.	106, 109
xxv. 2,	ii.	145
xxv. 3,	i.	37
xxv. 9,	iii.	298, 307
xxv. 12-18,	i.	383
xxv. 18,	ii.	142
xxvi.	iii.	277
xxvi. 1, 8,	i.	30
xxvi. 1, 8,	iii.	266
xxvi. 8,	i.	317
xxvi. 15, 16,	ii.	120
xxvi. 17-33,	iii.	278
xxvi. 34,	ii.	122
xxviii. 11-19,	iv.	26
xxviii. 19,	ii.	120
xxviii. 22,	iii.	183
xxix. 2, 3,	iii.	102
xxxi. 47,	ii.	120
xxxii. 7, 8,	ii.	228
xxxii. 22,	ii.	120, 228
xxxiii. 17,	ii.	120, 340, 341
xxxiii. 18,	ii.	350, 351
xxxiii. 18,	iv.	308

INDEX OF TEXTS.

Gen. xxxiii. 19, vol. ii. page 109,		123
xxxiv. 2,	ii. 117, 120,	123
xxxiv. 8,	ii.	120
xxxv. 1,	iv.	308
xxxv. 19,	iii.	331
xxxv. 27,	iii.	298
xxxv. 29,	iii.	307
xxxvi. 9,	ii.	135
xxxvi. 11, 12, 16,	ii.	144
xxxvi. 12,	ii.	141
xxxvi. 20-29,	ii.	134
xxxvi. 24,	iii.	69
xxxvi. 33,	i.	26
xxxvii. 12, 14, 28,	iv.	308
xxxvii. 14,	iii.	298
xxxvii. 17,	ii.	331
xxxvii. 27, 28,	i.	387
xl. 15,	ii.	106
xlv. 19, 21, 27,	iii.	328
xlviii. 7,	iii.	339
xlviii. 22,	ii.	125
xlviii. 22,	iv.	318
xlix. 9,	iii.	181
xlix. 11, 12,	iii.	198
xlix. 13,	iv.	280
xlix. 14,	iii.	189
xlix. 19,	iii.	188
l. 13,	iii. 48,	298
Exod. ii. 15,	i.	387
ii. 15-22,	ii.	144
iii. 1, 12, 18,	i.	319
iii. 22,	i.	81
v. 3,	i.	319
viii. 27, 28,	i. 320,	356
xi. 2,	i.	81
xii. 22,	i.	212
xii. 37,	ii.	340
xii. 38,	i.	291
xii. 38,	iii.	274
xiii. 17,	i.	316
xiii. 17,	iii. 203,	275
xv. 14,	iii.	276
xv. 22,	ii.	104
xv. 27,	i.	288
xvi. 1,	i.	322
xvi. 4,	i.	284
xvi. 14,	i. 277,	286
xvi. 15,	i.	285
xvi. 20,	i.	297
xvi. 31, 32, 34,	i.	285
xvi. 35,	i.	290
xvi. 35,	ii.	107
xvii. 1, 8,	i.	319
xvii. 3,	i.	291
xvii. 8,	iii.	269
xvii. 15,	i.	326
xviii. 14-23,	i.	383
xviii. 21-23,	i.	387
xix. 1, 2,	i.	199
Exod. xix. 2, 16-20, vol. i. page 217,		
	218,	222
xix. 16,	i. 222,	226
xix. 17,	i.	218
xx. 18, 21,	i.	218
xxi. 13,	i.	399
xxx. 23, 24,	i.	121
xxxii. 4,	i.	190
xxxii. 15-34,	i.	231
xxxiv. 11,	ii.	121
Lev. xiv. 4,	i.	212
Num. ix. 1,	i.	199
x. 12,	i. 22, 69, 199,	427
x. 29-33,	ii.	145
xi. 4,	iii.	274
xi. 7,	i.	121
xi. 8,	i.	273
xi. 8, 9,	i.	286
xi. 9,	i.	277
xi. 22, 31,	i.	68
xi. 35,	i.	67
xii. 14,	i.	67
xiii. 3, 21, 26,	i. 22,	69
xiii. 4, 27,	i.	427
xiii. 21,	ii.	207
xiii. 22,	iii.	292
xiii. 23,	iii.	258
xiii. 24,	iii. 294,	298
xiii. 29,	ii. 115,	128
xiii. 33,	ii.	131
xiv. 45,	i.	428
xiv. 45,	ii.	142
xix. 6,	i.	212
xx. 1,	i.	22
xx. 2,	i.	426
xx. 14-21,	i.	26
xx. 16,	ii.	135
xxi. 1,	i.	34
xxi. 3,	i.	426
xxi. 4,	i. 75,	291
xxi. 10, 11,	i.	38
xxi. 11,	ii.	148
xxi. 13, 34,	ii.	125
xxi. 14, 15,	ii.	150
xxi. 17, 18,	ii.	150
xxi. 21-26,	ii.	151
xxi. 24,	ii. 151,	157
xxi. 26,	ii.	126
xxi. 27-30,	ii.	150
xxi. 28,	i.	34
xxi. 30,	ii.	152
xxi. 31-35,	ii.	153
xxi. 33,	ii.	125
xxii. 4, 7,	ii.	147
xxii. 5,	ii.	154
xxii. 23, 24,	ii.	153
xxiii. 7,	ii.	154
xxiii. 44,	ii.	148
xxiv. 20,	ii.	141

INDEX OF TEXTS. 403

Num. xxiv. 20,	vol. iii.	page 268
xxv. 1,	ii.	153
xxv. 1,	iii.	2
xxv. 9-13,	i.	395
xxvii. 21,	i.	387
xxxii. 6, 16-18,	ii.	154
xxxii. 37,	ii.	152
xxxii. 33-38,	ii.	154
xxxii. 41,	iv.	338
xxxiii. 5,	ii.	340
xxxiii. 12-14,	i.	320, 322
xxxiii. 13, 14,	i.	8, 34
xxxiii. 17, 18,	i.	427
xxxiii. 17, 20,	i.	63, 67
xxxiii. 29,	i.	75
xxxiii. 38,	ii.	107
xxxiii. 41,	i.	36
xxxiii. 41-44,	ii.	148
xxxiii. 43, 44,	i.	38
xxxiii. 44,	ii.	148
xxxiii. 45-47,	ii.	150
xxxiii. 49,	ii.	153
xxxiii. 49,	iii.	2
xxxiii. 51,	ii.	107, 115
xxxiv.	iii.	175
xxxiv. 2-13,	ii.	107
xxxiv. 3, 4,	i.	427
xxxiv. 5,	ii.	113
xxxiv. 5,	iii.	203
xxxiv. 7,	iii.	175
xxxiv. 11,	ii.	235
Deut. i. 1,	i.	26, 37, 63
i. 2,	i.	427
i. 7, 19,	iii.	325
i. 19,	i.	427
i. 44,	i.	431
ii. 1, 5, 8, 12,	i.	429
ii. 1, 8,	ii.	136
ii. 4, 8,	ii.	135
ii. 6,	i.	291
ii. 9,	i.	34
ii. 10,	ii.	131, 149
ii. 10, 12, 19, 20,	ii.	109
ii. 12,	ii.	134
ii. 13,	ii.	149
ii. 13, 18,	iii.	78
ii. 14,	ii.	148
ii. 20,	ii.	131
ii. 23,	ii.	109, 113, 133
ii. 23,	iii.	210, 225, 265-267
ii. 24,	ii.	118
ii. 26-37,	ii.	151
iii. 4, 5,	ii.	153
iii. 5,	ii.	126
iii. 9,	ii.	160, 165
iii. 11,	ii.	132
iii. 14,	ii.	118
iii. 16,	ii.	150
iii. 17,	ii.	235
Deut. iv. 43,	vol. ii.	page 196
iv. 48,	ii.	160, 165
viii. 30,	iv.	303
xi. 26-28,	iv.	304
xi. 29,	iv.	303
xi. 30,	iv.	296
xxiii. 3,	ii.	158
xxiv. 3,	iii.	1
xxvii. 2,	iv.	303
xxxii. 17,	iii.	168
xxxii. 32,	iii.	21
xxxii. 43,	ii.	260
xxxiii. 2,	i.	22
xxxiii. 3,	ii.	158
xxxiii. 7,	iii.	181
xxxiii. 18,	iii.	189
xxxiii. 19,	iv.	280
xxxiv.	iii.	85
xxxiv. 3,	iii.	1
Josh. i. 4,	ii.	122
ii. 7,	iii.	4, 52
iii. 17,	iii.	41
iv. 13,	iii.	2
iv. 19, 20, 24,	iii.	46
v. 10,	iii.	2
v. 12,	ii.	107
v. 12,	iii.	46
vi. 26,	iii.	2, 32
vii. 2,	iv.	223
vii. 26,	iii.	45
viii. 1-35,	iv.	223
viii. 33, 34,	iv.	303
ix. 1,	ii.	124, 128, 235
ix. 3, 7, 15,	ii.	124
ix. 6,	iii.	47
ix. 17,	iii.	230
x.	iii.	269
x. 1, 2,	ii.	124
x. 1, 5,	ii.	128
x. 1-14,	ii.	126
x. 1-42,	iii.	325
x. 2,	iii.	230
x. 3,	iii.	258
x. 3-5,	iii.	247
x. 4,	i.	35
x. 5,	ii.	118
x. 6, 15,	iii.	47
x. 10, 11,	ii.	224
x. 16,	iii.	249
x. 43,	ii.	224
xi. 1-6,	ii.	118
xi. 1-12,	ii.	127
xi. 1-16,	iii.	325
xi. 1-20,	ii.	200
xi. 3,	ii.	115, 123
xi. 8, 10-13,	ii.	224
xi. 21,	iii.	257, 276, 324
xi. 21, 22,	ii.	133
xii. 2,	ii.	150, 153

Reference	Vol.	Page
Josh. xii. 3,	vol. ii.	page 235
xii. 5,	ii.	118, 125
xii. 14,	i.	35
xii. 16,	iv.	26
xii. 17,	ii.	345
xii. 19,	ii.	217
xii. 19-23,	iv.	334
xii. 21,	iv.	329
xii. 22,	iv.	336
xii. 23,	iv.	249
xii. 24,	ii.	351
xiii. 2,	iv.	334
xiii. 2-6,	iii.	178
xiii. 3,	ii.	133
xiii. 3,	iii.	203, 275
xiii. 4, 15-32,	iii.	178
xiii. 5,	ii.	215
xiii. 5,	iii.	186
xiii. 12,	ii.	118
xiii. 19,	iii.	68
xiii. 21-27,	i.	26
xiii. 27,	ii.	340
xiii. 32,	ii.	152
xiv. 14,	iii.	182
xiv. 15,	ii.	132
xiv. 15,	iii.	292
xv. 1,	ii.	35
xv. 1-5,	iii.	179
xv. 1-8,	i.	433
xv. 3,	i.	135
xv. 4, 47,	ii.	113
xv. 6, 7,	iii.	47
xv. 7,	ii.	10, 11, 45, 238
xv. 8,	ii.	128, 132
xv. 8,	iv.	147
xv. 9,	iii.	233
xv. 10,	iii.	241
xv. 11,	iii.	242, 243, 244
xv. 15,	iii.	257
xv. 16, 17,	iii.	182
xv. 17,	iii.	257
xv. 20-32,	iii.	284, 325
xv. 23,	i.	427
xv. 24,	iii.	103
xv. 25,	iii.	239
xv. 28,	i.	30
xv. 29,	i.	31
xv. 30,	i.	431
xv. 31,	iii.	247
xv. 32,	iii.	284
xv. 33-36,	iii.	240
xv. 34,	iii.	234
xv. 37-41,	iii.	247
xv. 39,	iii.	248
xv. 41,	ii.	352
xv. 41,	iii.	220
xv. 43,	iii.	239, 256
xv. 45-47,	iii.	243
Josh. xv. 46, 47,	vol. iii.	page 225, 257
xv. 47,	iii.	210
xv. 48,	iii.	108, 240, 284
xv. 48-51,	iii.	285
xv. 48-60,	iii.	283
xv. 49,	iii.	257
xv. 50,	iii.	107, 284, 285
xv. 52-54,	iii.	324
xv. 55,	iii.	103, 107, 286
xv. 55-57,	iii.	287
xv. 56,	iii.	239, 327
xv. 58, 59,	iii.	324
xv. 61, 62,	iii.	331
xvi. 1, 2,	iv.	222, 226
xvi. 5,	iv.	229, 242
xvi. 10,	ii.	126
xvi. 10,	iii.	183
xvii. 11,	ii.	327
xvii. 11,	iv.	329, 337
xvii. 11, 16,	ii.	323, 335
xvii. 14-18,	ii.	327
xvii. 15,	ii.	132
xvii. 21,	iv.	337
xviii. 1,	iii.	47, 182
xviii. 3,	iii.	182
xviii. 7,	iii.	183
xviii. 10,	iii.	182
xviii. 11-28,	iii.	7
xviii. 13,	iv.	222, 229, 242
xviii. 14, 15,	iii.	233
xviii. 16,	iv.	147
xviii. 17,	iii.	10
xviii. 21,	iii.	48
xviii. 22,	ii.	349
xviii. 23,	ii.	133
xviii. 23,	iii.	225
xviii. 23,	iv.	219, 225
xviii. 24,	ii.	294
xviii. 25,	iii.	231
xviii. 28,	ii.	128
xviii. 29,	iii.	183
xix. 1-9,	iii.	325
xix. 7,	iii.	284
xix. 9,	iii.	182
xix. 10-16,	iv.	336
xix. 10-40,	iv.	334
xix. 11, 26,	iv.	352
xix. 12,	ii.	314
xix. 12.	iv.	375
xix. 12, 22,	ii.	312
xix. 17-24,	iv.	337
xix. 21,	ii.	331
xix. 21,	iv.	345
xix. 24-32,	iv.	337, 361
xix. 27,	ii.	352
xix. 27,	iii.	188, 220
xix. 27,	iv.	334
xix. 28, 29,	iii.	183

INDEX OF TEXTS. 405

Reference	Vol.	Page
Josh. xix. 28, 29,	vol. iv.	page 380
xix. 32-40,	ii.	220
xix. 32-40,	iv.	338
xix. 35,	ii.	235, 257, 302
xix. 35-37,	ii.	219
xix. 36-38,	ii.	215
xix. 37,	ii.	217
xix. 38,	ii.	264
xix. 41-43,	iii.	242
xix. 42,	iv.	235
xix. 43,	iii.	243
xix. 45,	iii.	250
xix. 46,	iv.	254
xix. 47,	iii.	183
xix. 50,	iii.	325
xix. 51,	iii.	182
xx. 7,	ii.	217, 255
xx. 7,	iii.	191, 302
xx. 7,	iv.	308, 344
xx. 8,	ii.	196
xxi. 2,	iii.	183
xxi. 6,	iii.	264
xxi. 11,	iii.	257, 302
xxi. 14,	iii.	107, 284, 285
xxi. 16,	iii.	242
xxi. 24,	iii.	250
xxi. 27,	ii.	196
xxi. 32,	ii.	217, 255
xxi. 32,	iv.	334
xxi. 32, 33,	iii.	227
xxiv. 17, 18,	ii.	125
xxiv. 28, 29,	iv.	247
xxiv. 30,	iii.	325
xxiv. 32,	iv.	309, 315
xxiv. 33,	iv.	295
Judg. i. 4, 5,	ii.	121
i. 7,	ii.	117
i. 16,	i.	34
i. 16,	ii.	144
i. 17,	i.	431
i. 18,	iii.	210, 243, 275
i. 19,	ii.	127
i. 26,	ii.	123
i. 27,	ii.	327, 335
i. 27-33,	iv.	332
i. 31,	ii.	283
i. 31,	iii.	183, 187
i. 31,	iv.	361
i. 32,	iii.	187
i. 33,	iii.	189
i. 34, 36,	ii.	121, 126
ii. 8, 9,	iv.	247
iii. 1-4,	iii.	226
iii. 2,	iii.	276
iii. 3,	ii.	123
iii. 3,	iii.	210
iii. 5,	ii.	121
iii. 12-30,	ii.	155
iii. 13,	ii.	143
Judg. iii. 13,	vol. iii.	page 2
iii. 28,	iii.	52
iv. 2, 13, 16,	ii.	218, 224
iv. 3,	iii.	186
iv. 4,	iv.	230
iv. 6, 10,	ii.	217
iv. 6, 12,	ii.	317
iv. 6, 12,	iv.	353
iv. 11,	ii.	144
iv. 11,	iv.	338
v.	iv.	351
v. 8,	iii.	279
v. 12,	iv.	230
v. 17,	iii.	188
v. 21,	iv.	347
v. 23,	ii.	316
vi. 3,	ii.	147
vi. 33,	iv.	344
vii. 24,	iii.	424
vii. 25,	iii.	121
viii. 5-17,	ii.	341
viii. 10, 21-27,	i.	387
viii. 11,	ii.	152
viii. 33,	ii.	124
ix. 4,	ii.	124
ix. 46,	ii.	124
ix. 50-57,	ii.	341
x. 6,	ii.	157
x. 8,	ii.	125
x. 12,	i.	424
x. 12,	ii.	148
x. 12,	iii.	190
xi. 13,	ii.	125
xi. 18,	ii.	149
xi. 22,	ii.	125
xi. 33,	ii.	157
xii. 15,	ii.	144
xiii. 1,	iii.	239
xiv. 3,	iii.	278
xiv. 4,	iii.	225
xiv. 5,	iii.	239
xv. 18,	iii.	278
xvi. 1,	iii.	210
xviii. 7,	iii.	186
xviii. 7, 28,	ii.	204
xviii. 17,	ii.	196
xviii. 28,	ii.	207
xviii. 29,	iii.	183
xix. 13,	iv.	230
xx. 1,	i.	28
xx. 26,	iv.	26
xx. 45, 47,	iv.	219
xxi.	iv.	219
xxi. 12,	ii.	117
xxi. 24,	iv.	298
Ruth ii. 3-18,	iii.	249
1 Sam. ii. 3,	iii.	226
iii. 20, 21,	iv.	298
v. 4,	iii.	220

INDEX OF TEXTS.

Reference	Vol.	Page
1 Sam. v. 6, 7,	vol. iii.	page 226
v. 8,	iii.	250
vi. 17,	iii.	210, 225
vi. 21,	iii.	234
vii. 1,	iii.	234
vii. 1, 2,	iii.	233
vii. 5-7,	iii.	232
vii. 13,	iii.	279
vii. 16,	iii.	46
ix. 1-16,	iii.	235
x. 2-7,	iii.	235
x. 10,	iii.	46
xii. 9,	ii.	218
xiii. 5,	iii.	279
xiii. 15,	iii.	36
xiii. 16-18,	iv.	221
xiii. 17,	iv.	225
xiii. 19-22,	iii.	279
xiii. 23,	iv.	221
xiv. 1, 4, 5,	iv.	221
xiv. 4,	iv.	25
xiv. 6,	iii.	278
xiv. 31,	iv.	221
xv. 1-4,	iii.	241
xv. 2-7,	ii.	143, 144
xv. 6,	ii.	145
xv. 7,	ii.	142
xv. 12,	i.	30
xv. 13,	iii.	36
xvi. 11, 13,	iii.	339
xvii. 2, 19,	iii.	240
xvii. 4,	ii.	133
xvii. 4, 23,	iii.	250
xvii. 26,	iii.	278
xvii. 52,	iii.	251
xix. 18-24,	iii.	46, 231
xxi. 1,	iii.	97
xxi. 1,	iv.	218
xxii. 5,	iii.	326
xxii. 19,	iv.	218
xxiii. 14, 25,	iii.	104
xxv. 1,	iii.	231
xxv. 2,	i.	30
xxvii. 6,	iii.	247, 248
xxvii. 7,	iii.	226
xxvii. 8,	ii.	105, 118, 143
xxvii. 8,	iii.	268
xxvii. 10,	iii.	325
xxviii. 3,	iii.	231
xxviii. 7,	ii.	319
xxix. 1,	ii.	325
xxix. 1-11,	ii.	326
xxix. 1-11,	iii.	184
xxx. 1,	iii.	248
xxx. 1-22,	ii.	144
xxx. 14,	iii.	264, 325
xxx. 26,	iii.	235
xxx. 28,	i.	35
xxx. 29,	ii.	146
1 Sam. xxxi. 1-10,	vol. ii.	page 329
xxxi. 10,	ii.	335
2 Sam. i. 1,	iii.	248
ii. 2, 8, 9,	ii.	324
ii. 3,	iii.	248, 264
ii. 29,	iii.	52
iii. 32, 33,	iii.	296
iv. 2,	iv.	228
iv. 12,	iii.	295
v. 6, 7,	iii.	265
v. 6-9,	iv.	63
v. 18-25,	iv.	27
v. 19,	iii.	211
viii. 1,	iii.	211
viii. 2,	ii.	115
viii. 18,	iii.	265
x. 5,	iii.	2
x. 6,	ii.	207
x. 17,	iii.	52
xi. 3,	ii.	123
xv. 19,	iii.	265
xvi. 5-13,	iv.	213
xvii. 17,	iv.	147
xvii. 22-24,	iii.	52
xviii. 17, 18,	iv.	173
xix. 1,	iii.	52
xx. 14, 15,	ii.	213
xx. 19,	ii.	213
xxi. 2,	ii.	125
xxi. 14,	ii.	329
xxi. 15,	iii.	211
xxi. 15-22,	ii.	133
xxiii. 13,	iii.	97
xxiii. 15,	iii.	341
xxiii. 20,	ii.	155
xxiii. 37,	iv.	228
xxiii. 39,	ii.	123
xxiv. 7,	ii.	123
xxiv. 7,	iii.	325
xxiv. 16-25,	ii.	129
xxiv. 16-25,	iv.	64
1 Kings i. 9, 41,	iv.	146
ii. 10,	iv.	56
iii. 4,	iii.	231
iv. 9,	iii.	242
iv. 10,	ii.	345
iv. 11,	iv.	270
iv. 12,	ii.	325, 335, 341
iv. 12,	iv.	339
iv. 13,	ii.	334
iv. 24,	iii.	175, 211
v. 8,	i.	126
v. 11,	i.	120
v. 11,	iii.	188
v. 17, 18,	ii.	215
vii. 14,	iii.	190
ix. 11,	ii. 255, 256, iii. 190
ix. 13,	iii. 188, iv. 334
ix. 14,	i.	91, 137

INDEX OF TEXTS. 407

1 Kings	ix. 15,	vol. ii.	page 225
	ix. 20,	ii.	123, 129
	ix. 26,	i.	92
	ix. 26-28,	i.	64, 81, 89, 90
	x. 2,	i.	82, 145
	x. 11,	i.	82, 124
	x. 12,	i.	125
	x. 14,	i.	90
	x. 18,	i.	143
	x. 22,	i.	82, 143
	xi. 7,	iv.	24
	xi. 14-22,	ii.	138
	xii.	iii.	185
	xii. 14, 15,	iv.	309
	xii. 20,	ii.	205
	xiii. 32,	ii.	349
	xv. 20,	ii.	200, 213, 235
	xv. 22,	iii.	232
	xv. 29,	ii.	225
	xvi. 24,	ii.	349
	xvi. 32, 33,	iv.	352
	xvi. 34,	iii.	2
	xvii. 3, 5,	iii.	8
	xviii. 19-46,	iv.	352
	xix. 2, 13,	i.	4
	xix. 8,	iv.	353
	xx. 49,	i.	12
	xxii. 14,	i.	91
	xxii. 48,	i.	87
	xxii. 49,	i.	87, 89
	xxii. 51,	iv.	175
	xxxii. 49,	i.	64
2 Kings	i. 2,	iii.	280
	i. 16,	iii.	244
	ii. 19-22,	iii.	34
	iii. 9,	ii.	139
	iv. 8-37,	iv.	347
	iv. 38,	iii.	46
	v. 12,	iii.	52
	vi. 2-5,	iii.	52
	vi. 13,	ii.	331
	viii. 20-22,	i.	137
	viii. 20-22,	ii.	139
	ix. 21,	ii.	327
	x. 15,	ii.	327
	x. 15, 33,	ii.	146
	xiv. 7,	i.	419
	xiv. 7,	ii.	138, 139, 143
	xiv. 22,	i.	72, 138
	xiv. 25,	iv.	336
	xv. 20,	ii.	219, 223, 225, 236, 255
	xv. 29,	iv.	339
	xv. 35,	iv.	48
	xvi. 6,	i.	138, 317
	xvi. 12,	iv.	90
	xvii. 27-41,	iv.	288
	xx. 20,	iii.	334
	xx. 20,	iv.	70, 74, 90
2 Kings	xxiii. 6,	vol. iv.	page 162
	xxiii. 8,	iv.	221
1 Chron.	i. 43-54,	ii.	136
	vi. 71,	ii.	196
	vi. 76,	ii.	255
	vii. 21,	iii.	268
	ix. 12,	iv.	240
	xi. 5-8,	iv.	63
	xii. 15,	iii.	51
	xviii. 12,	ii.	138
	xxi. 18,	iv.	64
	xxiii. 2, 14,	i.	81
	xxvi. 4,	ii.	143
	xxviii. 29,	iv.	265
2 Chron.	ii. 3,	i.	126
	ii. 8,	i.	125
	ii. 10,	iii.	294
	ii. 10, 15,	i.	120
	ii. 15,	iii.	188
	ii. 16,	i.	126
	iii. 6,	i.	79
	viii. 5,	iv.	242
	viii. 18,	i.	81, 89, 90
	ix. 10,	i.	125
	xi. 5, 6,	iii.	99, 336
	xi. 8,	iii.	104, 251
	xi. 9,	iii.	259
	xii. 4, 19,	ii.	349
	xiii. 19,	iv.	225
	xiii. 21,	i.	83
	xvi. 4,	ii.	209, 213
	xx. 16,	iii.	109
	xx. 22,	ii.	125
	xx. 36, 37,	i.	64, 87
	xxiv. 16,	iv.	56
	xxiv. 21,	iv.	173
	xxv. 11, 14,	ii.	138, 143
	xxvi. 6,	iii.	251
	xxvii. 3,	iv.	48
	xxvii. 7,	ii.	148
	xxviii. 18,	iv.	235, 241
	xxviii. 27,	iv.	56
	xxxi. 14,	iv.	34
	xxxii. 2-6,	iv.	90
	xxxii. 3, 30,	iii.	334
	xxxii. 4,	iv.	95
	xxxii. 30,	iv.	70, 73, 90
	xxxv. 20-25,	iv.	330
	xxxv. 22,	ii.	23
Ezra	ii. 22,	iii.	239
	ii. 26,	iv.	230
	ii. 28,	iv.	223
	iii. 7,	iv.	254
	iv. 2, 4, 10,	iv.	289
	iv. 10,	iv.	287, 309
	vi. 2,	iii.	258
	ix. 1,	ii.	129, 156
	x. 18-44,	ii.	117
Neh.	ii. 13,	iv.	71, 145

Neh. ii. 14,	vol. iv.	page 152
ii. 19,	iv.	267
iii. 13, 14,	iv. 49, 53,	171
iii. 15,	iv. 52, 56, 148,	152
iii. 15-28,	iv.	47
iii. 16,	iv.	75
iii. 29,	iv.	34
iii. 30-32,	iv.	29
iii. 34,	iv. 267,	309
iv. 2,	iv.	267
iv. 7-18,	iii.	226
vii. 26,	iii.	239
vii. 28,	iv.	218
vii. 30,	iv.	230
xi. 31,	iv.	223
xi. 35,	iv.	240
xiii. 1,	ii.	156
xiii. 24,	iii.	226
Job xvi. 15,	ii.	188
xxii. 24,	i.	79
xxiv. 5-9,	ii.	134
xxviii. 16,	i.	81
xxx. 4,	i.	345
xxxix. 9-12,	ii.	210
Ps. i. 3,	ii.	247
xxix. 8,	i.	425
xlii. 6,	ii. 164,	319
xlv. 10,	i.	81
xlviii. 11-15,	iv.	58
lxvi. 6,	ii.	341
lxviii. 16,	iv.	353
lxx. 15,	i.	82
lxxii. 15,	i.	97
lxxvii. 19, 20,	i.	369
lxxvii. 68, 69,	iv.	20
lxxxiii. 5-9,	i.	403
lxxxiii. 9, 10,	ii.	319
lxxxiii. 12,	iii.	121
lxxxix. 12,	ii. 312, 313,	319
civ. 15,	iii.	297
civ. 18,	iii.	79
cxii. 9,	ii.	188
cxx. 4,	i.	345
cxxii. 3,	iv.	20
cxxv. 2,	iv.	18
cxxxii. 17,	ii.	188
cxxxiii. 3,	ii.	165
cxxxvii. 7-9,	ii.	139
cxlviii. 14,	ii.	188
Prov. xxx. 24-29,	iii.	79
Eccles. ii. 5, 6,	iii.	337
Song of Sol. i. 10,	iv.	328
i. 12,	i.	121
i. 14,	iii.	25
ii. 11, 13,	iii.	338
iv. 3,	iii.	25
iv. 12,	iii.	334
vii. 5,	iv.	352
vii. 9,	iii.	297
Isa. v. 24,	vol. ii.	page 252
vii. 3,	iv.	30
viii. 6, 7,	iv.	148
ix. 1,	ii.	230
ix. 1, 2,	ii. 255,	261
ix. 1, 2,	iv.	342
x. 26,	iii.	121
x. 31,	iv.	295
xiii. 12,	i.	81
xv. 7,	iii.	78
xvii. 1, 10,	i.	419
xvii. 5,	iv.	28
xx. 1,	iii.	226
xxi. 13, 14,	i.	37
xxii. 9,	iv.	71, 74
xxii. 9-11,	iv. 70, 73,	91
xxii. 11,	iv.	95
xxii. 15-17,	iv.	162
xxiii. 6, 10,	i.	89
xxvii. 12,	i.	41
xxviii. 27,	iii.	27
xxxii. 14,	iv.	48
xxxiii. 19,	iv.	352
xxxv. 2,	iv.	352
xxxvi. 2,	iv.	91
xliii. 3-5,	iv.	170
xliii. 14,	i. 110,	138
xlviii. 17,	iv.	70
liii. 5, 11,	iv.	170
lxiii. 1,	i.	37
lxiii. 6, 10,	i.	89
Jer. i. 1,	iv.	217
ii. 13,	iv.	142
ii. 21,	iii.	297
iv. 5, 6,	ii.	335
vi. 1,	iii.	96
vii. 12, 14,	iv.	208
x. 9,	i.	79
x. 28-33,	iv.	217
xii. 5,	iii.	52
xix. 1, 2, 11,	iv.	165
xix. 2, 8,	iv.	71
xxv. 23,	i.	37
xxvi. 18,	iv.	58
xxvii. 3,	ii.	139
xxxi. 38-40,	iv.	135
xxxii. 8,	iv.	217
xxxv. 6, 7,	ii.	146
xxxvii. 36,	iv.	217
xli. 5,	iv.	309
xlvi. 18,	ii.	317
xlvii. 1,	iii.	210
xlvii. 4,	iii. 263,	281
xlvii. 5,	iii.	211
xlviii. 1,	iii.	73
xlviii. 24, 41,	iii.	74
xlix. 7, 8,	i.	36, 37
xlix. 13, 22,	i.	26, 37
xlix. 19,	iii.	52

INDEX OF TEXTS.

Reference	Vol.	Page
Lam. iv. 21,	vol. ii.	page 134
Ezek. xxv. 8-14,	ii.	139, 156
xxv. 13,	i.	36, 37
xxv. 16,	iii.	264
xxvii. 6,	ii.	123
xxvii. 9,	ii.	215
xxvii. 12,	i.	82
xxvii. 15,	i. 118, 122, 135, 146
xxvii. 17,	i.	120
xxvii. 22,	i.	97, 135
xxxviii. 13,	i.	89
xlvii.	iii.	175
xlvii. 1-12,	iv.	96
xlvii. 19,	i.	35
xlviii. 28,	i.	35
Hosea ix. 13,	iv.	321
x. 14,	ii.	266
Joel iii. 12,	iii.	82
Amos i. 2,	iv.	352
i. 12,	i.	26, 37
ii. 2,	iii.	74
iii. 15,	i.	143
v. 4,	iv.	227
vi. 14,	iii.	78
ix. 3,	iv.	354
ix. 7,	iii.	263
Obad. 3,	ii.	133
3, 4,	i.	442
Micah i. 6,	iv.	324
iii. 12,	iv.	58
iv. 4,	ii.	188
vii. 14,	iv.	353
Zeph. ii. 5,	iii.	264
Zech. ix. 2-6,	iii.	281
ix. 9, 10,	ii.	328
ix. 9, 10,	iii.	197
xii. 11,	iv.	350
xiv. 8,	iv.	96
xiv. 16-18,	iv.	2
Matt. ii. 11,	iii.	346
iii. 1,	iii.	42
iii. 11,	iii.	41
iii. 12-14,	ii.	261
iii. 15,	ii.	256
iv. 8,	iii.	8
iv. 13,	ii. 229, 264, 275
iv. 15, 16,	iv.	342
iv. 18,	ii.	236
iv. 25,	ii.	335
v. 14,	ii.	220
viii. 23,	ii.	251
viii. 28,	ii.	127, 302
ix. 9,	ii.	230
xi. 21,	ii.	270
xi. 21-23,	ii.	274
xii. 1-6,	iii.	104
xiii. 25,	ii.	192
xiv. 1-22,	ii.	258
xiv. 13,	ii.	276
Matt. xiv. 34,	vol. ii.	page 235
xv. 29, 39,	ii.	263
xix. 1,	iv.	225
xx. 29,	iii.	32
xxi. 8,	iv.	33
xxiii. 27,	iii.	245
xxiii. 29,	iv.	177
xxiv. 1, 2,	iv.	112
xxvi. 3, 4,	iv.	28
xxvi. 36,	iv.	170
xxvi. 69,	ii.	255
xxvii. 7, 8,	iv.	165
xxvii. 25,	iii.	117
xxvii. 32,	iv.	127
xxvii. 57,	iv.	261
xxviii. 2,	iv.	167
Mark i. 4,	iii.	42
ii. 14,	ii.	230
v. 1,	ii.	302
vi. 45, 53,	ii.	270
vi. 33,	ii.	276
vi. 53,	ii.	235
viii. 10,	ii.	263
viii. 22,	ii.	234
ix. 2,	ii.	312
x. 1,	iv.	225
xiii. 1, 2,	iv.	112
xiv. 32,	iv.	170
xv. 40,	ii.	263
Luke ii. 4, 11,	iii.	339
ii. 7,	iii.	346
ii. 22,	iv.	152
iii. 1,	ii.	233
v. 27,	ii.	230
vi. 1,	iii.	104
viii. 2,	ii.	263
viii. 26,	ii.	302
viii. 27,	ii.	307
ix. 10,	ii.	234, 276
x. 13,	ii.	270
xi. 47,	iv.	176
xiii. 4,	iv.	149
xvii. 18,	iv.	289
xix. 1, 28,	iii.	32
xix. 1, 29,	iii.	5
xix. 4,	iii.	23
xxi. 37,	iv.	170
xxii. 39,	iv.	170
xxiii. 6,	ii.	255
xxiii. 33,	iv.	134
xxiv. 13-35,	iv.	236
John i. 44,	ii.	270
i. 46,	ii.	256
i. 47,	iv.	380
ii. 11,	iv.	380
iii. 23,	ii.	345
iv. 5,	iv.	318
iv. 12,	iv.	288
iv. 20,	iv.	290

INDEX OF TEXTS.

John iv. 25,	vol. iv. page	307
iv. 39,	iv.	309
iv. 48,	iv.	380
v. 2-7,	iv.	156
v. 2-9,	iv.	144
vi. 3,	ii.	276
vi. 17-24,	ii.	276
vi. 18,	ii.	251
vii. 52,	ii.	256
ix. 7,	iv.	148
xi. 31, 38,	iv.	214
xi. 47,	iv.	28
xi. 54,	iv.	225
xii. 13,	iv.	33
xii. 21,	ii.	233, 270
xv. 1,	iii.	297
xix. 17, 20, 41, 42,	iv.	127
xix. 38,	iv.	261
xxi. 2,	iv.	380
Acts i. 11,	ii.	254
i. 19,	iv.	165
ii. 5,	iv.	2
Acts ii. 7,	vol. ii. page	254
ii. 29,	iv.	57
vii. 58,	iv.	32
viii. 5-25,	iv.	310, 326
viii. 20,	iv.	272
viii. 26,	iii.	211
viii. 26, 28,	iii.	329
ix. 1,	iv.	272
ix. 10, 11,	iv.	257
ix. 30,	iv.	272
ix. 31,	iii.	192
xi. 1,	iv.	272
xii. 19-24,	iv.	272
xviii. 22,	iv.	272
xxi. 8,	iv.	272
xxi. 34-37,	iv.	110
xxiii. 23-35,	iv.	243
Heb. xi. 38,	iii.	7
xiii. 12,	iv.	127
Jas. ii. 23,	iii.	294
2 Pet. i. 18,	ii.	313

END OF VOL. IV.

www.ingramcontent.com/pod-product-compliance
Lightning Source LLC
Chambersburg PA
CBHW030553300426
44111CB00009B/964